Crop Production: Principles and Practices

Crop Production: Principles and Practices

Editor: Shirley Doy

www.callistoreference.com

Callisto Reference,
118-35 Queens Blvd., Suite 400,
Forest Hills, NY 11375, USA

Visit us on the World Wide Web at:
www.callistoreference.com

ISBN: 978-1-63239-783-6 (Hardback)

The publisher's policy is to use permanent paper from mills that operate a sustainable forestry policy. Furthermore, the publisher ensures that the text paper and cover boards used have met acceptable environmental accreditation standards.

Printed in the United States of America.

Cataloging-in-publication Data

Crop production : principles and practices / edited by Shirley Doy.
  p. cm.
Includes bibliographical references and index.
ISBN 978-1-63239-783-6
1. Crop science. 2. Agriculture. 3. Crops. 4. Crop yields. I. Doy, Shirley.
SB91 .C76 2017
633--dc23

Table of Contents

Preface..IX

Chapter 1   **Impact of Interspecific Hybridization between Crops and Weedy Relatives
on the Evolution of Flowering Time in Weedy Phenotypes**............................1
Corinne Vacher, Tanya M. Kossler, Michael E. Hochberg, Arthur E. Weis

Chapter 2   **Non-Uniform Distribution Pattern for Differentially Expressed Genes of
Transgenic Rice Huahui 1 at Different Developmental Stages and
Environments**...8
Zhi Liu, Jie Zhao, Yunhe Li, Wenwei Zhang, Guiliang Jian, Yufa Peng,
Fangjun Qi

Chapter 3   **Estimation of Agricultural Water Consumption from Meteorological and
Yield Data**..16
Zaijian Yuan, Yanjun Shen

Chapter 4   **A Novel Universal Primer-Multiplex-PCR Method with Sequencing Gel
Electrophoresis Analysis**..25
Wentao Xu, Zhifang Zhai, Kunlun Huang, Nan Zhang, Yanfang Yuan,
Ying Shang, Yunbo Luo

Chapter 5   **Recent Weather Extremes and Impacts on Agricultural Production and
Vector-Borne Disease Outbreak Patterns**...35
Assaf Anyamba, Jennifer L. Small, Seth C. Britch, Compton J. Tucker,
Edwin W. Pak, Curt A. Reynolds, James Crutchfield, Kenneth J. Linthicum

Chapter 6   **Impact of Single and Stacked Insect-Resistant Bt-Cotton on the Honey Bee
and Silkworm**...44
Lin Niu, Yan Ma, Amani Mannakkara, Yao Zhao, Weihua Ma, Chaoliang Lei,
Lizhen Chen

Chapter 7   **Genetically Modified Crops and Food Security**..53
Matin Qaim, Shahzad Kouser

Chapter 8   **Bacterial Communities Associated with the Surfaces of Fresh Fruits and
Vegetables**..60
Jonathan W. Leff, Noah Fierer

Chapter 9   **Root Interactions in a Maize/Soybean Intercropping System Control
Soybean Soil-Borne Disease, Red Crown Rot**...69
Xiang Gao, Man Wu, Ruineng Xu, Xiurong Wang, Ruqian Pan, Hye-Ji Kim,
Hong Liao

Chapter 10   **Sustainable Management in Crop Monocultures: The Impact of Retaining
Forest on Oil Palm Yield**...78
Felicity A. Edwards, David P. Edwards, Sean Sloan, Keith C. Hamer

Chapter 11   **Early Root Overproduction not Triggered by Nutrients Decisive for Competitive Success Belowground**..86
Francisco M. Padilla, Liesje Mommer, Hannie de Caluwe, Annemiek
E. Smit-Tiekstra, Cornelis A. M. Wagemaker, N. Joop Ouborg, Hans de Kroon

Chapter 12   **A Meta-Analysis of the Impacts of Genetically Modified Crops**.........................95
Wilhelm Klümper, Matin Qaim

Chapter 13   **Alfalfa (*Medicago sativa* L.)/Maize (*Zea mays* L.) Intercropping Provides a Feasible way to Improve Yield and Economic Incomes in Farming and Pastoral Areas of Northeast China**...102
Baoru Sun, Yi Peng, Hongyu Yang, Zhijian Li, Yingzhi Gao, Chao Wang,
Yuli Yan, Yanmei Liu

Chapter 14   **Expression of Cry1Ab and Cry2Ab by a Polycistronic Transgene with a Self-Cleavage Peptide in Rice**..114
Qichao Zhao, Minghong Liu, Miaomiao Tan, Jianhua Gao, Zhicheng Shen

Chapter 15   **Soil Chemical Property Changes in Eggplant/Garlic Relay Intercropping Systems under Continuous Cropping**..120
Mengyi Wang, Cuinan Wu, Zhihui Cheng, Huanwen Meng, Mengru Zhang,
Hongjing Zhang

Chapter 16   **Reduced Levels of Membrane-Bound Alkaline Phosphatase are Common to Lepidopteran Strains Resistant to Cry Toxins from *Bacillus thuringiensis***...................133
Juan Luis Jurat-Fuentes, Lohitash Karumbaiah, Siva Rama Krishna Jakka,
Changming Ning, Chenxi Liu, Kongming Wu, Jerreme Jackson, Fred Gould,
Carlo Blanco, Maribel Portilla, Omaththage Perera, Michael Adang

Chapter 17   **Movement of Soil-Applied Imidacloprid and Thiamethoxam into Nectar and Pollen of Squash (*Cucurbita pepo*)**..141
Kimberly A. Stoner, Brian D. Eitzer

Chapter 18   **Effects of Dominance and Diversity on Productivity along Ellenberg's Experimental Water Table Gradients**..146
Andy Hector, Stefanie von Felten, Yann Hautier, Maja Weilenmann,
Helge Bruelheide

Chapter 19   **Intercropping Competition between Apple Trees and Crops in Agroforestry Systems on the Loess Plateau of China**..154
Lubo Gao, Huasen Xu, Huaxing Bi, Weimin Xi, Biao Bao, Xiaoyan Wang,
Chao Bi, Yifang Chang

Chapter 20   **How to Design a Targeted Agricultural Subsidy System: Efficiency or Equity?**.........................162
Rong-Gang Cong, Mark Brady

Chapter 21   **Environmental Fate of Soil Applied Neonicotinoid Insecticides in an Irrigated Potato Agroecosystem**..174
Anders S. Huseth, Russell L. Groves

Chapter 22   **Effects of Transgenic Cry1Ac + CpTI Cotton on Non-Target Mealybug Pest**
Ferrisia virgata **and its Predator** *Cryptolaemus montrouzieri*.. **185**
Hongsheng Wu, Yuhong Zhang, Ping Liu, Jiaqin Xie, Yunyu He,
Congshuang Deng, Patrick De Clercq, Hong Pang

Permissions

List of Contributors

Index

Preface

This book is a valuable compilation of topics, ranging from the basic to the most complex advancements in the field of crop production. This book presents the complex subject of crop production in the most comprehensible and easy to understand language. The book covers the fundamentals of the topic and critically addresses the advancements in the field of crop production. The objective of this book is to give a general view of the different areas of crop production. As this field is emerging at a rapid pace, the contents of this book will help the readers understand the modern concepts and applications of the subject.

In my initial years as a student, I used to run to the library at every possible instance to grab a book and learn something new. Books were my primary source of knowledge and I would not have come such a long way without all that I learnt from them. Thus, when I was approached to edit this book; I became understandably nostalgic. It was an absolute honor to be considered worthy of guiding the current generation as well as those to come. I put all my knowledge and hard work into making this book most beneficial for its readers.

I wish to thank my publisher for supporting me at every step. I would also like to thank all the authors who have contributed their researches in this book. I hope this book will be a valuable contribution to the progress of the field.

Editor

Impact of Interspecific Hybridization between Crops and Weedy Relatives on the Evolution of Flowering Time in Weedy Phenotypes

Corinne Vacher[1,2]*, **Tanya M. Kossler**[3,4], **Michael E. Hochberg**[2], **Arthur E. Weis**[3,5]

1 INRA, UMR1202 Biodiversité Gènes et Communautés, Cestas, France, **2** Université Montpellier II, UMR5554 Institut des Sciences de l'Evolution, Montpellier, France, **3** Department of Ecology and Evolutionary Biology, University of California Irvine, Irvine, California, United States of America, **4** Department of Biology, Duke University, Durham, North Carolina, United States of America, **5** Department of Ecology and Evolutionary Biology, University of Toronto, Toronto, Canada

Abstract

Background: Like conventional crops, some GM cultivars may readily hybridize with their wild or weedy relatives. The progressive introgression of transgenes into wild or weedy populations thus appears inevitable, and we are now faced with the challenge of determining the possible evolutionary effects of these transgenes. The aim of this study was to gain insight into the impact of interspecific hybridization between transgenic plants and weedy relatives on the evolution of the weedy phenotype.

Methodology/Principal Findings: Experimental populations of weedy birdseed rape (*Brassica rapa*) and transgenic rapeseed (*B. napus*) were grown under glasshouse conditions. Hybridization opportunities with transgenic plants and phenotypic traits (including phenological, morphological and reproductive traits) were measured for each weedy individual. We show that weedy individuals that flowered later and for longer periods were more likely to receive transgenic pollen from crops and weed×crop hybrids. Because stem diameter is correlated with flowering time, plants with wider stems were also more likely to be pollinated by transgenic plants. We also show that the weedy plants with the highest probability of hybridization had the lowest fecundity.

Conclusion/Significance: Our results suggest that weeds flowering late and for long periods are less fit because they have a higher probability of hybridizing with crops or weed×crop hybrids. This may result in counter-selection against this subset of weed phenotypes, and a shorter earlier flowering period. It is noteworthy that this potential evolution in flowering time does not depend on the presence of the transgene in the crop. Evolution in flowering time may even be counter-balanced by positive selection acting on the transgene if the latter was positively associated with maternal genes promoting late flowering and long flowering periods. Unfortunately, we could not verify this association in the present experiment.

Editor: Hans Henrik Bruun, University Copenhagen, Denmark

Funding: Financial support came from the French Ministère de la Recherche et de l'Enseignement Supérieur (Program: Impact of Biotechnologies in Agro-Ecosystems), and grants from the United States National Science Foundation (Grant: DEB-034530) and the National Science and Engineering Research Council of Canada to AEW. The funders had no role in study design, data collection and analysis, decision to publish, or preparation of the manuscript.

Competing Interests: The authors have declared that no competing interests exist.

* E-mail: corinne.vacher@pierroton.inra.fr

Introduction

When transgenic plants were initially developed, most plant evolutionary biologists and geneticists considered spontaneous hybridization between species to be rare and of little importance in terms of evolution. This view extended to both crops and their wild or weedy relatives [1], but has now radically changed. More than twenty years of gene-flow research has shown that interspecific hybridization is very common in some groups of vascular plants [2,3] and may be of considerable evolutionary significance. Hybridization may occasionally result in the extinction of a population [1,4], may trigger the evolution of plant invasiveness [5], or initiate speciation [6,7]. A substantial body of evidence [8,9] has now accumulated, demonstrating the high potential for interspecific hybridization between agricultural crops and their wild or weedy relatives. Transgenic crops are no

exception, and empirical studies have provided evidence of transgene dispersal from GM crops to their weedy relatives [10,11,12,13,14].

Many factors have been shown to influence the rate of hybrid formation between crops and their wild or weedy relatives. Population effects such as the local densities of the parental types and their relative frequencies, have been demonstrated in several cases [12,15,16,17,18,19]. Mating system differences at the individual level due to, for example, selfing rates and apomixis, have also been found to affect hybridization rates [20]. Moreover, several studies have shown that overlap in the flowering periods of crop and weed plants affect opportunities for hybridization [18,21].

The aim of this study is to gain insight into the impact of hybridization with transgenic crops on the evolution of the weedy relatives by (1) verifying that hybridization opportunities for weedy plants depend on their phenotypic traits (including flowering phenology), (2) measuring the relative fitness of hybridizing weeds, and (3) searching for associations between the transgenic trait and the phenotypic traits increasing hybridization opportunities in the offspring of weedy plants.

We studied hybridization opportunities, phenotypic traits (including phenological, morphological and reproductive traits) and offspring phenotype of weedy individuals (Table 1) in experimental plant populations cultivated under glasshouse conditions. Experimental populations were composed of weeds (birdseed rape, *Brassica rapa* L., AA, 2n = 20) and transgenic plants in a 1:1 ratio. Transgenic plants were crop plants of the *Brassica* genus (rapeseed, *Brassica napus* L. *ssp oleifera*, AACC, 2n = 38), F1 hybrids between *B. rapa* and *B. napus*, or first-generation backcrosses. Crop plants were all homozygous for the *Btcry1Ac* transgene from *Bacillus thuringiensis* (*Bt*) [22], F1 hybrids were all hemizygous and first-generation backcrosses and consisted of an equal mixture of hemizygotes and null homozygotes. Hybridization opportunities for each weedy individual was calculated as the expected proportion of pollen received from transgenic plants (*PPR*) based on the observed flowering schedules.

This experimental system was ideal for addressing the question of interest in this study, for three reasons. First, despite barriers to interspecific mating such as apomixis [20] or preferential exclusion of hybrid zygotes [23], numerous studies [24] have shown that *B. napus* and *B. rapa* readily hybridize under controlled conditions, but also in the field. Spontaneous hybridization has, for instance, been reported in weedy populations of *B. rapa* growing in agricultural crops [12,25,26] and in natural populations of *B. rapa* occurring near waterways [27]. Second, flowering time has been extensively studied in *B. rapa* [18,28,29], and temporal clines in phenotypic traits have been observed. For example, time to first flowering has been shown to be positively correlated with stem height and stem diameter [28,30,31]. Third, transgenic lines of *B. napus* containing a green fluorescent protein (GFP) gene associated with the *Bt* transgene have been constructed [32,33]. The presence of the *Bt* transgene in the offspring of weedy plants can therefore be inferred by exposing the plants to UV light [32,34].

Results

(1) Relationship between hybridization opportunities for weedy individuals, their flowering phenology, and their morphology

As expected from previous results [18], weeds flowered earlier than transgenic crops and hybrids (Fig. 1), with the F1 hybrids flowering the latest. Correspondingly, the expected proportion of crosses between weeds and F1 plants was lower than that for crop

Table 1. Phenotypic traits studied in weedy mother plants (M) and their offspring (O).

Trait	Generation
Time to first flower	M, O
Flowering duration	M
Stem diameter on the day of first flower	M, O
Stem height on the day of first flower	M
Total number of filled seeds	M
Expression of the *Bt*-transgene	O

or backcross plants (Table 2A). Moreover, *PPR* (log transformed) increased with the time to first flower and the duration of flowering in weedy individuals (see overall slopes in Table 2B). The overall slope for the interaction between the two phenological traits (Table 2B) was close to zero and did not qualitatively modify these effects. However, significant interactions (Table 3) indicated that the effects of phenology of weedy plants on *PPR* depended on transgenic type (crop, F1 hybrid or first-generation backcross). The regression coefficients and their 95% confidence limits indicated that a longer time to flowering and a longer flowering duration increased *PPR* more for F1 hybrids than for crops or first-generation backcrosses (see within-type slopes in Table 2B). Thus, weedy individuals flowering later and for longer periods were more likely to receive transgenic pollen, particularly if the transgenic donors were first-generation crop x weed F1 hybrids.

As expected from the results of previous studies [28,30], we observed temporal clines in the morphological traits under study. Time to first flower was positively correlated with stem diameter ($r_s = 0.31$, $P<0.001$) and stem height ($r_s = 0.18$, $P<0.05$). These correlations indicate that the opportunity for hybridization may not be random, and may instead depend on the morphology of the weed. We found a significant, single effect of stem diameter on *PPR* ($F_{1,105} = 5.0$, $P<0.05$). The overall slope was positive and its 95% confidence interval did not include zero (slope = 0.05, CL = (0.01, 0.09)), indicating that plants with large stems on the day of the first flower were more likely to hybridize with transgenic plants. No such effect was detected for stem height, either as a single effect ($F_{1,104} = 0.05$, $P = 0.82$) or in interaction with transgenic type ($F_{2,104} = 0.87$, $P = 0.42$).

(2) Relative fitness of hybridizing weeds

For any given weedy plant in the experimental populations, the total number of filled seeds decreased significantly with *PPR* (see overall slopes in Table 4B and the significant effect of *PPR* in Table 5). We observed no significant interaction between *PPR* and transgenic type (Table 5), indicating that this decrease in fecundity with *PPR* was not dependant on transgenic type (crop, F1 hybrid or first-generation backcross). This decrease in fecundity was observed despite the positive correlation between *PPR* and total flower production within weedy plants ($r_s = 0.37$, $P<0.001$). An alternative analysis (not shown), including transgenic type as fixed effect and phenological traits of weeds (time to first flower or flowering duration) as covariates also predicted the total number of filled seeds. We found a significant effect of the time to first flower on the number of seeds, in interaction with transgenic type ($F_{2,107} = 3.66$, $P<0.05$). However, all the 95% confidence intervals of the regression coefficients for each transgenic type included zero, making further interpretation impossible. Flowering duration was significant as a single effect ($F_{1,105} = 37.4$, $P<0.001$). The overall slope was negative and its 95% confidence interval did not include zero (slope = −21.3, CL = (−28.77, −14.51)), indicating that weedy plants with longer flowering times produced fewer seeds. Thus, the weedy plants with the highest probability of being pollinated by *Bt*-transgenic plants were those with the lowest fecundity (Fig. 2).

(3) Associations between the transgenic trait and the phenotypic traits increasing hybridization opportunities in the offspring of weedy individuals

An analysis of offspring phenotype showed that time to first flower in weedy mother plants had a significant effect on the average time to first flower of their offspring ($F_{1,104} = 7.48$, $P<0.05$). Transgenic type (crop, F1 hybrid or first-generation

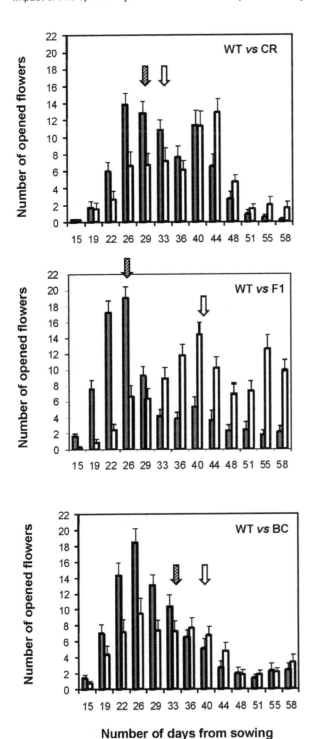

Figure 1. Phenology of transgenic and weedy plants. Phenology of weedy plants (WT; *hatched bars*) and their *Bt*-transgenic relatives (CR, F1 or BC; *white bars*). For each combination, three mixed populations of 30 plants were monitored. Bars represent are the mean numbers of opened flowers per population for each day of observation, with standard errors. WT: weedy plants of *B. rapa*; CR: *Bt*-crop plants of *B. napus*; F1: F1 hybrids between WT and CR; BC: *Bt*-plants from the backcross of F1 on WT. Arrows indicate the date at which 50% of the flowers had been produced.

backcross) did not affect time to first flowering in the offspring, either as a main effect ($F_{2,71.3} = 0.20$, $P = 0.82$) or in interaction with maternal time to first flower ($F_{2,97.4} = 0.40$, $P = 0.66$). In contrast, offspring stem diameter was not affected by maternal diameter ($F_{1,104} = 1.80$, $P = 0.18$) or maternal time to first flower ($F_{1,104} = 2.03$, $P = 0.15$). These results confirm that late-flowering plants tend to produce late-flowering offspring [28]. Because late-flowering plants were also more likely to receive transgenic pollen, we therefore expected to find more transgenic offspring in the offspring of late-flowering weedy mothers and an association between the transgenic trait and time to first flower in the offspring generation.

Contrary to expectation, we found no evidence to suggest that weedy individuals with higher *PPR* produced more transgenic offspring. A total of 1648 seedlings, obtained from 126 weedy plants, were scored under UV light for the *Bt*-GFP construct. Only 38 seedlings, produced by 17 weedy mothers, scored positively. None of them was sired by the pollen of F1 hybrids. The proportions of fluorescent seedlings were equal to 0.04±0.01 for populations with crop plants, 0.0±0.0 for populations with F1 hybrids and 0.02±0.01 for populations with first-generation backcrosses ($\chi^2 = 17.55$, df = 2, P<0.001). Significant differences were observed between replicates for the proportion of positive scores ($\chi^2 = 15.76$, df = 2, P<0.001). The proportion of *Bt*-GFP+ seedlings was not correlated with *PPR* in populations with crop plants ($r_s = -0.16$, $P = 0.33$) nor with backcross plants ($r_s = 0.23$, $P = 0.17$). Correlations with the proportion of *Bt*-GFP+ seedlings were also weak and non significant for all other maternal traits measured. Thus, variation in the probability of weedy mother plants being pollinated by transgenic donors did not translate into variation in the proportion of *Bt*-seedlings in their offspring.

Because of the very low proportions of *Bt*-GFP+ seedlings, we could not study the associations between the transgenic trait and the phenotypic traits increasing hybridization opportunities in the offspring of weedy plants. Among the 1654 seedlings scored under UV light, 1048 reached the first flower stage and were measured. Unfortunately, only nine of these plants were *Bt*-GFP+, and seven of these nine plants were half sibs (the nine plants were produced by only three weedy mothers). The 31 remaining *Bt*-GFP+ seedlings did not reach the first flower stage. There were, therefore, clearly too few *Bt*-GFP+ plants to compare the phenotypic characteristics of *Bt*-GFP+ and *Bt*-GFP- offspring.

Discussion

The aim of our experiment was to assess the impact of interspecific hybridization between weedy *B. rapa* and transgenic *B. napus* on the evolution of the weedy phenotype. This was done by identifying the phenotypic traits increasing hybridization opportunities for weedy individuals, searching for associations between thesephenotypic traits and the transgenic trait in the offspring of weedy mothers and evaluating the relative fitness of hybridizing weeds. Our results show that weedy individuals that flowered later and for longer periods were more likely to receive transgenic pollen from crops and weed×crop hybrids. Because stem diameter is correlated with flowering time [28,30], plants with wider stems were also more likely to be pollinated by transgenic plants. Our results suggest that the transgene and maternal genes promoting late flowering, long flowering periods and stem thickening may be preferentially associated in the offspring of weedy mothers. However, although time to first flower is a heritable trait in *B. rapa* [28], our experiment did not confirm the gametic association between the transgene and genes promoting late-flowering in the offspring of hybridized weedy

Table 2. Effects of phenological traits on the expected proportion of pollen received from transgenic plants (*PPR*) of weedy *B. rapa* in mixed populations including transgenic *B. napus* crop and crop-weed hybrids.

A. Means	Overall	Transgenic type		
		Crop	F1	Backcross
Expected proportion of pollinations by transgenic plants (*PPR*)	0.39	0.39	0.36	0.41
± Standard error	±0.01	±0.01	±0.02	±0.02
B. Slopes				
Time to first flower	**0.06**	0.01	**0.21**	0.11
95% C.L.	(0.03; 0.09)	(−0.05; 0.06)	(0.06; 0.36)	(−0.03; 0.25)
Flowering duration	**0.16**	0.02	**0.46**	0.24
95% C.L.	(0.07; 0.26)	(−0.15; 0.20)	(0.02; 0.90)	(−0.18; 0.65)
Time x duration	**−0.005**	0.00	**−0.02**	−0.01
95% C.L.	(−0.01; 0.00)	(−0.01; 0.01)	(−0.04; 0.00)	(−0.03; 0.01)

A. The mean *PPR* for the transgenic type treatments. **B.** The influence of weedy traits. "Slopes" are the coefficients for the effect of each trait on *PPR*. The "overall" slope indicates the effect across all transgenic types. The within-type slopes were obtained from the mixed linear model presented in Table 3, they indicate the relationships for crop, F1 and backcross migrants. Coefficients that do not include zero in their 95% confidence interval are shown in bold typeface.

plants. Indeed, given the very small numbers of *Bt*-GFP+ seedlings recovered from the experimental populations, we could not study the association between the transgenic trait and other phenotypic traits in weed plant offspring.

We also found that the weedy plants with the highest probability of hybridization produced fewer seeds, despite producing larger numbers of flowers. The most straightforward interpretation of this result is that fecundity was reduced by hybrid crosses. Controlled crosses between the weedy and transgenic plants used in the experiment (unpublished results) and several previous studies [35,36] have indeed shown that crops and weed×crop hybrids have lower siring success than do weeds. Therefore, our experiment suggests that maternal weeds that flowered late and for long periods are less fit, because they have a higher probability of hybridizing with GM crop plants or hybrids. This may result in counter-selection against this subset of weed phenotypes, and a shorter earlier flowering period. It is noteworthy that this potential evolution in flowering time does not depend on the presence of the *Bt* transgene in the crop, and may even be counter-balanced by positive selection acting on the transgene if the latter was positively associated with maternal genes promoting late flowering and long

flowering periods. Recent experiments indeed indicate that the *Bt* transgene does not induce any fitness costs in hybrids between transgenic *B. napus* and weedy relatives [37,38]. It may therefore convey a selective advantage under insect herbivore pressure [39].

In conclusion, our analyses show that phenological differences between weedy birdseed rape and transgenic rapeseed are likely to alter the phenotypic structure of weed populations, by promoting interspecific hybridization in only a subset of weedy plants with specific phenotypes and by altering the fitness of hybridizing weeds. Unfortunately, we could not verify the non-random association between the transgenic trait and other phenotypic traits in the offspring of weedy populations because of the very low rate of transgene introgression.

Materials and Methods

Experimental design

Nine populations, each composed of 15 *Brassica rapa* plants and 15 of one of three types of transgenic plants (see below) were sown as seeds and then grown from germination until death in a glasshouse at the University of California, Irvine. The nine populations were divided into three blocks, with each transgenic type replicated once per block. Plants were grown in individual Conetainer® (3.8×21 cm) pots filled with a 75/25 mixture of potting soil and sand. Before planting, seeds were vernalized on wet filter paper at 4°C for 5 days. Pots were spaced 7.6 cm apart and were watered every day until 90% stopped producing flowers. An equal amount of 10:10:10 NKP liquid fertilizer was applied to each pot on the sowing date.

The three types of transgenic plants were: *Bt*-transgenic *B. napus* crop plants, *Bt*-transgenic *B. napus* × *B. rapa* F1 hybrids, and first-generation backcrosses (*B. rapa* ×F1 hybrids). Over 20 unique seed and 20 unique pollen parents were used to produce each of the three types. *B. rapa* plants served as seed parents for the F1 and backcross types. *B. napus* were all homozygous for the *Bt*-GFP insertion, whereas the F1 plants were all hemizygous. The backcross generation was expected to consist of an equal mixture of hemizygotes and null homozygotes for the insertion.

B. rapa seeds were obtained from over 400 mature plants in a population at Back Bay, near Irvine, California [40]. Transgenic *B. napus* plants were derived from spring rapeseed lines (variety

Table 3. Mixed linear model for the effects of transgenic type and weedy plant phenological traits on the expected proportion of pollen received from transgenic plants (*PPR*).

Source	df (numerator)	df (denominator)	F value	P
Transgenic type	2	105	8.28	0.0005
Time to first flower	1	99.2	21.25	<.0001
Flowering duration	1	99.5	27.63	<.0001
Type × time	2	99.2	5.41	0.0059
Type × duration	2	99.4	8.52	0.0004
Time × duration	1	99.3	14.07	0.0003
Type × time × duration	2	99.3	5.47	0.0056

−2 residual log likelihood = −35.1.
Akaike's information criterion = −31.1.

Table 4. Effect of the expected proportion of pollen received from transgenic plants (*PPR*) on the total number of filled seeds produced by weedy *B. rapa* in mixed populations including transgenic *B. napus* crop and crop-weed hybrids.

A. Means	Overall	Transgenic type		
		Crop	F1	Backcross
Total number of seeds	151.68	140.23	140.26	176.44
± standard error	±7.63	±12.58	±12.00	±14.60
B. Slopes				
PPR	**−303.15**	**−372.1**	−319.21	−290.42
95% C.L.	(−466.73; −139.57)	(−729.13; −15.07)	(−1098.72; 460.30)	(−1160.18; 579.34)

A. Mean seed production for the transgenic type treatments. **B.** The influence of *PPR*. "Slopes" are the coefficients for the effect of each trait on seed production. The "overall" slope indicates the effect across all transgenic types. The within-type slopes were obtained from the mixed linera model presented in Table 5, they indicate the relationships for crop, F1 and backcross plants. Coefficients that do not include zero in their 95% confidence interval are in shown in bold typeface.

Westar, supplied by Dr. Neal Stewart, University of Tennessee). In addition to the *Btcry1Ac* gene from *Bacillus thuringiensis* (*Bt*) [22], these lines contained a green fluorescent protein (GFP) gene (mGFP5er) under the control of the cauliflower mosaic virus 35S promoter and a nopaline synthase terminator cassette [32,33]. The fate of the *Bt* transgene could therefore be inferred by exposing the offspring to UV light [32,34].

Flowering schedules were constructed for each individual plant by recording the time to first flower (i.e., the number of days between sowing and the first observed flower) and the number of opened flowers on every fourth day until the end of the flowering period. The lifetime of a flower is about three days (Weis A., pers. obs.), so this procedure made it possible to estimate the total number of flowers produced by each plant over the flowering period. The length of the flowering period was defined as the number of scoring days on which the plant had opened flowers.

Every fourth day, all open flowers on all plants were hand pollinated in each of the nine experimental populations (there were no natural pollinators in the experimental glasshouse). Each experimental population was composed of 30 plants which were numbered from 1 to 30. On each pollination day, a random sequence of 30 numbers (without repetition) was generated for each population. For a given population, a pollination session consisted of brushing all the flowers of the first plant in the sequence, and then brushing all of the flowers of the next plant. This was continued until the brush from the 30th plant was used to transfer pollen to the first plant. Each plant was brushed up and down several times to deposit the pollen from the previous plant in the sequence and collect the maximum amount of pollen. A given plant was only brushed if it was alive and had one or more open

flowers. Otherwise the next plant in the sequence was considered. Each of the nine populations had its own brush, and new brushes were used for each pollination session. This hand-pollination procedure was chosen to approximate the behaviour of a bumble bee in a patch of oilseed rape. Bumblebees tend to visit many plants successively and rarely revisit the plants [41]. They deposit most of the pollen from a source plant on immediate neighbours [42].

We did not keep track of the random sequences of plants generated for each experimental population on each pollination day so we used observed flowering schedules to calculate the expected proportion of pollen received from transgenic plants (*PPR*) for each weedy plant. On each pollination day, the probability of a weedy plant receiving pollen from a transgenic plant was assumed to be proportional to the number of transgenic plants in flower in the experimental population. Over the entire flowering period:

$$PPR_{ij} = \sum_d \theta_{ijd} X_{jd}$$

where PPR_{ij} is the expected proportion of flowers crossed with a transgenic plant for weedy plant i from population j, θ_{ijd} is the proportion of flowers open on pollination day d for the weedy plant i from population j, and X_{jd} is the proportion of plants in flower that were of the transgenic type on pollination day d in population j. The proportion X_{jd} was calculated by excluding the focal plant i, since *B. rapa* is known to be largely self-incompatible [43].

In addition to phenological traits, several morphological and reproductive traits were assessed. On the day of first flower, we recorded basal stem diameter and stem height. Dry siliques were collected once the plants had died. The aggregate mass of filled seeds was determined for each plant by separating these seeds from the lighter, aborted seeds, using an air-flow system. We selected five seeds per plant at random and weighed them, to estimate the total number of seeds per plant. We confirmed the accuracy of these measures by counting and weighing all the seeds for 47 plants spanning the range of seed masses.

Finally, for each weedy plant of the nine experimental populations described above, 14 randomly chosen seeds were sown and grown until the day of the first flower. If a mother plant had less than 14 seeds in total, all were sown. Growing conditions were identical to those for the parental generation. Each seedling was scored for fluorescence under high-intensity UV light, at the four-leaf stage. At this stage, the petioles and main nerves of the

Table 5. Mixed linear model for the effects of transgenic type and expected proportion of pollen received from transgenic plants (*PPR*) on the total number of filled seeds.

Source	df (numerator)	df (denominator)	F value	P
Transgenic type	2	28.7	0.07	0.929
PPR	1	74.3	12.19	0.0008
Transgenic type x PPR	2	76.2	0.05	0.948

−2 residual log likelihood = 1240.8.
Akaike's information criterion = 1244.8.

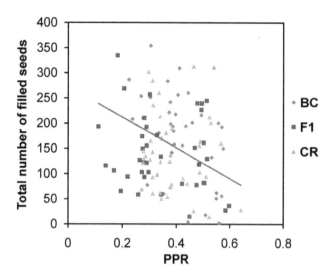

Figure 2. Decrease in fecundity with *PPR*. Total number of filled seeds (*TNS*) produced by weedy individuals as a function of the expected proportion of pollen (*PPR*) received from transgenic plants (CR: *Bt*-crop plants of *B. napus*; F1: F1 hybrids between weedy plants and CR; BC: *Bt*-plants from the backcross of F1 on weedy plants). The grey line corresponds to the regression line across all transgenic types (*TNS* = −303.15 *PPR* +272.69); its slope is the overall slope given in Table 5.

leaves of transgenic plants displayed fluorescence [34]. This made it possible to determine the proportion of *Bt*-GFP+ seedlings for each mother plant. To investigate the association between the transgenic trait and phenotypic traits in the offspring, time to first flower was recorded for each seedling and, on the day of the first flower, basal stem diameter was measured.

Statistical analysis

We performed all statistical analyses with SAS/STAT® software [44]. Plants that died during the experiment were excluded from the analysis and the final data set contained 117 weedy plants.

We first investigated how phenological traits affected the chances of interspecific hybridization between *Bt*-trangenic plants and weeds. We used a mixed linear model (SAS, Procedure MIXED), with transgenic type (crop, F1 hybrid or first-generation backcross) as the fixed treatment effect, phenological traits of weeds (time to first flower, flowering duration and total number of flowers) as covariates, and block and treatment×block interaction as random effects. The response variable was the proportion of flowers receiving pollen from *Bt*-transgenic plants (*PPR*). The response variable was log-transformed to increase its normality (Kolmogorov-Smirnov goodness-of-fit; SAS, Procedure UNIVARIATE). If a factor was not significant as a single effect or in interaction with other factors, it was eliminated from the model and the analysis was rerun. We continued until there was no further improvement in residual maximum likelihood.

We then investigated how morphological traits affected the chances of hybridization. Temporal phenotypic clines were assessed by correlating morphological traits of weeds (with time to first flower (Spearman's rank correlation test; SAS, Procedure CORR). A mixed linear approach (SAS, Procedure MIXED) was then used to determine whether the morphological traits changing with time to first flower had a significant effect on *PPR*. As above, transgenic type (crop, F1 hybrid or first-generation backcross) was treated as a fixed treatment effect, morphological traits were covariates and block and treatment×block interaction were treated as random effects.

We used the mixed linear approach (SAS, Procedure MIXED) with block and treatment x block interactions as random effects, to investigate whether the phenological and morphological traits which were found to favour hybridization of weedy mothers were transmitted to their offspring. In this model, transgenic type (crop, F1 hybrid or first-generation backcross) was treated as a fixed effect, the maternal trait as a covariate and the average offspring phenotypic trait as the response variable. The normality of the response variables was checked (Kolmogorov-Smirnov goodness-of-fit; SAS, Procedure UNIVARIATE), and data was transformed as necessary.

Finally we investigated the relationship between opportunities for hybridization and fecundity in weeds. We used the mixed linear approach (SAS, Procedure MIXED) with transgenic type (crop, F1 hybrid or first-generation backcross) as the fixed treatment effect, *PPR* as the covariate and block and treatment× block interaction as random effects. The response variable was the total number of filled seeds. Its normality was checked with a Kolmogorov-Smirnov goodness-of-fit test (SAS, Procedure UNIVARIATE).

We then checked that the mother plants with the highest expected probability of receiving transgenic pollen (*PPR*) also had the highest proportion of *Bt*-GFP+ seedlings. The proportion of *Bt*-GFP+ seedlings did not follow a normal distribution (Kolmogorov-Smirnov goodness-of-fit; SAS, Procedure UNIVARIATE) and could not be transformed. We therefore checked the effects of transgenic type, *PPR* and block separately, in non parametric one-way ANOVAs (SAS, Proc NPAR1WAY, Kruskal-Wallis test). The correlation between *PPR* and the proportion of *Bt*-GFP+ seedlings was assessed using Spearman's rank correlation test (SAS, Proc CORR).

Acknowledgments

We thank Donald Hermann for carrying out the preliminary experiments for this study.

Author Contributions

Conceived and designed the experiments: CV MH AEW. Performed the experiments: CV TMK AEW. Analyzed the data: CV AEW. Wrote the paper: CV TMK MH AEW.

References

1. Ellstrand NC (2003) Current knowledge of gene flow in plants: implications for transgene flow. Philosophical Transactions of the Royal Society of London Series B-Biological Sciences 358: 1163–1170.
2. Ellstrand NC, Whitkus R, Rieseberg LH (1996) Distribution of spontaneous plant hybrids. Proceedings of the National Academy of Sciences of the United States of America 93: 5090–5093.
3. Rieseberg LH, Carney SE (1998) Plant hybridization. New Phytologist 140: 599–624.
4. Hedge SG, Nason JD, Clegg JM, Ellstrand NC (2006) The evolution of California's wild radish has resulted in the extinction of its progenitors. Evolution 60: 1187–1197.
5. Ellstrand NC, Schierenbeck KA (2000) Hybridization as a stimulus for the evolution of invasiveness in plants? Proceedings of the National Academy of Sciences of the United States of America 97: 7043–7050.
6. Abbott RJ (1992) Plant Invasions, Interspecific Hybridization and the Evolution of New Plant Taxa. Trends in Ecology & Evolution 7: 401–405.

7. Rieseberg LH (1997) Hybrid origins of plant species. Annual Review of Ecology and Systematics 28: 359–389.

8. Ellstrand NC, Prentice HC, Hancock JF (1999) Gene flow and introgression from domesticated plants into their wild relatives. Annual Review of Ecology and Systematics 30: 539–563.

9. Stewart CN, Halfhill MD, Warwick SI (2003) Transgene introgression from genetically modified crops to their wild relatives. Nature Reviews Genetics 4: 806–817.

10. Hall L, Topinka K, Huffman J, Davis L, Good A (2000) Pollen flow between herbicide-resistant Brassica napus is the cause of multiple-resistant B-napus volunteers. Weed Science 48: 688–694.

11. Quist D, Chapela IH (2001) Transgenic DNA introgressed into traditional maize landraces in Oaxaca, Mexico. Nature 414: 541–543.

12. Simard MJ, Legere A, Warwick SI (2006) Transgenic Brassica napus fields and Brassica rapa weeds in Quebec: sympatry and weed-crop in situ hybridization. Canadian Journal of Botany-Revue Canadienne De Botanique 84: 1842–1851.

13. Snow A (2009) Unwanted Transgenes Re-Discovered in Oaxacan Maize. Molecular Ecology 18: 569–571.

14. Pineyro-Nelson A, Van Heerwaarden J, Perales HR, Serratos-Hernandez JA, Rangel A, et al. (2009) Transgenes in Mexican maize: molecular evidence and methodological considerations for GMO detection in landrace populations. Molecular Ecology 18: 750–761.

15. Johannessen MM, Andersen BA, Jorgensen RB (2006) Competition affects gene flow from oilseed rape (female) to Brassica rapa (male). Heredity 96: 360–367.

16. Johannessen MM, Damgaard C, Andersen BA, Jorgensen RB (2006) Competition affects the production of first backcross offspring on F-1-hybrids, Brassica napus x B-Rapa. Euphytica 150: 17–25.

17. Linder CR, Taha I, Seiler GJ, Snow AA, Rieseberg LH (1998) Long-term introgression of crop genes into wild sunflower populations. Theoretical and Applied Genetics 96: 339–347.

18. Pertl M, Hauser TP, Damgaard C, Jorgensen RB (2002) Male fitness of oilseed rape (Brassica napus), weedy B-rapa and their F-1 hybrids when pollinating B-rapa seeds. Heredity 89: 212–218.

19. Vacher C, Weis AE, Hermann D, Kossler T, Young C, et al. (2004) Impact of ecological factors on the initial invasion of Bt transgenes into wild populations of birdseed rape (Brassica rapa). Theoretical and Applied Genetics 109: 806–814.

20. Pallett DW, Huang L, Cooper JI, Wang H (2006) Within-population variation in hybridisation and transgene transfer between wild Brassica rapa and Brassica napus in the UK. Annals of Applied Biology 148: 147–155.

21. Cummings CL, Alexander HM, Snow AA, Rieseberg LH, Kim MJ, et al. (2002) Fecundity selection in a sunflower crop-wild study: Can ecological data predict crop allele changes? Ecological Applications 12: 1661–1671.

22. Maagd RA, Bravo A, Crickmore N (2001) How Bacillus thuringiensis has evolved specific toxins to colonize the insect world. Trends in Genetics 17: 193–199.

23. Hauser TP, Jorgensen RB, Ostergard H (1997) Preferential exclusion of hybrids in mixed pollinations between oilseed rape (Brassica napus) and weedy B-campestris (Brassicaceae). American Journal of Botany 84: 756–762.

24. FitzJohn RG, Armstrong TT, Newstrom-Lloyd LE, Wilton AD, Cochrane M (2007) Hybridisation within Brassica and allied genera: evaluation of potential for transgene escape. Euphytica 158: 209–230.

25. Warwick SI, Simard MJ, Legere A, Beckie HJ, Braun L, et al. (2003) Hybridization between transgenic Brassica napus L. and its wild relatives: Brassica rapa L., Raphanus raphanistrum L., Sinapis arvensis L., and Erucastrum gallicum (Willd.) OE Schulz. Theoretical and Applied Genetics 107: 528–539.

26. Warwick SI, Legere A, Simard MJ, James T (2008) Do escaped transgenes persist in nature? The case of an herbicide resistance transgene in a weedy Brassica rapa population. Molecular Ecology 17: 1387–1395.

27. Wilkinson MJ, Elliott LJ, Allainguillaume J, Shaw MW, Norris C, et al. (2003) Hybridization between Brassica napus and B-rapa on a national scale in the United Kingdom. Science 302: 457–459.

28. Weis AE, Kossler TM (2004) Genetic variation in flowering time induces phenological assortative mating: Quantitative genetic methods applied to Brassica rapa. American Journal of Botany 91: 825–836.

29. Weis AE, Winterer J, Vacher C, Kossler TM, Young CA, et al. (2005) Phenological assortative mating in flowering plants: the nature and consequences of its frequency dependence. Evolutionary Ecology Research 7: 161–181.

30. Dorn LA, Mitchell-Olds T (1991) Genetics of Brassica campestris. 1. Genetic constraints on evolution of life-history characters. Evolution 45: 371–379.

31. Franks SJ, Sim S, Weis AE (2007) Rapid evolution of flowering time by an annual plant in response to a climate fluctuation. Proceedings of the National Academy of Sciences of the United States of America 104: 1278–1282.

32. Harper BK, Mabon SA, Leffel SM, Halfhill MD, Richards HA, et al. (1999) Green fluorescent protein as a marker for expression of a second gene in transgenic plants. Nature Biotechnology 17: 1125–1129.

33. Haseloff J, Siemering KR, Prasher DC, Hodge S (1997) Removal of a cryptic intron and subcellular localization of green fluorescent protein are required to mark transgenic Arabidopsis plants brightly. Proceedings of the National Academy of Sciences of the United States of America 94: 2122–2127.

34. Halfhill MD, Richards HA, Mabon SA, Stewart CN (2001) Expression of GFP and Bt transgenes in Brassica napus and hybridization with Brassica rapa. Theoretical and Applied Genetics 103: 659–667.

35. Allainguillaume J, Alexander M, Bullock JM, Saunders M, Allender CJ, et al. (2006) Fitness of hybrids between rapeseed (Brassica napus) and wild Brassica rapa in natural habitats. Molecular Ecology 15: 1175–1184.

36. Hauser TP, Shaw RG, Ostergard H (1998) Fitness of F-1 hybrids between weedy Brassica rapa and oilseed rape (B-napus). Heredity 81: 429–435.

37. Halfhill MD, Sutherland JP, Moon HS, Poppy GM, Warwick SI, et al. (2005) Growth, productivity, and competitiveness of introgressed weedy Brassica rapa hybrids selected for the presence of Bt cry1Ac and gfp transgenes. Molecular Ecology 14: 3177–3189.

38. Kun DC, Neal Stewart J, Wei W, Bao-cheng S, Zhi-Xi T, et al. (2009) Fitness and maternal effects in hybrids formed between transgenic oilseed rape (Brassica napus L.) and wild brown mustard [B. juncea (L.) Czern et Coss.] in the field. Pest Management Science 65: 753–760.

39. Stewart CN, All JN, Raymer PL, Ramachandran S (1997) Increased fitness of transgenic insecticidal rapeseed under insect selection pressure. Molecular Ecology 6: 773–779.

40. Franke DM, Ellis AG, Dharjwa M, Freshwater M, Fujikawa M, et al. (2006) A steep cline in flowering time for Brassica rapa in southern California: Population-level variation in the field and the greenhouse. International Journal of Plant Sciences 167: 83–92.

41. Cresswell JE (2000) A comparison of bumblebees' movements in uniform and aggregated distributions of their forage plant. Ecological Entomology 25: 19–25.

42. Cresswell JE, Bassom AP, Bell SA, Collins SJ, Kelly TB (1995) Predicted pollen dispersal by honey-bees and three species of bumble-bees foraging on oil-seed rape: A comparison of three models. Functional Ecology 9: 829–841.

43. Ellstrand NC (2003) Dangerous liaisons? When cultivated plants mate with their wild relatives. Dangerous liaisons? When cultivated plants mate with their wild relatives. xx+244 p.

44. SAS (1999) SAS/STAT® software, Version 8 of the SAS System for Unix, Copyright © 1999–2000, SAS Institute Inc., Cary, (North Carolina).

Non-Uniform Distribution Pattern for Differentially Expressed Genes of Transgenic Rice Huahui 1 at Different Developmental Stages and Environments

Zhi Liu, Jie Zhao, Yunhe Li, Wenwei Zhang, Guiliang Jian, Yufa Peng*, Fangjun Qi*

State Key Laboratory for Biology of Plant Diseases and Insect Pests, Institute of Plant Protection, Chinese Academy of Agricultural Sciences, Beijing, People's Republic of China

Abstract

DNA microarray analysis is an effective method to detect unintended effects by detecting differentially expressed genes (DEG) in safety assessment of genetically modified (GM) crops. With the aim to reveal the distribution of DEG of GM crops under different conditions, we performed DNA microarray analysis using transgenic rice Huahui 1 (HH1) and its non-transgenic parent Minghui 63 (MH63) at different developmental stages and environmental conditions. Considerable DEG were selected in each group of HH1 under different conditions. For each group of HH1, the number of DEG was different; however, considerable common DEG were shared between different groups of HH1. These findings suggested that both DEG and common DEG were adequate for investigation of unintended effects. Furthermore, a number of significantly changed pathways were found in all groups of HH1, indicating genetic modification caused everlasting changes to plants. To our knowledge, our study for the first time provided the non-uniformly distributed pattern for DEG of GM crops at different developmental stages and environments. Our result also suggested that DEG selected in GM plants at specific developmental stage and environment could act as useful clues for further evaluation of unintended effects of GM plants.

Editor: Sunghun Park, Kansas State University, United States of America

Funding: This work was supported by the National Genetically Modified Organisms Breeding Major Project (2008ZX08011-001) and Special Funding of State Key Laboratory of BPDIP, China (SKL2010SR01). The funders had no role in study design, data collection and analysis, decision to publish, or preparation of the manuscript.

Competing Interests: The authors have declared that no competing interests exist.

* E-mail: fjqi@ippcaas.cn (FQ); yfpeng@ippcaas.cn (YP)

Introduction

With the development of transgenic technology, GM crops have increased the farm income dramatically during the past years [1]. However, there have been, and would continue to be, considerable public concerns for the commercialization of GM crops. Such concerns focus on whether random insertion of transgenes into host plant genomes would result in unpredicted changes in expression pattern of other intrinsic genes, leading to unintended effects on GM crops and their products [2]. It is generally agreed that unintended effects should be paid particular attention in the process of safety assessment of GM crops and their products, especially in regard to some long-term and potential food safety issues [3].

The use of profiling technologies, such as DNA microarray analysis, has been proved to be an effective way to detect differentially expressed genes (DEG) and investigate unintended effects in a number of transgenic plant systems. For example, Gregerson *et al.* compared the gene expression profiles of wild type wheat seeds and GM wheat seeds at three developmental phases using a 9K unigene cDNA microarray and found only slight differences in gene expression profiles [4]. Affymetrix *Arabidopsis* ATH1 GeneChip was used to search for transcriptome changes in *Arabidopsis* and the result turned out that the insertion and expression of the marker genes, *uidA* and *nptII*, did not induce changes to the expression profiles under optimal growth conditions and under physiological stress imposed by low temperatures [5].

Also, Affymetrix *Arabidopsis* ATH1 GeneChip was used to study the pleiotropic effects of the *bar* gene and glufosinate on the *Arabidopsis* transcriptome by detecting DEG [6]. Microarray analysis was performed on *Arabidopsis* plants overexpressing transcription factor ABF3, and no unintended effects were discovered [7].

However, the majority of researches investigating DEG and unintended effects of GM crops [4–14], were carried out using GM plants at specific developmental phases and/or particular environments. As a consequence, the results of such investigations might be invalid unless DEG and unintended effects of GM crops at specific developmental phases and/or environmental conditions could be representative for GM plants in all conditions.

The distribution of DEG under different conditions (developmental stages or environments), however, still remains unclear. Apparently, it is possible that the distribution pattern of DEG might vary under different conditions. Theoretically, there are three possible distribution patterns of DEG: (I) uniform distribution; the amount of DEG remain more or less constant regardless of developmental phases or environmental factors, (II) extreme distribution; the number of DEG differ dramatically in different conditions, with extremely huge amount of DEG in some conditions and a nominal sum of DEG in other conditions, (III) non-uniform distribution; DEG distribute randomly, with various considerable amount of DEG in different conditions. If DEG were uniformly distributed, then the DEG detected under any condition

Table 1. Number of DEP on each rice chromosome in HH1 at different developmental stages and environments.

	Chr1	Chr2	Chr3	Chr4	Chr5	Chr6	Chr7	Chr8	Chr9	Chr10	Chr11	Chr12
30-day	49	28	26	20	27	16	22	13	18	7	33	10
HT	13	9	6	5	10	3	17	4	15	4	34	4
LT	45	19	24	16	24	21	36	19	31	9	30	8
60-day	27	15	16	8	21	8	23	2	14	4	30	8
75-day	67	39	49	32	34	26	51	17	41	23	43	18
90-day	24	15	17	14	21	6	29	15	16	8	31	5
Jxol	19	17	19	14	14	8	27	10	23	8	37	7
Pxo99	54	34	51	36	34	19	37	19	34	10	45	13
Rs105	14	12	17	12	12	4	22	6	19	2	31	5
Xv5	13	14	15	7	18	6	27	4	22	3	34	3

would be representative and valid for investigations of unintended effects. If DEG were non-uniformly distributed, then the number and distribution of DEG under different conditions might vary, but considerable DEG could still be detected, if there were, and unintended effects based on DEG were valid. If DEG were extremely distributed, however, the DEG were not representative and invalid for investigations of unintended effect, since extremely huge or nominal number of DEG might be detected under different conditions. So it is crucial to clarify the distribution of DEG before investigating unintended effects and assessing safety of GM plants.

Transgenic rice Huahui 1 (HH1) and its corresponding non-transgenic parent rice Minghui 63 (MH63) were used in this study. HH1 was an insect-resistant rice expressing *BT* fusion protein derived from *Cry1Ab* and *Cry1Ac*. HH1 was created by micro projectile bombardment with two plasmids, pFHBT1 and pGL2RC7, into the elite Chinese cytoplasmic male sterile restorer line, MH63. The plasmid pFHBT1 harbored a hybrid *Cry1Ab/Ac* gene regulated by the rice *actinI* gene promoter and the nopaline synthase (NOS) terminator; plasmid pGL2RC7 carried a Chitinase gene (*RC7*) and a selectable marker gene (*Hph*). The selectable marker gene *Hph* was further removed from the gene of interest by self-segregation [15,16]. Field tests showed that production efficiency of HH1 was increased through resistance against yellow stem borers and leaf folders [17].

In this paper, with the aim to define the distribution pattern of DEG, we performed DNA microarray analysis on HH1 and MH63 at 4 different developmental stages and in 6 different environments (high temperature, low temperature and pathogen inoculations). DEG and significantly changed pathways of HH1 at different developmental stages and environments were analyzed. The results suggested that DEG were non-uniformly distributed in HH1 at different developmental stages and/or environments. Thus DEG detected by comparative transcriptome microarray analysis under certain conditions would be representative for DEG of GM plants under other conditions, and would act as valid clues for further investigation of unintended effects of GM plants.

Results

The insertion of *Cry1Ab/1Ac* did not cause differential expression of genes in insertional positions in HH1

To study whether or not there are DEG near the insertion site in HH1, we performed microarray analysis using HH1 and MH63 at different developmental stages (30-day, 60-day, 75-day and 90-

day) and environmental conditions (temperature and pathogen stress) and investigated the expression level of genes located within 100 kb up-stream and 100 kb down-stream of the insertion site. According to the reported 3'- and 5'-franking sequences of the hybrid *Cry1Ab/1Ac* gene [15], BLAST analysis was performed. The result indicated that the hybrid *Cry1Ab/1Ac* gene was inserted into chromosome 10, between 5378530 and 5378531. There were 8 genes within 100 kb up- and down-stream of the insertion site. The expression levels of all these 8 genes were not obviously changed (with fold change between 0.5 and 2.0).

Number of DEP on each rice chromosome in HH1 at different developmental stages and environments

In order to determine the global distribution pattern of DEG, we performed microarray analysis using HH1 and MH63 at different developmental stages and environmental conditions. Table 1 shows the distribution pattern of differentially expressed probe sets (DEP, with fold change ≥2.0 or ≤0.5) on each chromosome. In each case, the numbers of DEP on each chromosome were different: on chromosome 10, where *Cry1Ab/1Ac* was inserted, there were only a few DEP; on chromosome 12, there were also a small number of DEP; on chromosome 1, 5, 7, 9, etc, there were a large number of DEP. This result indicated that DEP were non-uniformly distributed on chromosome in HH1 at different developmental stages and environments.

Identification of DEG responding to developmental stages and environmental conditions

To determine numbers of DEG at different developmental stages, we performed microarray analysis using HH1 and MH63 at the age of 30-day, 60-day, 75-day and 90-day, respectively. Considerable DEP were detected (Table 1, Table S1). Since some genes were represented by more than one probe, the corresponding numbers of DEG of HH1 at the four developmental stages were 261, 167, 422 and 195, respectively (Table 2). To explore numbers of DEG at different environmental conditions, we treated HH1 and MH63 with high-temperature (HT) and low-temperature (LT) at the age of 30-day and inoculated HH1 and MH63 with pathogens (JxoI, Pxo99, Rs105, Xv5) at the age of 75-day old, respectively, and performed microarray analysis. There were 116 and 271 DEG in HH1 treated with HT and LT, and 194, 372, 148 and 157 DEG in HH1 inoculated with JxoI, Pxo99, Rs105 and Xv5, respectively (Table 2). Furthermore, as shown in volcano plots (Figure 1), there were more down-regulated probes (with fold

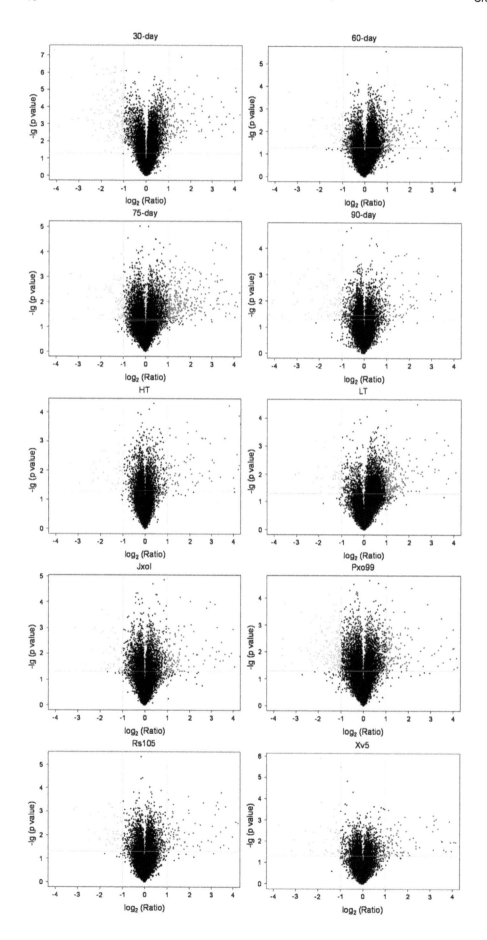

Figure 1. Volcano plots for differentially expressed genes in HH1. Each point represents a gene detected in microarray analysis. Red spots represent differentially expressed genes with fold change ≥2.0; green spots represent differentially expressed genes with fold change ≤0.5. The $\log_2$ (ratio) of expression (HH1/MH63) is shown on the X-axis and the –lg (p-value) is shown on the Y-axis. The vertical lines represent 2-fold change ratio and the horizontal line represents statistical-significance level where p = 0.05. 30-day, 60-day, 75-day, 90-day: HH1 and MH63 at developmental stage of 30-day, 60-day, 75-day, 90-day, respectively; HT: HH1 and MH63 treated with high-temperature at 45°C for 6 hours; LT: HH1 and MH63 treated with low-temperature at 12°C for 6 hours; Jxol, Pxo99: HH1 and MH63 inoculated with *X. oryzae* pv. *oryzae* Jxol and Pxo99 strain; Rs105: HH1 and MH63 inoculated with *X. oryzae* pv. *oryzicola* Rs105 strain; Xv5: HH1 and MH63 inoculated with non-host pathogen *X. compestris* pv. *vesicatoria* Xv5 strain.

change ≤0.5) than up-regulated probes (with fold change ≥2.0) in the group of 30-day, and in all the other groups, there were more up-regulated probes than down-regulated probes, indicating that the number of up-regulated probes and down-regulated probes varied in HH1 at different conditions. These results indicated that considerable differentially expressed DEP and DEG could be detected in HH1 at different stages and environmental conditions.

Common DEP among HH1 at different developmental stages and environmental conditions

In order to clarify whether there were common DEP among HH1 at different developmental stages and environmental conditions, we performed pairwise comparisons between each group of DEP. It turned out that there were considerable common DEP between each group of HH1 (Figure 2, Table S2). The ratio of number of common DEP to number of the smaller group of DEP in the comparison was calculated and represented by different boxes (Figure 2). The numbers of common DEP were not proportional to the numbers of DEP in each group of HH1. Furthermore, numbers of common DEP ranked from 59 (HT and LT) to 149 (Pxo99 and Jxol). These results suggested that common DEP between each group of HH1 were neither uniformly distributed nor extremely distributed; instead, they were non-uniformly distributed and the amount of common DEP in each group of HH1 was adequate for investigating unintended effects.

Significantly changed pathways among HH1 at different developmental stages and environmental conditions

To further explore influences of DEG on HH1, we analyzed changes in pathways of HH1 using Plant MetGenMAP system. A number of significantly changed pathways were selected. Among these significantly changed pathways, 16 pathways were found in all groups of HH1, 8 pathways were found in the majority of groups of HH1 (Table 3), and the other significantly changed pathways dispersed in each group of HH1 (Table S3). This finding indicated that a certain number of common significantly changed pathways were shared among HH1 under different conditions. These changes, with the mere differences at expression level, were everlasting existed in HH1, suggesting that they were probably caused by genetic modification rather than differences in developmental stages and/or environments.

Discussion

GM plants, first planted in 1996, have occupied 148 million hectares cropland in 2010, nearly 10% of all 1.5 billion hectares cropland in the world [18]. Compared to traditional breeding approaches, transgenic approach is direct and breaks the reproductive isolation, with which scientists can transfer any gene-of-interest from any species into chosen crops [19]. Despite the many benefits of the GM crops [18], people are concerned about safety of GM crops and products derived from them. Since random insertion of exogenous specific DNA sequences into plant genome may cause disruption, modification or silencing of active genes and/or activation of silenced genes, resulting in unintended effects [19].

Detecting unintended effects is an important task in safety assessment of GM crops. Traditional methods to detect unintended effects, such as comparing agronomic characters, evaluating environmental adaptability, and analyzing the chemical compositions between GM and non-GM plants [20], are considered as targeted approaches; the limitations of these methods are obvious, especially in the aspects of time and economic consuming and lack of objectivity and impartiality [21,22]. With technical breakthroughs in recent years, DNA microarray has emerged as an indispensable methodology for large-scale and high-throughput analysis of genes in the crops. DNA microarray is a non-targeted approach, and has been proved to be an effective and comprehensive method to detect DEG and investigate unintended effects in GM crops [4–7]. Most of these studies focus on DEG and unintended effects of GM plants at specific developmental stage and/or particular environmental condition and neglect the fact that certain factors, such as developmental stages and environmental factors, may influence distribution pattern of DEG, which further may influence the occurrences of unintended effects. Without detailed evidence on the distribution pattern of DEG of GM plants at different developmental stages and/or environments, it might be questionable to investigate unintended effects using GM plants at specific developmental stages and/or particular environments.

Thus, it is necessary to understand the distribution pattern of DEG in GM plants at different developmental stages and/or environments before investigating unintended effects. As discussed above, there are three possible distribution patterns of DEG: (I) uniform distribution, (II) extreme distribution and (III) non-uniform distribution. DEG detected from microarray analysis were valid for further predicting unintended effects if they were

Table 2. Numbers of DEP and DEG in HH1 compared with MH63 at different developmental stages and environments.

Treatments	Developmental stage				Temperature stress		Pathogen inoculation			
	30 d	60 d	75 d	90 d	HT	LT	Jxol	Pxo99	Rs105	Xv5
DEP Number	271	177	442	203	125	283	207	389	157	168
DEG Number	261	167	422	195	116	271	194	372	148	157

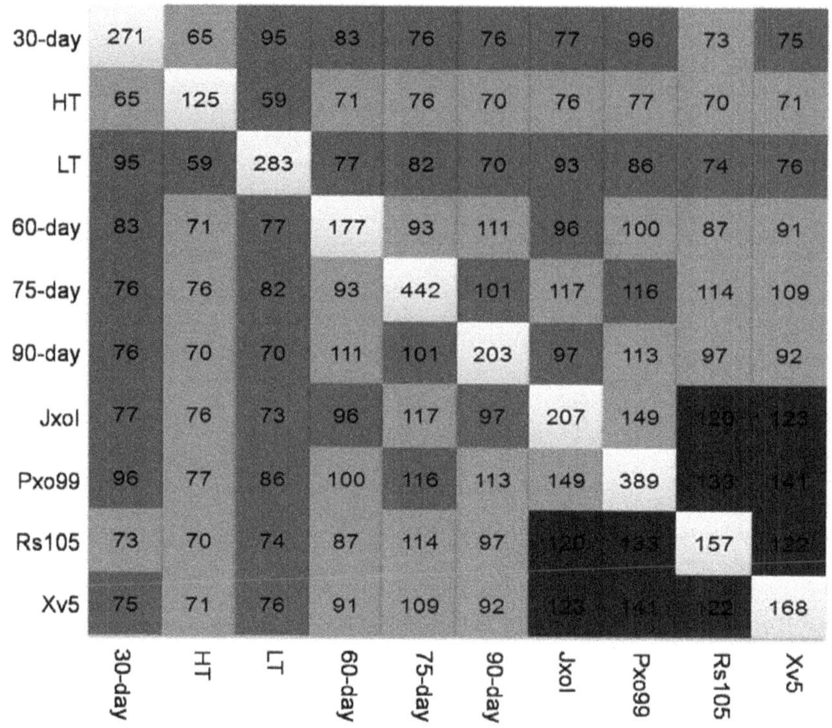

Figure 2. Distribution patterns of common DEP of HH1 in pairwise comparison. Pairwise comparisons were carried out between groups of DEP of HH1 at different developmental stages and environments. Ratio of number of common DEP to number of the smaller group of DEP in the pairwise comparison was calculated and represented by different colors: ■- 25~50%, ■- 50~75%, ■- 75~100%. ■ represents numbers of DEP in HH1. 30-day, 60-day, 75-day, 90-day: HH1 and MH63 at developmental stage of 30-day, 60-day, 75-day, 90-day, respectively; HT: HH1 and MH63 treated with high-temperature at 45°C for 6 hours; LT: HH1 and MH63 treated with low-temperature at 12°C for 6 hours; JxoI, Pxo99: HH1 and MH63 inoculated with *X. oryzae* pv. *oryzae* JxoI and Pxo99 strain; Rs105: HH1 and MH63 inoculated with *X. oryzae* pv. *oryzicola* Rs105 strain; Xv5: HH1 and MH63 inoculated with non-host pathogen *X. compestris* pv. *vesicatoria* Xv5 strain.

uniformly distributed or non-uniformly distributed; if DEG were extremely distributed, however, DEG detected from microarray analysis were invalid for assessment of unintended effects, since DEG were not representative.

In this study, with the purpose of revealing the distribution pattern of DEG in GM plants, we performed microarray analysis with groups of HH1 and MH63 at different developmental stages and environments. No DEG were found near the insertion site (100 kb up- and down-stream of the insertion site), suggesting the transgene event did not cause changes on expression level of intrinsic genes near the insertion site. In each case, the numbers of DEP, detected in microarray analysis, on each chromosome were difference, indicating DEG on each chromosome was non-uniformly distributed. Considerable DEG were found in each group of HH1, and the numbers of DEG varied with changes in developmental stages and/or environments (Table 2). We found that distribution pattern of DEG in HH1 was closest to the non-uniform distribution pattern discussed above, so we conclude that DEG in HH1 was non-uniformly distributed. In addition, we investigate the relationship between DEP and growing conditions (developmental stages and/or environmental conditions in which GM plants are growing) of GM plants. If the number of DEP detected in different cases is relevant to growing conditions, then the numbers of DEP detected in HH1 at growing conditions would be about the same and the numbers of DEP detected in HH1 at different growing conditions would be significantly different. So we carried out pairwise comparisons using DEP detected in each case. The growing conditions of HH1 in our

study could be classified into three types: different developmental stages (30-day, 60-day, 75-day and 90-day), temperature stress (HT and LT) and pathogen stress (JxoI, Pxo99, Rs105 and Xv5). As shown in Figure 2, the numbers of DEP in all the three types of growing conditions were around 200 (except for the case of 75-day and the case of Pxo99). The numbers of DEP in all the three types of growing conditions were about the same and no significant differences in the numbers of DEP were found between HH1 at different types of growing conditions, indicating that DEP had no relevance to growing conditions of HH1. This irrelevance is especially obvious in the case of pathogen stress. JxoI, Pxo99 and Rs105 are pathogenic pathogens that cause diseases on rice plants, and Xv5 is a non-host pathogen that does not cause any diseases on rice plants; the stresses caused by pathogenic pathogens and non-host pathogen are totally different. So the growing conditions in these four cases could be subdivided into two types: pathogenic stress and non-pathogenic stress. The numbers of DEP in HH1 under pathogenic stresses and non-pathogenic stress, however, remained about the same. So we concluded that the number of DEP in HH1 was not relevant to growing conditions. For the same reason, we got the conclusion that the number of common DEP was also not relevant to growing conditions of HH1. Moreover, Figure 2 showed that the number of common DEP in each case was large enough to be valid for assessment of unintended effects and were not proportional to the numbers of DEP. Based on these findings, we concluded that both DEG and common DEG were non-uniformly distributed, and the numbers of DEG and common DEG detected in HH1 had no relevance to growing conditions

Table 3. Common significantly changed pathways of HH1 at different developmental stages and environments.

Pathway name	p value									
	30-day	HT	LT	60-day	75-day	90-day	Jxol	Pxo99	Rs105	Xv5
jasmonic acid biosynthesis	√	√	√	√	√	√	√	√	√	√
enterobactin biosynthesis	√	√	√	√	√	√	√	√	√	√
sucrose degradation to ethanol and lactate (anaerobic)	√	√	√	√	√	√	√	√	√	√
oxidative ethanol degradation I	√	√	√	√	√	√	√	√	√	√
phenylalanine degradation III	√	√	√	√	√	√	√	√	√	√
methionine degradation III	√	√	√	√	√	√	√	√	√	√
ethanol fermentation to acetate	√	√	√	√	√	√	√	√	√	√
valine degradation II	√	√	√	√	√	√	√	√	√	√
tetrapyrrole biosynthesis I	√	√	√	√	√	√	√	√	√	√
leucine degradation III	√	√	√	√	√	√	√	√	√	√
mixed acid fermentation	√	√	√	√	√	√	√	√	√	√
isoleucine degradation II	√	√	√	√	√	√	√	√	√	√
cytokinins 7-N-glucoside biosynthesis	√	√	√	√	√	√	√	√	√	√
cytokinins 9-N-glucoside biosynthesis	√	√	√	√	√	√	√	√	√	√
betanidin degradation	√	√	√	√	√	√	√	√	√	√
cytokinins-O-glucoside biosynthesis	√	√	√	√	√	√	√	√	√	√
aerobic respiration – electron donor II	√	√	√	√	X	√	√	√	√	√
photorespiration	√	X	√	√	√	√	√	√	√	√
NAD salvage pathway II	√	√	√	√	√	X	√	√	X	√
brassinosteroid biosynthesis II	√	√	√	√	√	√	X	√	X	√
medicarpin biosynthesis	√	X	X	√	√	√	X	√	X	√
maackiain biosynthesis	√	X	X	√	√	√	X	√	X	√
cellulose biosynthesis	X	√	X	X	√	√	√	√	√	X
starch degradation	X	X	X	X	√	√	√	√	√	√

30-day, 60-day, 75-day, 90-day: HH1 and MH63 at developmental stage of 30-day, 60-day, 75-day, 90-day, respectively; HT: HH1 and MH63 treated with high-temperature at 45°C for 6 hours; LT: HH1 and MH63 treated with low-temperature at 12°C for 6 hours; Jxol, Pxo99: HH1 and MH63 inoculated with *X. oryzae* pv. *oryzae* Jxol and Pxo99 strain; Rs105: HH1 and MH63 inoculated with *X. oryzae* pv. *oryzicola* Rs105 strain; Xv5: HH1 and MH63 inoculated with non-host pathogen *X. compestris* pv. *vesicatoria* Xv5 strain. √: detected; X: not found (For detailed information, please refer to Table S3).

and the numbers of DEG and common DEG detected in HH1 at specific conditions were large enough to be representative and valid for investigating unintended effects.

Furthermore, we analyzed changes in expression level of pathways of HH1, and selected a number of significantly changed pathways. Among these significantly changed pathways, 16 pathways were found in all groups of HH1 (Table 3), 8 pathways were found in the majority of groups of HH1 (Table 3). These common DEG and significantly changed pathways in HH1 were probably to be caused by insertion of exogenous DNA fragment and had nothing to do with other factors, such as developmental stages and/or environmental factors. Among these common significantly changed pathways, jasmonic acid biosynthesis [23], medicarpin biosynthesis [24], and maackiain biosynthesis [24] were associated with response to biotic and abiotic stress. These changes were possibly intended effects of HH1, since HH1 were genetically modified to be resistant to pest insects. However, five common significantly changed pathways (Table 3), phenylalanine degradation III, methionine degradation III, valine degradation II, leucine degradation III, isoleucine degradation II, were associated with amino acid degradation. So it was necessary to carry out further research to determine whether these changes were intended effects or unintended effects.

Our finding provided evidences on the non-uniform distribution pattern of DEG in GM plants. So we could use DEG, especially common DEG and common significantly changed pathways, as a clue to investigate unintended effects of GM plants in future safety assessment of GM plants. However, DEG do not always mean unintended effects, since some DEG are directly associated with the transgenes introduced or with the desired new characteristics of GM plants. Further works should focus on distinguishing whether these DEG are associated with intended effects or unintended effects.

Materials and Methods

Plant Materials

Transgenic rice line Huahui 1 (HH1) and its corresponding non-transgenic line Minghui 63 (MH63) were used for microarray analysis. HH1 was genetically engineered to be insect-resistant through expressing fused insect-resistant gene of *Cry1Ac/Cry1Ab* by Huazhong Agricultural University, and obtained the first security certificate for genetically modified rice in China from Hubei Province in 2009 [25].

Rice sample preparation

Rice seeds were surface-disinfected and then soaked in distilled sterile water for germination at 28°C for 2 days. Rice seedlings were grown in pots fertilized with half-strength of basal macro- and micro-salt nutrition components of Murashige and Skoog medium [26] in controlled climate chambers at 16-h-light (30°C)/ 8-h-dark (26°C) cycle. At the age of 30-day, 60-day, 75-day and 90-day old, seedling samples were collected, frozen in liquid nitrogen and kept at −80°C. Seedlings, at the age of 30-day old, were treated with high-temperature (45°C) and low-temperature (12°C) respectively at climate chambers, and samples were collected 6 hours after treatment. Seedlings were inoculated respectively with compatibility pathogen *Xanthomonas* oryzae pv. *oryzae* (rice leaf blight disease pathogen) JxoI and Pxo99 strains, *Xanthomonas* oryzae pv. *oryzicola* (rice leaf streak disease pathogen) Rs105 strain and rice non-host pathogen of *Xanthomonas compestris* pv. *vesicatoria* Xv5 strain at the age of 75-day old according to leaf rubbing inoculation method [27], and samples were collected 2 days after inoculation. All samples were kept at −80°C until RNA extraction. Each treatment was performed with three replicates, and more than 20 whole seedlings were collected for each sample.

RNA extraction and microarray hybridization

Total RNA was extracted from rice samples according to the manufacturer's instructions of TRIzol reagent (Invitrogen Life Technologies, Carlsbad, CA, USA). The integrity of extracted RNA was checked and then sent to CapitalBio Corporation (an Affymetrix platform service facility at Beijing) for further quality and quantity examination and microarray hybridization. 1 μg of RNA samples was used for hybridization with Affymetrix GeneChip® Rice Genome Arrays according to the manufacturer's instructions. The array was designed mainly based on the annotation of TIGR version 2.0 and contained 55, 515 probe sets to query 48,564 transcripts of rice japonica subspecies and 1,260 transcripts of rice *indica* subspecies. Microarray hybridization was performed at 45°C with rotation lasting for 16 h using an Affymetrix GeneChip Hybridization Oven 640. Following hybridization, the arrays were washed and stained at Affymetrix GeneChip Fluidics Station 450 and then scanned with Affymetrix GeneChip® Scanner 3000 7G.

Data analysis

The scanned images were analyzed with Affymetrix GeneChip® Command Console™ (AGCC) software. The expression flags (indicators of expressed genes) were determined using the Affymetrix® Expression Console™ software application MAS 5.0 algorithm as "present", "marginal" and "absent" calls. Then normalization and expression analysis were performed with .CEL files and .mas5.CHP files by DNA-chip analyzer (dChip). All these data were deposited in NCBI GEO database with accession number GES33204. DEP and their corresponding DEG in HH1 were selected using the Significance Analysis of Microarrays (SAM version 3.02) software by two class unpaired method with q value ≤5% and fold change ≥2.0 or ≤0.5 when compared with control samples (MH63).

Analysis of significantly changed pathways

Significantly changed pathways of HH1, in comparison with MH63, were analyzed by the Plant MetGenMAP system [28]. All changed pathways were selected by the raw p value with the threshold 0.05. Significantly changed pathways were selected by the FDR (False Discovery Rate) corrected p value with threshold 0.05 [29].

Supporting Information

Table S1 Differentially expressed probe sets in HH1. Differentially expressed probe sets (DEP) in HH1 were selected using the Significance Analysis of Microarrays (SAM version 3.02) software by two class unpaired method with q value ≤5% and fold change ≥2.0 or ≤0.5 when compared with MH63. DEP with fold change ≥2.0 were represented in green color, and DEP with fold change ≤0.5 were represented in red color. 30-day, 60-day, 75-day, 90-day: HH1 and MH63 at developmental stage of 30-day, 60-day, 75-day, 90-day, respectively; HT: HH1 and MH63 treated with high-temperature at 45°C for 6 hours; LT: HH1 and MH63 treated with low-temperature at 12°C for 6 hours; JxoI, Pxo99: HH1 and MH63 inoculated with X. oryzae pv. oryzae JxoI and Pxo99 strain; Rs105: HH1 and MH63 inoculated with X. oryzae pv. oryzicola Rs105 strain; Xv5: HH1 and MH63 inoculated with non-host pathogen X. compestris pv. vesicatoria Xv5 strain.

Table S2 Commom Differentially expressed probe sets (DEP) among HH1 at different developmental stages and enviromental conditions. Differentially expressed probe sets (DEP) in HH1 were selected using the Significance Analysis of Microarrays (SAM version 3.02). Commen DEP Counted reprecented the number of DEP among HH1 at different developmental stages and enviromental conditions. "1" represent the absense of DEP. 30-day, 60-day, 75-day, 90-day: HH1 and MH63 at developmental stage of 30-day, 60-day, 75-day, 90-day, respectively; HT: HH1 and MH63 treated with high-temperature at 45°C for 6 hours; LT: HH1 and MH63 treated with low-temperature at 12°C for 6 hours; JxoI, Pxo99: HH1 and MH63 inoculated with X. oryzae pv. oryzae JxoI and Pxo99 strain; Rs105: HH1 and MH63 inoculated with X. oryzae pv. oryzicola Rs105 strain; Xv5: HH1 and MH63 inoculated with non-host pathogen X. compestris pv. vesicatoria Xv5 strain.

Table S3 Significantly changed pathways in HH1 at different developmental stages and environmental conditions. 30-day, 60-day, 75-day, 90-day: HH1 and MH63 at developmental stage of 30-day, 60-day, 75-day, 90-day, respectively; HT: HH1 and MH63 treated with high-temperature at 45°C for 6 hours; LT: HH1 and MH63 treated with low-temperature at 12°C for 6 hours; JxoI, Pxo99: HH1 and MH63 inoculated with X. oryzae pv. oryzae JxoI and Pxo99 strain; Rs105: HH1 and MH63 inoculated with X. oryzae pv. oryzicola Rs105 strain; Xv5: HH1 and MH63 inoculated with non-host pathogen X. compestris pv. vesicatoria Xv5 strain. ××: not found.

Acknowledgments

We are grateful to Professor Lin Yong-Jun from Huazhong Agricultural University for kindly providing seeds of HH1 and MH63.

Author Contributions

Conceived and designed the experiments: FJQ YFP YHL. Performed the experiments: ZL JZ FJQ. Analyzed the data: ZL WWZ GLJ. Contributed reagents/materials/analysis tools: YHL GLJ. Wrote the paper: ZL FJQ.

References

1. Brooks G, Barfoot P (2011) GM crops: global socio-economic and environmental impacts 1996–2009. PG Economics Ltd, UK.

2. Baudo MM, Lyons R, Powers S, Pastori GM, Edwards KJ, et al. (2006) Transgenesis has less impact on the transcriptome of wheat grain than conventional breeding. Plant Biotechnol J 4: 369–380.

3. Deng PJ, Zhou XY, Yang DY, Hou HL, Yang XK, et al. (2008) The definition, source, manifestation and assessment of unintended effects in genetically modified plants. J Sci Food Agr 88: 2401–2413.

4. Gregersen PL, Brinch-Pedersen H, Holm PB (2005) A microarray-based comparative analysis of gene expression profiles during grain development in transgenic and wild type wheat. Transgenic Res 14: 887–905.

5. Ouakfaoui SE, Miki B (2004) The stability of the *Arabidopsis* transcriptome in transgenic plants expressing the marker genes *nptII* and *uidA*. Plant J 41: 791–800.

6. Abdeen A, Miki B (2009) The pleiotropic effects of the bar gene and glufosinate on the *Arabidopsis* transcriptome. Plant Biotechnol J 7: 266–282.

7. Abdeen A, Schnell J, Miki B (2010) Transcriptome analysis reveals absence of unintended effects in drought-tolerant transgenic plants overexpressing the transcription factor *ABF3*. BMC Genomics 11: 69.

8. Montero M, Coll A, Nadal A, Messeguer J, Pla M (2011) Only half the transcriptomic differences between resistant genetically modified and conventional rice are associated with the transgene. Plant Biotechnol 9: 693–702.

9. Dubouzet JG, Ishihara A, Matsuda F, Miyagawa H, Iwata H, et al. (2007) Integrated metabolomic and transcriptomic analyses of high-tryptophan rice expressing a mutant anthranilate synthase alpha subunit. J Exp Bot 58: 3309–3321.

10. Baudo MM, Powers SJ, Mitchell RA, Shewry PR (2009) Establishing substantial equivalence: transcriptomics. Methods Mol Biol 478: 247–272.

11. Coll A, Nadal A, Palaudelmas M, Messeguer J, Mele E, et al. (2008) Lack of repeatable differential expression patterns between MON810 and comparable commercial varieties of maize. Plant Mol Biol 68: 105–117.

12. Coll A, Nadal A, Collado R, Capellades G, Kubista M, et al. (2010) Natural variation explains most transcriptomic changes among maize plants of MON810 and comparable non-GM varieties subjected to two N-fertilization farming practices. Plant Mol Biol 7: 349–362.

13. Barros E, Lezar S, Anttonen MJ, Van Dijk JP, Rohlig RM, et al. (2010) Comparison of two GM maize varieties with a near-isogenic non-GM variety using transcriptomics, proteomics and metabolomics. Plant Biotechnol J 8: 436–451.

14. Kogel KH, Voll LM, Schäfer P, Jansen C, Wu YC, et al. (2010) Transcriptome and metabolome profiling of field-grown transgenic barley lack induced differences but show cultivar-specific variances. Proc Natl Acad Sci USA 107: 6198–6203.

15. Tu GM, Zhang QF, Huang HQ, Pan G, He YQ, et al. (2006) Transgenic rice culture method. CN 1840655A.

16. Tu JM, Datta K, Oliva N, Zhang GA, Xu CG, et al. (2003) Site-independently integrated transgenes in the elite restorer rice line Minghui 63 allow removal of a selectable marker from the gene of interest by self-segregation. Plant Biotechnol J 1: 155–165.

17. Tu JM, Zhang GA, Datta K, Xu CG, He YQ, et al. (2000) Field performance of transgenic elite commercial hybrid rice expressing *Bacillus thuringiensis* δ-endotoxin. Nat Biotechnol 18: 1101–1104.

18. James C (2010) Global Status of Commercialized Biotech/GM Crops: 2010. ISAAA: Ithaca, NY. ISBN: 978-1-892456-49-4.

19. Ren YF, Lv J, Wang H, Li LC, Peng YF, et al. (2009) A comparative proteomics approach to detect unintended effects in transgenic *Arabidopsis*. J genet genomics. 36: 629–639.

20. Cellini F, Chesson A, Colquhoun I, Constable A, Davies HV, et al. (2004) Unintended effects and their detection in genetically modified crops. Food Chem Toxicol 42: 1089–1125.

21. FAO/WHO (2009) Food derived from modern biotechnology, second edition. ISBN 978-92-5-105914-2.

22. Kuiper HA, Kok EJ, Engel KH (2003) Exploitation of molecular profiling techniques for GM food safety assessment. Curr Opin Biotech 14: 238–243.

23. Creelman RA, Mullet EJ (1995) Jasmonic acid distribution and action in plants: Regulation during development and response to biotic and abiotic stress. Proc Natl Acad Sci USA 92: 4114–4119.

24. VanEtten HD, Matthews PS, Mercer EH (1983) (+)-Maackiain and (+)-medicarpin as phytoalexins in *Sophora Japonica* and identification of the (−) isomers by biotransformation. Phytochemistry 22: 2291–2295.

25. China MoAotPsRo The second list of approved security certificates of agricultural gmos (http://www.stee.agri.gov.cn/biosafety/spxx/).

26. Murashige T, Skoog F (1962) A revised medium for rapid growth and bioassays with tobacco tissue cultures. Physiol Plant 15: 473–497.

27. Vera Cruz CM, Gosselé F, Kersters K, Segers P, Van Den Mooter M, et al. (1984) Differentiation between *Xanthomonas campestris* pv. *oryzae*, *Xanthomonas campestris* pv. *oryzicola* and the Bacterial 'Brown Blotch' Pathogen on Rice by Numerical Analysis of Phenotypic Features and Protein Gel Electrophoregrams. Microbiology 130: 2983–2999.

28. Joung JG, Corbett AM, Fellman SM, Tieman DM, Klee HJ, et al. (2009) Plant MetGenMAP: An integrative analysis system for plant systems biology. Plant Physiol 151: 1758–1768.

29. Benjamini Y, Hochberg Y (1995) Controlling the false discovery rate-a practical and powerful approach to multiple testing. J R Stat Soc B: Methodol 57: 289–300.

3

Estimation of Agricultural Water Consumption from Meteorological and Yield Data: A Case Study of Hebei, North China

Zaijian Yuan[1,2], Yanjun Shen[2]*

1 School of Economics & Management, Hebei University of Science and Technology, Shijiazhuang, China, 2 Center for Agricultural Resources Research, Institute of Genetics and Developmental Biology, Chinese Academy of Sciences, Shijiazhuang, China

Abstract

Over-exploitation of groundwater resources for irrigated grain production in Hebei province threatens national grain food security. The objective of this study was to quantify agricultural water consumption (AWC) and irrigation water consumption in this region. A methodology to estimate AWC was developed based on Penman-Monteith method using meteorological station data (1984–2008) and existing actual ET (2002–2008) data which estimated from MODIS satellite data through a remote sensing ET model. The validation of the model using the experimental plots (50 m^2) data observed from the Luancheng Agro-ecosystem Experimental Station, Chinese Academy of Sciences, showed the average deviation of the model was −3.7% for non-rainfed plots. The total AWC and irrigation water (mainly groundwater) consumption for Hebei province from 1984–2008 were then estimated as 864 km^3 and 139 km^3, respectively. In addition, we found the AWC has significantly increased during the past 25 years except for a few counties located in mountainous regions. Estimations of net groundwater consumption for grain food production within the plain area of Hebei province in the past 25 years accounted for 113 km^3 which could cause average groundwater decrease of 7.4 m over the plain. The integration of meteorological and satellite data allows us to extend estimation of actual ET beyond the record available from satellite data, and the approach could be applicable in other regions globally where similar data are available.

Editor: Zhi Zhou, National University of Singapore, Singapore

Funding: The Natural Science Foundation of China (NSFC) and Chinese Academy of Sciences (CAS) supported this work through grants 40901130, 40871021, and KSCX2-EW-J-5. The funders had no role in study design, data collection and analysis, decision to publish, or preparation of the manuscript.

Competing Interests: The authors have declared that no competing interests exist.

* E-mail: yjshen@sjziam.ac.cn

Introduction

The most critical resource for agroecosystems in China is water [1]. The total annual water resources available in China are around 2,800 km^3. With a population of 1.3×10^9, the available water per capita is only 2,100 m^3/y. Thus, China is a nation with high water scarcity compared to a global average of 6,466 m^3/y [2]. Water resources in the northern parts of China account for less than 20% of the national total, whereas arable land accounts for 65% of the total [3], and the grain production in the North has exceeded to 50% since 2005. As 80% of China's food is produced on irrigated farmland, irrigation water plays an important role in feeding the large population [4,5]. The North China Plain (NCP) is one mostly important granary of China. It has 140,000 km^2 of arable land and produces about 20% of the nation's grain food.

The natural rainfall cannot meet crop water requirements in NCP, supplementary irrigation is therefore widely applied to increase yields and to secure the food supply for the nation [3]. However, excessive use of diverted river ows and groundwater has caused severe environmental problems. For example, since 1972 the lower reaches of the Yellow River has frequently dried up during the dry seasons for several years. During the droughts of 1997 it didn't reach the sea for even 228 days. However, it must be mentioned that since the beginning of the 2000s, after a river basin

management plan approach was adopted in Yellow River Basin, no drying up has occurred so far [2].

On the other hand, in most places of the NCP, such as Hebei province, groundwater is the primary source of water for irrigation. Grain production in Hebei province totaled 2.9×10^{10} kg in 2008, accounting for 5.5% of the country's total, while the production of wheat and corn shared for 10.9% and 8.7% of the national total, respectively. Due to continually over-pumping, groundwater resource has been greatly depleted and facing to great challenges in sustainability. The water table at the piedmont plain for example has declined rapidly from ~10 m below ground surface in the 1970s to ~30 m in 2001 [6], and to ~40 m in 2010 [7].

It is extremely important for a sustainable agricultural water management to explicitly estimate the groundwater consumption for agriculture in recent decades in NCP. FAO Penman-Monteith equation combined with crop coefficient was widely used for estimation crop water requirement over the world. For the NCP region, Liu et al. [8] calculated the crop water requirement for winter wheat and summer maize in North China in the past 50 years and found a widely decreasing trend of $-0.9 \sim -19.2$ mm per decade for wheat and $-8.3 \sim -24.3$ mm per decade for maize, respectively. Li et al. [9] successfully estimated the water consumption and crop water productivity of winter wheat in NCP using remote sensing for a growing season in 2003–2004. Their calculation suggested the average water consumption (i.e. ET) by winter wheat in 83 counties was 424 mm, which was 118 mm higher than the precipitation. Yang et al. [10] estimated that the crop water requirement for five major crops (wheat, maize, cotton, fruit trees, vegetables) in NCP using crop models DSSAT and COTTON2K, and found wheat accounted for over 40% of total irrigation water requirement in the plain, while maize and cotton together accounted for 24% of the total irrigation water requirement. They also estimated that the annual averaged irrigation requirement for grain crops was 6.16 km^3 during the period of 1986–2006. This estimation is of great importance to make regional water resources planning. Though the crop model with careful calibrations can provide relatively accurate estimation of crop water consumption, the difficulties in collecting huge amount of information on soil profiles and crops biology and phenology together with the complicated parameterization restrict the wide application of crop model to regional water resources management, especially for the regions with limited data. Moreover, even in some developed countries, actual evapotranspiration (ET) has been observed only in recent 1–2 decades, mostly at field scale. Simple methods to estimate agricultural water consumption (AWC) at larger spatial and temporal scales are urgently needed for water resources assessment and planning. In the present study, we attempt to propose a simple method to calculate long-term regional AWC by using limited meteorological and census data.

Therefore, the main objectives of the present study are to estimate 1) the AWC changes over past decades in Hebei province; and 2) irrigation water consumption for agriculture and related groundwater depletion. The results from this study will provide critical information for the future development of sustainable agricultural water resources management practices for local governments.

Materials and Methods

Hebei province (36°05′N-42°40′N, 113°27′E-119°50′E, Figure 1) is 190,000 km^2 in area with a population of 69 million (2009), and is divided into 11 prefectures (including 138 counties). The topography consists of mountains, hills, and plateaus in the north and west part, and a broad plain in the central and southeastern region. 34% of the provincial land area is cultivated with grain crops such as wheat, maize, rice, soybean, potato and millet, and among them the yield of winter wheat and summer maize account for 85% of the total grain yield (winter wheat is cultivated from early October to early June, summer maize grows from mid-June to late September). In plain area, most arable lands are irrigated except for the eastern part where the saline shallow groundwater restrains the irrigation but irrigation increased gradually in recent 3 decades due to technology evolution.

The study area is located in a temperate and continental monsoon climate zone with a mean annual precipitation of 500 mm (1984–2008), 70% of which occurs between June and September. Mean annual temperature is 10°C (1984–2008). Precipitation and temperature decrease from southeast to northwest.

Data

The meteorological data for 1984–2008, including daily average temperature, relative humidity, precipitation, sunshine duration, atmospheric pressure, vapor pressure, wind speed, were obtained from 55 national weather stations (Figure 1). The economic statistics data for each county from 1984 to 2008, including grain yield, sowing area and effective irrigation area, were obtained from Hebei economic statistical yearbooks. The meteorological data were used to calculate reference evapotranspiration, and economic data were employed to estimate the actual evapotranspiration.

An independent remote sensing ET dataset was employed to analyze the relationship between grain yield and ET and to calibrate a key parameter, i.e. K_f, of the model we proposed. The remote sensing ET data were produced based on moderate-resolution imaging spectroradiometer (MODIS) data by combining meteorological records and an scheme called ETWatch. There are 7 years (2002–2008) ET data available for Hebei province with a 1 km spatial resolution. Wu et al. [11] presented the details of the algorithm of ETWatch and its validation.

Validation data are collected from five years (2007–2011) field experiments on irrigation and water productivity at Luancheng Agro-ecosystem Experimental Station (35°53′N, 114°41′E), the Chinese Academy of Sciences, which is located at the piedmont, with an elevation of 50 m above sea level. The experiments have been conducted in 16 water balance plots with an area of 50 m^2 each. Irrigation was applied as five treatments to control the soil moisture at different levels (see Sun et al. [12] for details). The data of annual irrigation amount, annual total yield of the double crops wheat and maize, actual ET calculated from soil water balance for each treatment were collected as well as the daily meteorological data and groundwater depth monitoring data. The meteorological data was used to calculate the reference ET at this station, other data were employed to validate and evaluate the model's applicability.

Reference Evapotranspiration

Reference evapotranspiration was estimated through FAO56-PM model [13],

$$ET_0 = \frac{0.408\Delta(R_n - G) + \gamma\frac{900}{T+273}u_2(e_s - e_a)}{\Delta + \gamma(1 + 0.34u_2)} \tag{1}$$

where ET_0 is reference evapotranspiration (mm d^{-1}) and annual ET_0 was accumulated from daily ET_0; R_n is the net radiation at the crop surface (MJ m^{-2}d^{-1}); G is the soil heat flux density (MJ m^{-2}d^{-1}); T is daily average temperature (°C); u_2 is the wind speed at 2 m height (m s^{-1}); e_s is the vapor pressure of the air at saturation (kPa); e_a is the actual vapour pressure (kPa); Δ is the slope of the vapor pressure curve (kPa °C^{-1}) and γ is the psychrometric constant (kPa °C^{-1}). A complete set of equations is proposed by Allen et al. [13] to compute the parameters of Eq. (1) according to the available weather data and the time step computation, which constitute the so-called FAO-PM method. G can be ignored for daily time step computation. Using Eq. (1), we firstly calculated ET_0 for the 55 weather stations based on conventional meteorological observation data, and then estimated ET_0 for 138 counties through Kriging interpolation.

Actual evapotranspiration (ET) of croplands. Actual evapotranspiration of croplands was calculated by using the following equation,

$$ET = K_c \times K_f \times ET_0 \tag{2}$$

where ET is actual evapotranspiration (mm); K_c is crop coefficient; K_f is soil moisture correction coefficient. The crop coefficient is largely dependent on crop varieties and planting patterns such as sowing density, fertilizer management, etc. So it varies largely in space and time and difficult to be collected, especially for the past, because information on grain varieties and growing observation

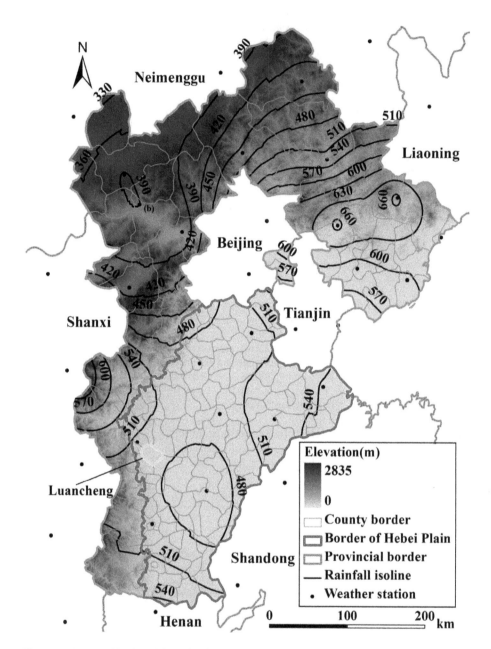

Figure 1. Geographical position of Hebei province. The contour lines and the points indicate average precipitation (1984–2008) and locations of weather stations used in this study, respectively.

data are not available. Alternatively, we assume that the temporal change of crop coefficient can be reflected by the grain yield coefficient (GY_c) without distinguishing crop species in this study.

The GY_c is based on our analysis of the relationship between grain yield (GY) and water consumption, i.e. ET, in 121 counties of the all 138 counties by using the statistical yield for 2002–2008 and independent source of remote sensing derived ET data (thereafter, ET_{rs}) for the same period. There are 17 counties, where the cultivated croplands mostly grow cotton and the grain croplands shares little to their total cultivated land, were removed from the correlation analysis of observed GY and ET_{rs} at county level. Figure 2 illustrated that the annual ET from the remote sensing ET products is significantly linearly correlated to the grain yield.

The grain yield coefficient GY_c is calculated as follows,

$$GY_{ci} = \frac{GY_i}{\overline{GY}} = \frac{GY_i}{\sum\limits_{i=1984}^{2008} GY_i/25} \tag{3}$$

where GY_{ci} is grain yield coefficient of year i, GY_i is grain yield of year i (t/ha), and $\overline{GY}$ is the mean yield from 1984 to 2008 (t/ha). Therefore, Eq. (2) can be modified as,

$$ET = GY_c \times K_f \times ET_0 \tag{4}$$

Then, the soil moisture correction coefficient K_f can be expressed

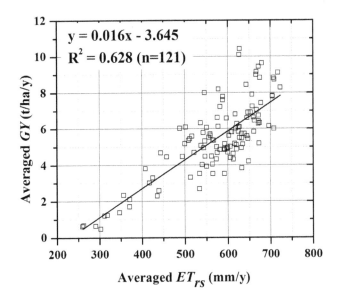

Figure 2. Relationship between remote sensing derived annual ET (ET_{rs}) and grain yield (GY) from 2002 to 2008 of 121 major grain production counties.

as below,

$$K_f = ET/(GY_c \times ET_0) \qquad (5)$$

Determination of K_f

The annual K_f of farmland was calculated using Eq. (5) on the basis of annual remote sensing ET_{rs} products (2002–2008) for each county, combined with annual ET_0 calculated for the same period. In this calculation, we used the areal weighted ET_{rs} from grain crop lands for each county since the grain yield coefficient GY_c only presents the water consumption and productivity from grain lands. Then we got the K_f parameter for all the counties during the period 2002–2008.

There are no long-term soil moisture data available for the region, but it should reflect the annual precipitation and irrigation. So, we analyzed the correlations of annual K_f with annual precipitation (P) and annual irrigation rate (Irr_{rate}) of the 121 counties during the 7 years (totally, 847 samples), and resulted in an empirical equations below,

$$K_f = 0.00033P + 0.2754\,Irr_{rate} + 0.0818$$
$$(R^2 = 0.54, n = 847, F = 489.4) \qquad (6)$$

where P is annual precipitation for each county (mm), Irr_{rate} is annual irrigation rate for each county which was defined as the effective irrigation area (EIA, km^2) of a county to its total cultivated area (A, km^2),

$$Irr_{rate} = EIA/A \qquad (7)$$

With assumption of the regression coefficients in Eq. (6) keep stationary during the study period, we can get K_f for each year in

each county by using Eq. (6) from the mean annual precipitation and irrigation rates from 1984 through 2008.

Agricultural Water Consumption and Irrigation Water Consumption

Agricultural water consumption (WC_{ag}, km^3) for each county was estimated as follows,

$$WC_{ag} = ET \times A/1000000 \qquad (8)$$

According to the water balance equation, total ET for the study period also can be estimated by the following formula,

$$ET = P + Irr_n + (SM_0 - SM_1) - (R_o - R_i) \qquad (9)$$

where Irr_n is effective irrigation water (mm); SM_0 is initial and SM_1 is final soil moisture (mm), when water balance for a relatively long period were calculated, $SM_0 - SM_1 \approx 0$; R_o is outflow runoff (mm), R_i is inflow runoff (mm) of croplands. In counties on the plain R_i is basically equal to R_o, and in mountainous counties, we estimate the difference between annual R_o and R_i using the method proposed by Ji et al. [14].

$$R_o - R_i = 15.782\,e^{0.0035P} \qquad (10)$$

According to the above analysis, the annual agricultural irrigation water of a county on plain area can be estimated as follows,

$$Irr_n = ET - P \qquad (11)$$

while for the counties in mountainous region, it can be expressed as,

$$Irr_n = ET + 15.782\,e^{0.0035P} - P \qquad (12)$$

The total irrigation water consumption Irr_{nc} (km^3), or net groundwater mining for each county can be estimated as follows,

$$Irr_{nc} = Irr_n \times EIA/1000000 \qquad (13)$$

So, the annual decline rate of the groundwater table affected by agricultural irrigation can be simply estimated through,

$$D_g = (Irr_{nc} \times 1000/LA)/P_e \qquad (14)$$

where D_g is the decline depth of groundwater (m); LA is land area (km^2); P_e is effective porosity, which ranges from 10 to 30% in the piedmont area, 5 to 20% in the middle alluvial plain, and 5 to 7% in the coastal plain, respectively [15]. In this study, uniform effective porosity of 25% was used across the plain area of the province.

Validation and Sensitivity Analysis

The 5 years experimental data from Luancheng Agro-ecosystem Experimental Station as introduced earlier were employed to validate and evaluate the model's performance. We assumed the five different irrigation treatments for the five years as different irrigation rate. Firstly, according to the different irrigation levels, such as rainfed, fully irrigation, 80% irrigation, 75% irrigation and 70% irrigation, we set the irrigation rates (Irr_{rate}) of the 5 treatments as 0, 1.0, 0.8, 0.75 and 0.7, respectively. Then, the key parameters of GY_c and K_f were calculated according to Eq. (3) and (6). GY_c ranged from 0.44 to 1.30 and K_f from 0.20 to 0.55, respectively. Finally, we applied all the yield data, P, and ET_0 to the model and calculated the ET for different treatments in each year.

The comparison of calculated ET with field observed ET through soil water balance demonstrated a quite good consistency (Figure 3) except for 3 rainfed treatments in dry years. The relative bias for the 22 samples is only -3.7% and RMSE is 78.9 mm. But for the rainfed treatment in dry years, without supplementary irrigation the grain yield will be largely dependent the occurrence of rainfall on by both amount and timing, which induces uncertainty of the grain yield response to rainfall. In our studies, the main purpose is to give a good projection of the groundwater depletion because of irrigation pumping in past decades. We judge that the bias happened in rainfed cropland will have minor effects on this objective.

In order to evaluate the effectiveness of parameter K_f, we conducted a sensitivity analysis of the estimated ET to the key variables in Eq. 6, i.e. precipitation (P) and irrigation rate (Irr_{rate}). Figure 4 illustrated the responses of ET change to changes in P under different irrigation rates (Figure 4a) and to changes in irrigation rate under different annual precipitations (Figure 4b). $\Delta ET/\Delta P$ varies from 0.32–0.67 when irrigation rate varies from 100% to zero (Figure 4a); the dependence of ET on P decreases as irrigation rate increases. While, the dependence of ET on irrigation rate shows a smaller range, $\Delta ET/\Delta Irr_{rate}$ varies from 0.35–0.53 when annual precipitation decreases from 700 mm to 200 mm. So, the soil wetness parameter K_f is sensitive enough to annual P and irrigation rate, and the model can reflect good responses of estimated ET to precipitation and irrigation at an annual base.

Results

ET_0, ET and AWC

Mean annual ET_0 (1984–2008) ranged from 1,294 mm to 1,365 mm, decreasing gradually from southeast to northwest and showing a similar spatial pattern to air temperature (Figure 5a). And mean annual ET of croplands in each county ranged from 286 to 674 mm (Figure 5b). ET has significantly increased during the past 25 years except for a few counties located in mountainous regions. Increasing ET from croplands is attributed mainly to intensified agricultural activity, such as changes in sowing density, irrigation rate, etc., especially in the plain areas, and partly to increasing temperature (Figure 5c). Decreased ET was detected in some mountainous counties as was shown in Figure 5c, this phenomenon may reflect the effects of the state policy so-called 'Grain to Green', which was launched in the end of 1990s to prevent the land desertification and sand storm through returning cropland to forest or grassland. Mean annual agricultural water consumption (AWC) for the counties ranged from 50 million m^3 to 550 million m^3 in Hebei province during the period from 1984 to 2008. The total water consumption for agricultural grain production was estimated as much as 864 km^3 in Hebei province during the 25 years.

Net consumption of Irrigation Water

Mean annual net irrigation water (mainly groundwater, Irr_n) for each county ranged from 16 to 214 mm (Figure 6a) during the study period, in other words, the groundwater table changes would response to these numbers. The counties at the southeast part of the low plain region showed large increase in irrigation water consumption during the period of 1999–2008 compared with that in 1984–1993 (Figure 6b). That region used to be saline soil and shallow groundwater. The grain productivity has increase greatly during past 3 decades due to the efforts in drainage system construction and irrigation technology evolution. The total net groundwater consumption for irrigation (Irr_{nc}), calculated through Eq. (11 & 12) for the plain area during the study period was projected as 113 km^3 with the mean annual value for each county ranging from 2.7 million m^3 to 140 million m^3.

Discussion and Conclusions

Water shortage has become a major limiting factor for the sustainable development of agriculture in Hebei. The estimation of agricultural water and groundwater irrigation net consumption will provide scientific information for developing efficient irrigation practices to improve crop water productivity. During the study period from 1984 to 2008, the 138 counties in Hebei province produced a total of 6.1×10^8 Mg of grain, and consumed 864 km^3 of water (with an average of 34.6 km^3/y), including 139 km^3 of groundwater. The AWC estimating result was close to the fresh water footprint of agriculture, which was calculated by using of Gini coefficient and Theil index accounting for 33.4 km^3 in Hebei province in 2007 [16]. Figure 7 shows the variations of annual grain yield (GY), water from precipitation (Q_p) and groundwater irrigation net consumption (Irr_{nc}) in Hebei from 1984 to 2008. In terms of spatial distribution, the grain yield and AWC in the southeastern part of the province are significantly higher than those in the northwest.

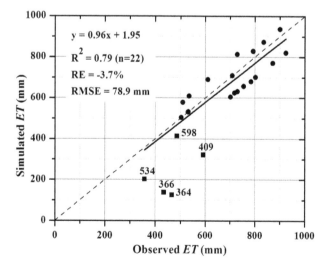

Figure 3. Validation of the model by using field experimental data from Luancheng Agro-ecosystem Experimental Station, Chinese Academy of Sciences. The numbers associated with the 5 points, representing rain-fed treatments, at the lower part refer to annual precipitation. The data for 3 filled blocks (409 mm, 366 mm, and 364 mm) are not included in the regression line because of the large deviations to the observed ET, indicating the model will underestimate actual ET for the non-irrigated lands when annual P less than around 400 mm.

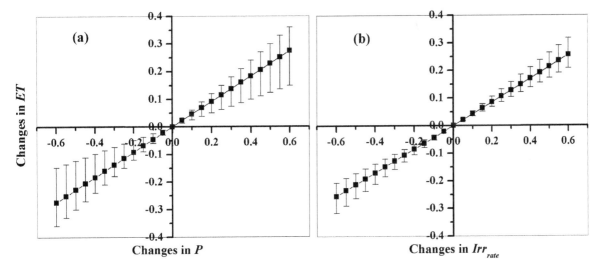

Figure 4. Sensitivities of estimated ET to the changes in annual precipitation (a), and to the changes in irrigation rates (b). The error bars indicates the range of different irrigation rates in (a), and range of different annual precipitation in (b), respectively.

Based on the linear correlation of ET and grain yield of each county (Figure 2), we estimated the grain yield without groundwater consumption, and subtracted this rain water fed yield from the actual statistical yield to obtain the grain yield gain (GYG) benefited from groundwater irrigation. It is found that the accumulated grain yield gain in the 25 years would be 1.9×10^8 Mg, which accounts for 31% of the province's total grain production during the same period.

In addition, we took Luancheng County (location shown in Figure 1) as a typical example to analyze the trade-off between groundwater consumption and grain yield gain. The irrigation rate of Luancheng County has reached to more than 90% since the beginning of 1980s. Although exploitation of groundwater ensured a stable increase in grain production, the groundwater table in

Luancheng fell 20.82 m from 1984 to 2008 due to continual over-pumping (Figure 8). The total groundwater consumption in Luancheng County estimated by the model accounted for 1.2 km^3 in the 25 years, which could cause the underground water table falling of 13.5 m in Luancheng area during the same period. Our estimation attributes the agricultural irrigation for grain production contributed 65% of the groundwater depletion in this county.

Large-scale mining of groundwater in Hebei Province began in the 1970s, the rapid socio-economic development consumed a large amount of groundwater in recent decades, the consumption of agricultural irrigation accounted for 77% of the total. Due to over-exploitation of groundwater, the underground water level was steadily declining. The total groundwater consumption in the

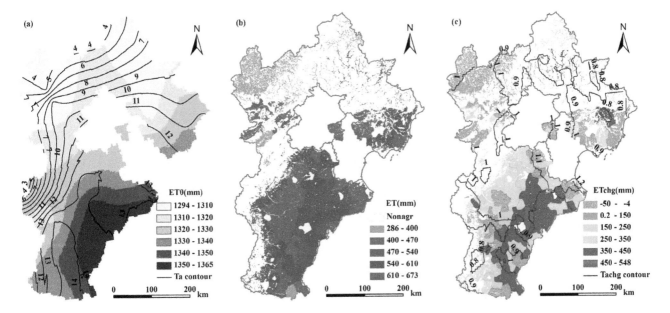

Figure 5. Distribution of annual mean ET_0 (a), actual ET (b), and changes in averaged ET for the period of 1984–1993 to 1999–2008 (c). The contour lines in (a) indicate the distribution of annual mean temperature (Ta); contour lines in (c) indicate the change of annual air temperature (Tachg) for the same time slices.

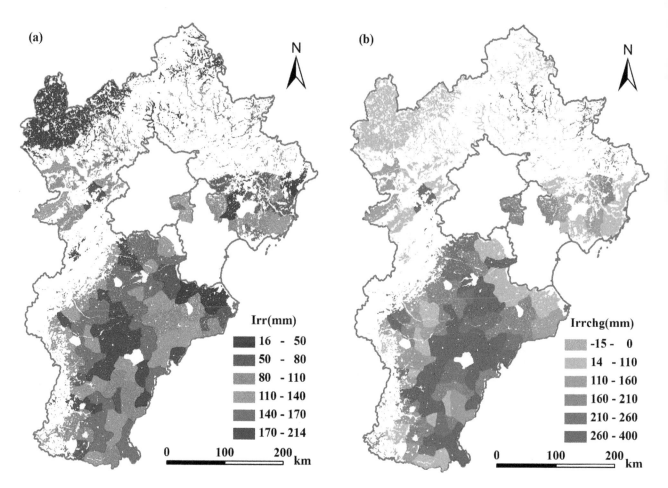

Figure 6. Annual mean net groundwater irrigation in 1984–2008 (a), and its change (b) from the periods of 1984–1993 to 1999–2008.

plain area during the study period was estimated as 113 km^3, which could cause an average groundwater falling of 7.4 m over the plain. This estimation is greatly agree with the results reported by Cao [17], who used a numerical groundwater flow model to simulate the groundwater pumping and water table decline over the Hebei plain.

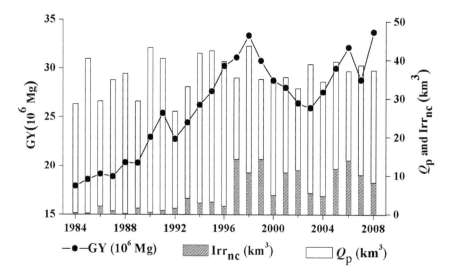

Figure 7. Inter-annual variations of precipitation (Q_p), net groundwater consumption (Irr_{nc}), and grain yield (GY) in Hebei Province (1984–2008).

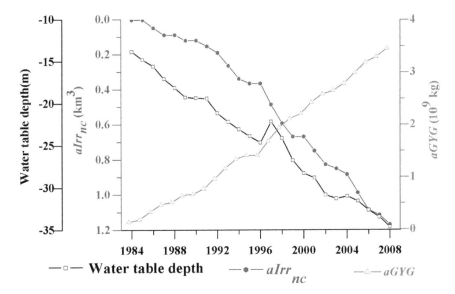

Figure 8. Accumulated net groundwater consumption (a/rr_nc), grain yield gain (aGYG), and observed groundwater table depth change for Luancheng county (1984–2008).

In this study, we aimed to estimate the AWC and groundwater irrigation consumption of Hebei province in recent decades using a simple model. The model proposed in this study need only the basic meteorological data and annual grain yield data. Based on some important assumptions the model can give good estimates of agricultural water consumption and net groundwater consumption for grain food production, and meet the study objectives well. However, we would like to call the audience attention to the uncertainties included in this study. First of all, we used a grain yield coefficient to substitute the crop coefficient in calculation of actual ET. This assumption ignored the differences in crop varieties, planting and field managements, irrigation methods, etc. and might cause some deviations of the results. Second, soil moisture of each year is different, but for any region it remains basically unchanged over the long term. The ET calculated by Eq. (1) for each year therefore varies, but it is reasonable to use Eq. (1) to calculate the sum and mean annual ET over the 25 years period. The ET products derived from remote sensing data and economic statistics data also contain some uncertainties [18], these sources of uncertainty may affect the accuracy of this study.

However, through comparing our estimations with the observed ET and groundwater depth at Luancheng county and further with the independent simulation over Hebei Plain [17] we have great confidence to believe the method proposed in this study could be extrapolated and applied to other regions where limited data such as meteorological and yield census data are available. Also it may be used in those regions for assessing the water footprint or aiding better water management for sustainable development.

Acknowledgments

Hongwei Pei helped calculating water balance for validation using the experimental data from Luancheng Agro-ecosystem Experimental Station, Chinese Academy of Sciences. We are also grateful to the constructive comments from 3 anonymous reviewers and editors.

Author Contributions

Conceived and designed the experiments: YJS. Performed the experiments: ZJY. Analyzed the data: ZJY. Wrote the paper: ZJY YJS.

References

1. Heilig GK (1999) China food: Can China feed itself? IIASA, Laxenburg (CD-ROM Vers. 1.1).
2. FAO's Information System on Water and Agriculture (2011) Available: http://www.fao.org/nr/water/aquastat/countries_regions/CHN/index.stm.Accessed 2013 Feb 9.
3. Deng XP, Shan L, Zhang HP, Turner NC (2006) Improving agricultural water use efficiency in arid and semiarid areas of China. Agricultural Water Management 80: 23–40.
4. Zhang H, Wang X, You M, Liu C (1999) Water-yield relations and water use efficiency of winter wheat in the North China Plain. Irrigation Science 19: 37–45.
5. Yang H, Zhang XH, Zehnder JB (2003) Water scarcity, pricing mechanism and institutional reform in northern China irrigated agriculture. Agricultural Water Management 61: 143–161.
6. Shen YJ, Kondoh A, Tang C, Zhang Y, Chen J, et al. (2002) Measurement and analysis of evapotranspiration and surface conductance of a winter wheat canopy. Hydrological Processes 16: 2173–2187.
7. Zhang YC, Shen YJ, Sun HY, Gates J (2011) Evapotranspiration and its partitioning in an irrigated winter wheat field: A combined isotopic and micrometeorologic approach. Journal of Hydrology 408: 203–211.
8. Liu XY, Li YZ, Hao WP (2005) Trend and causes of water requirement of main crops in North China in recent 50 years. Transactions of the CSAE 21: 155–159 (in Chinese with English abstract).
9. Li HJ, Zheng L, Lei YP, Li CQ, Liu ZJ, et al. (2008) Estimation of water consumption and crop water productivity of winter wheat in North China Plain using remote sensing technology. Agricultural Water Management 95: 1271–1278.
10. Yang YM, Yang YH, Moiwo JP, Hu YK (2010) Estimation of irrigation requirement for sustainable water resources reallocation in North China. Agricultural Water Management 97: 1711–1721.
11. Wu BF, Yan NN, Xiong J, Bastiaanssen WGM, Zhu WW, et al. (2012) Validation of ETWatch using field measurements at diverse landscapes: A case study in Hai Basin of China. Journal of Hydrology 436–437: 67–80.
12. Sun HY, Shen YJ, Yu Q, Flerchinger GN, Zhang YQ, et al. (2010) Effect of precipitation change on water balance and WUE of the winter wheat-summer maize rotation in the North China Plain. Agricultural Water Management 97: 1139–1145.
13. Allen RG, Pereira LS, Raes D, Smith M (1998) Crop evapotranspiration-Guidelines for computing crop water requirements. FAO Irrigation and drainage paper 56. FAO, Rome, Italy.
14. Ji ZH, Yang CX, Qiao GJ (2010) Reason analysis and calculation of surface runoff rapid decrease in the northern branch of Daqing River. South-to-North

Water Transfers and Water Science & Technology 8: 73–75, 79 (in Chinese with English abstract).

15. Chen W (1999) Groundwater in Hebei Province. Beijing: Seismological Press (in Chinese).

16. Sun CZ, Liu YY, Chen LX, Zhang L (2010) The spatial-temporal disparities of water footprints intensity based on Gini coefficient and Theil index in China. Acta Ecologica Sinica 30: 1312–1321 (in Chinese with English abstract).

17. Cao GL (2011) Recharge estimation and sustainability assessment of groundwater resources in the North China Plain. Ph.D. Thesis, Tuscaloosa: the University of Alabama.

18. Long D, Sing VP (2011) How sensitivity is SEBAL to changes in input variables, domain size and satellite sensor? Journal of Geophysical Research- Atmospheres 116: D21107.

4

A Novel Universal Primer-Multiplex-PCR Method with Sequencing Gel Electrophoresis Analysis

Wentao Xu[1,2❂], Zhifang Zhai[1❂], Kunlun Huang[1,2]*, Nan Zhang[2], Yanfang Yuan[1], Ying Shang[2], Yunbo Luo[1,2]*

1 Laboratory of Food Safety, College of Food Science and Nutritional Engineering, China Agricultural University, Beijing, China, 2 The Supervision, Inspection and Testing Center of Genetically Modified Food Safety, Ministry of Agriculture, Beijing, China

Abstract

In this study, a novel universal primer-multiplex-PCR (UP-M-PCR) method adding a universal primer (UP) in the multiplex PCR reaction system was described. A universal adapter was designed in the 5′-end of each specific primer pairs which matched with the specific DNA sequences for each template and also used as the universal primer (UP). PCR products were analyzed on sequencing gel electrophoresis (SGE) which had the advantage of exhibiting extraordinary resolution. This method overcame the disadvantages rooted deeply in conventional multiplex PCR such as complex manipulation, lower sensitivity, self-inhibition and amplification disparity resulting from different primers, and it got a high specificity and had a low detection limit of 0.1 ng for single kind of crops when screening the presence of genetically modified (GM) crops in mixture samples. The novel developed multiplex PCR assay with sequencing gel electrophoresis analysis will be useful in many fields, such as verifying the GM status of a sample irrespective of the crop and GM trait and so on.

Editor: Anil Kumar Tyagi, University of Delhi, India

Funding: The study was supported by the National GMO Cultivation Major Project of New Varieties (No. 2008ZX08012-001) and National Natural Science Foundation of China (No. 30800770). And the authors express their gratitude to the Ministry of Science and Technology and the Ministry of Agriculture of China for financial support respectively. The funders had no role in study design, data collection and analysis, decision to publish, or preparation of the manuscript.

Competing Interests: The authors have declared that no competing interests exist.

* E-mail: caufoodsafety@gmail.com (KH); lyb@yahoo.cn (YL)

❂ These authors contributed equally to this work.

Introduction

Nucleic acid analysis has become increasingly important in a variety of applications, such as the genotyping of individuals, the detection of infectious diseases, tissue typing for histocompatability, identifying individuals in forensic diagnosis, paternity testing, and monitoring the genetic make-up of plants and animals in agricultural breeding programs [1]. Techniques based on polymerase chain reaction (PCR) provide a powerful tool for the amplification of minute amounts of initial target sequences. Most PCR protocols involve reactions that amplify a single target. Multiplex PCR is a variation of the conventional technique in which two or more targets are simultaneously amplified in the same reaction. This approach has the potential for greater reliability, flexibility, and cost reduction. As far as we know, nine-target multiplex PCR method has been reported to simultaneously detect eight maize lines as well as the endogenous *Zein* gene in a single reaction tube [2], which contains the most targets in reported multiplex-PCR methods.

Multiplex PCR is an essential cost-saving technique for large scale scientific, clinical, and commercial applications, such as infectious microorganisms detection [3], gene expression [4,5], whole-genome sequencing [6], forensic analysis including human identification and paternity testing [7], the diagnosis of infectious diseases [8], and pharmacogenomic studies aimed at understanding the connection between individual genetic traits, drug response and disease susceptibility [9,10]. In recent years, multiplex PCR

has emerged as a core enabling technology for high-throughput SNP genotyping [7,10].

With the rapid development of GM crops, more and more studies have recently described the use of multiplex PCR as a rapid and convenient screening assay for the detection of GMOs. In GM crops such as soybean, maize, and canola, a multiplex PCR system has been developed to detect multiple target sequences using simultaneous amplification profiling [11]. A sensitive and specific triplex nested PCR assay was developed for the detection of housekeeping gene (lectin) and inserted elements of Roundup Ready soybean, i.e., constitutively expressed *CaMV 35S* promoter, *Cp4 epsps* gene encoding for 5-enol-pyruvyl-shikimate-3-phosphate for herbicide tolerance, *nos* terminator, and a chloroplast transit peptide (ctp) facilitating transport of epsps protein, in highly processed products [12]. Multiplex PCR simultaneously detecting eight lines of GM maize by employing sequence-specific primers and the maize endogenous *Zein* gene was developed [2], which is also a most targets multiplex PCR system nowadays. Recently, multiplex PCR assays simultaneously amplifying the commonly used selectable marker genes, i.e., *aadA*, *bar*, *hpt*, *nptII*, *pat*, and a reporter gene *uidA* were developed as a reliable tool for qualitative screening of GM crops [13]. What's more, multiplex PCR-based assays have also been developed to simultaneously detect functional transgenes, control elements and housekeeping genes, such as *cry1Ac* gene for insect resistance, *CaMV 35S* promoter and endogenous *SRK* (S-locus Receptor Kinase) gene in Bt cauliflower [14]; *osmotin* gene for salinity and drought tolerance, *CaMV35S* promoter and endogenous *LAT52* (late anther tomato) gene in GM tomato[15]; *cry1Ab* gene for insect resistance, *CaMV 35S* promoter; *npt II* marker gene and endogenous *UGPase* (uridine diphosphate glucose pyrophosphorylase) gene in Bt potato[16]. These studies demonstrate that the multiplex PCR system is also a convenient, cost-effective, and efficient assay for GM detection.

Although multiplex PCR has so many advantages, it has several disadvantages that can not be ignored, mainly including the self-inhibition among different sets of primers, low amplification efficiency and no identical efficiency on different templates, which restricts its further development and broad application. Even the reported nine-target multiplex PCR method cannot avoid the disadvantage of worse reproducibility and stability. A novel universal primer-multiplex PCR (UP-M-PCR) method was devised at the basis of the problem originating from conventional multiplex PCR, and up to now, it has been used to simultaneously detect four meat species including chicken, cattle, pig and horse [17], to simultaneously detect *Escherichia coli*, *Listeria monocytogenes*, and *Salmonella* spp. in food samples [18], and used in the event-specific detection of stacked genetically modified maize Bt11×GA21 [19]. In the present study, we also use this novel universal primer-multiplex PCR method to simultaneously detect nine commonly used selectable marker and reporter genes, including *hpt*, *gus*, *nptII*, *aadA*, *pat*, *35s*, *bar*, *nos*, *uidA*, as well as six endogenous genes *Pa*, *Ivr*, *Lec*, *sps*, *sad1* and *FatA* of six most common GM crops papaya, maize, soybean, rice, cotton and canola, respectively, which could overcome the shortcomings of conventional multiplex PCR described previously, and significantly simplify the procedure of identification for GMOs. For the analysis of PCR products, sequencing gel electrophoresis (SGE) was chosen as its large separation scale and extraordinary resolution, which could not restrict the increase of detecting targets in multiplex PCR. The fifteen-target UP-M-PCR we developed not only contains the most targets in all the multiplex PCR methods reported before, but also covers most selectable marker and reporter genes commonly employed in GM crops, which will totally meet the demand of verifying the GM status of a sample irrespective of the crop and GM trait.

Results

Specificity of Compound Specific Primers

The new designed compound primer pairs originated from specific primers have been tested to get equivalent intensities of bands on gels with the same template concentration (Figure 1A), which showed that the set of compound specific primers worked efficiently and had the same specificity as the specific primers from reference. Because the compound specific primers contained a common sequence (20 bp) at the 5′-end, as a result they got a higher annealing temperature and generated amplicons larger of 40 bp than the products amplified by corresponding specific primers, so the bands on gel were a little higher too. In fifteen-plex PCR, all the primers mixed together with the optimized concentration, while the template of each primer pair was added separately in every single reaction, and only one expected PCR amplicon was achieved corresponding to a certain template DNA (Figure 1B). That there were no unexpected bands showed there was no unexpected reaction, which also proved that the specificity of compound specific primers was high.

Feasibility of Universal Primer (UP)

Keeping the concentration of templates at 50 ng, with the amount of the specific primers for *Lec* gene decreasing (500 nmol L^{-1}, 50 nmol L^{-1}, 25 nmol L^{-1}, 5 nmol L^{-1}), the

intensity of band fell down markedly (25 nmol L^{-1}) until to nothing (5 nmol L^{-1}) in conventional singlet PCR (Figure 2, lanes 1, 2, 3, 4), which showed that the concentration of amplified fragments became lower and lower. While in the novel singlet PCR, for the addition of universal primer (500 nmol L^{-1}), though there is a down gradient concentration of compound specfic primers Lec-118-F/R from 500 nmol L^{-1} to 5 nmol L^{-1}, the PCR system above worked efficiently and got an equivalent amount of amplified products (Figure 2, lanes 5, 6, 7, 8). Similar results were achieved from other compound specific primers (data not shown) with UP in novel singlet PCR. The sharp contrast showed that the universal primer was well designed to work efficiently for the PCR amplification and had a high feasibility to amplify the amplicons produced by compound specific primers.

Optimization of the UP-M-PCR

The concentrations of primers strongly influence the efficiency and disparity of PCR reaction, which is very important for the PCR reaction, especially in multiplex PCR. The final optimized concentration of universal primer (UP) was 500 nmol L^{-1} in both singlet PCR and UP-M-PCR, which is the same as in normal singlet PCR, while the compound specific primers were 25 nmol L^{-1} (about 1/20 of UP) in singlet PCR that can ensure an efficient amplification for all these primers, but in fifteen-plex UP-M-PCR, because of the interaction and the difference of work efficiency among primers, all compound specific primers at 25 nmol L^{-1} could not get an equivalent amount of amplified products, thus there need an adjustment. The process of the adjustment was showed in Figure S2. The final optimized concentration of the compound specific primers were 10 nmol L^{-1} for 35s-195-F/R and Nos-F/R (about 1/50 of UP), 16 nmol L^{-1} for nptII-508-F/R and Pa-363-F/R(about 1/30 of UP), 50 nmol L^{-1} for hpt-839-F/R, Ivr-262-F/R and Lec-110-F/R(about 1/10 of UP), and 25 nmol L^{-1} for all other primers (including gus-565-F/R, aadA-406-F/R, pat-262-F/R, bar-177-F/R, sps-110-F/R, sad1-91-F/R, uidA-82-F/R and FatA-76-F/R). To test the efficiency of Taq Polymerase to be employed in PCR assays, comparative tests were made with several Taq polymerases, such as Phire™ Hot Start DNA polymerase, iProof™ High-Fidelity DNA polymerase, and *TaKaRa Taq*™. The Phire™ HotStart DNA polymerase, coupled with a preoptimized primer mix for different multiplex reactions, gave the best results both in terms of reproducibility and robustness. To find the best annealing temperature, a gradient temperature PCR from 56 to 64°C has been performed. At last the optimum annealing temperature was chosen at 60°C. Similarly the best extension temperature was chosen at 70°C The time for annealing and extension was chosen at 50 s. Figure 3 showed the amplification results by UP-M-PCR of the optimizing process on 5.0% polyacrylamide gel (3.5 M urea included). All the reactions were performed with the same amount of template (50 ng for each). Each compound specific primer pair in the mixture was sensitive and specific enough to amplify the corresponding sequence and generated the expected length of amplicons the same as in the singlet PCR and no unexpected PCR products were detected. There was less or even no disparity between various primers as in UP-M-PCR. Similar duplex, triplex, four-plex, even thirteen-plex or fourteen-plex PCR results were achieved with arbitrary combination of compound specific primer pairs (Figure 4).

The sensitivity of UP-M-PCR

The sensitivity of UP-M-PCR method was assayed in fifteen-plex PCR with only one DNA template. A set of UP-M-PCR

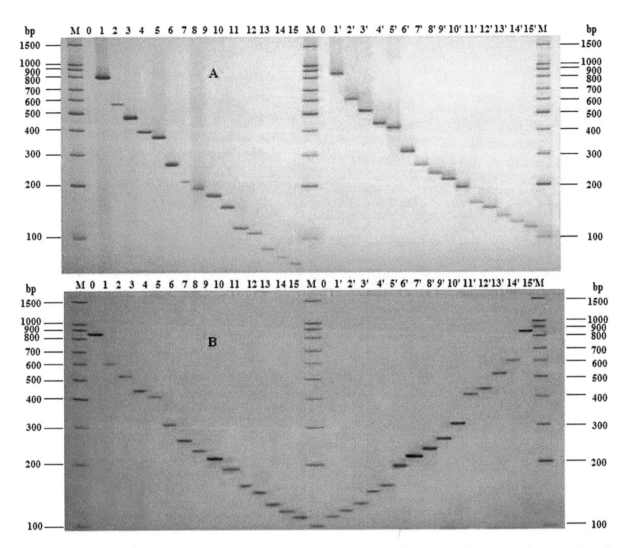

Figure 1. Detection of the specificity of UP-M-PCR. (A) Comparison of specificity between specific primers and compound specific primers. Lanes 1~15, amplicon fragments of specific primer; lanes 1'~15', amplicon fragments of UP and compound specific primer; lane 0/0', negative control without template; lane M, 100 bp DNA Marker. (B) Lanes 1~15, amplicon fragments of one specific primer in singlet PCR; lanes 1'~15', amplicon fragments of fifteen-plex PCR with only one certain template DNA; lane M, 100 bp DNA Marker.

reactions were performed with a series of change of template concentration from 50, 5, 0.5 and 0.1 ng to 0.05 ng, which resulted in significantly disproportionate amplification of target DNA by polyacrylamide gel electrophoresis assay. The amount of amplified products fell down along with the decrease of content of corresponding template, so did the intensity of bands. When the target template's amount was as low as 0.05 ng, there was no corresponding amplicon, so the detection limit of UP-M-PCR for target gene was 0.1 ng target DNA per reaction (Figure 5). Compared with the published conventional multiplex PCR system detecting eight GM maize lines at the same time [2], the limit of detection of which is 0.25% GM in the total 100 ng template, equaling to 0.25 ng GM content in the mixed samples, the novel PCR system has a much higher sensitivity.

Application of UP-M-PCR in Detection as a Rapid Screening Method

For most incidents, the detection samples are complicated by more than one GM material accompanied with multiple characteristics. Multiplexing provides a cost-effective diagnostic

assay for GM detection with higher through-put and less consumption of samples and reagents as compared to simplex assays. The novel UP-M-PCR method was applied to test common GM and non-GM crops including cotton, papaya, maize, rice, canola and soybean, and there was significant difference between GMO and their non-GM parents (Figure 6).

Discussion

The present paper has described the development and application of a novel UP-M-PCR method with an additional universal primer (UP) in the multiplex PCR reaction system. The designing of primers was very important on multiplex PCR techniques, because primer specificity and melting temperature (Tm) were more critical than conventional PCR. Comparing genomic DNA sequences of cotton, papaya, maize, rice, canola and soybean with the universal adapter sequence using DNA-MAN, the universal adapter sequence were not matched with genomic DNA sequences. The process of selecting universal primer was showed in Table S1 and Figure S1. In optimized UP-M-PCR system, the universal primer has a concentration of

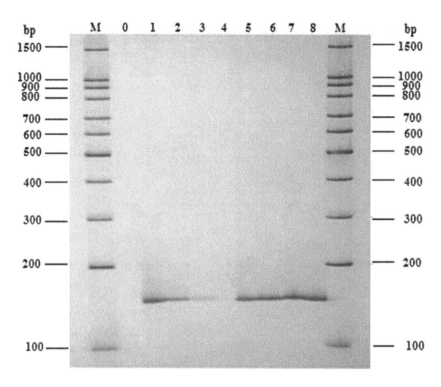

Figure 2. Impact of concentration of universal primer on singlet PCR. Lane 0, negative control without template; lanes 1, 2, 3, 4, amplicon fragments by compound specific primer pair Lec-118-F/R at the concentration of 500 nmol L^{-1}, 50 nmol L^{-1}, 25 nmol L^{-1}, 5 nmol L^{-1} respectively; lanes 5, 6, 7, 8, amplicon fragments by UP (500 nmol L^{-1}) and compound specific primer Lec-118-F/R at a series concentrations of 500 nmol L^{-1}, 50 nmol L^{-1}, 25 nmol L^{-1},5 nmol L^{-1}; lane M, 100 bp DNA Marker.

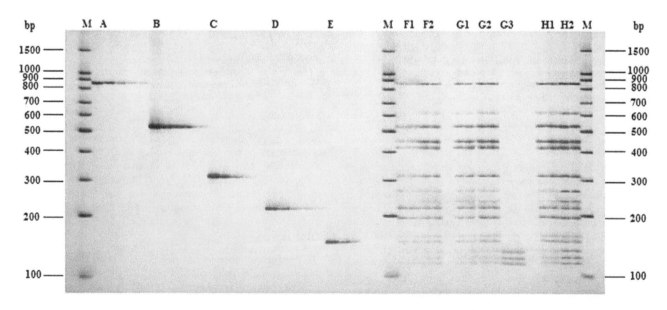

Figure 3. Optimization of the UP-M-PCR. Lane A, B, C, D, E, amplicon fragments by UP (500 nmol L^{-1}) and compound specific primer hpt-839, nptII-508, pat-262, bar-226 and sps-110 at a series concentrations of 500 nmol L^{-1}, 50 nmol L^{-1}, 25 nmol L^{-1}, 5 nmol L^{-1}, 0.5 nmol L^{-1}; lane F1, amplicon fragments by UP at 500 nmol L^{-1} and all compound specific primers at 25 nmol L^{-1}; lane F2, amplicon fragments by UP at 500 nmol L^{-1} and all compound specific primers at the optimized concentration; lane G1,G2,G3, amplicon fragments by all primers at the optimized concentration with *TaKaRa Taq*^{TM}, Phire^{TM} Hot Start DNA polymerase, iProof^{TM} High-Fidelity DNA polymerase; lane H1, amplicon fragments by all primers at the optimized concentration with Phire^{TM} Hot Start DNA polymerase under the common amplification conditions; lane H2, amplicon fragments by all primers at the optimized concentration with Phire^{TM} Hot Start DNA polymerase under the optimized amplification conditions; lane M, 100 bp DNA Marker.

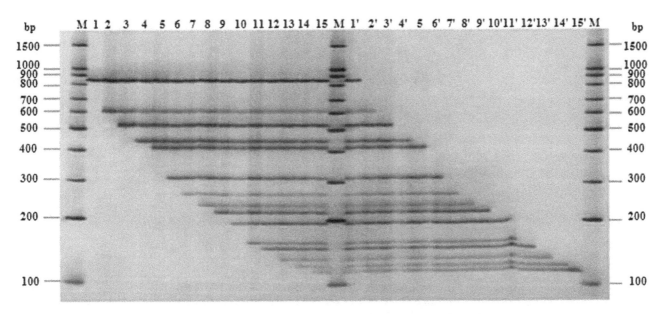

Figure 4. Multiplex PCR assay for testing of primer interference. Using equivalent DNA mix of six different GM events including Bt11 of maize, MON 15985 of cotton, GM rice with *bar* gene, Huanong No. 1 of papaya, RRS of soybean and GM canola with *hpt* gene. Lanes 1~15, PCR runs starting with the largest amplicon (by *hpt* gene-specific primer, amplicon size 879 bp), followed by the addition of a second primer pair, until the fifteenth primer pair; lanes1'~15', PCR runs starting with the fifteen-plex amplicon, followed by the elimination of the largest amplicon primer pair (by *hpt* gene-specific primer, amplicon size 879 bp), until only the smallest amplicon (by *FatA* gene-specific primer, amplicon size 116 bp) remained; lane M, 100 bp DNA Marker.

500 nmol L^{-1} at normal degree, while the concentration of every compound specific primer was as low as 1/50 to 1/10 of UP, therefore, the total amount of all the primers was almost equal to that of conventional singlet PCR and far less than that in conventional multiplex PCR, in which all the primers are mixed with a normal concentration about 500 nmol L^{-1}. In a word, it really simplified the multiplex PCR reaction system, which was also the reason why it could circumvent the amplification disparity resulting from different primers in traditional multiplex PCR.

The choice of DNA polymerase is very important for the optimum performance of the PCR. The Phire^{TM} Hot Start DNA polymerase, coupled with a preoptimized primer mix for different multiplex reactions, gave the best results both in terms of reproducibility and robustness. The use of hot start DNA polymerase prevents the formation of misprimed products and reduces primer-dimer formation. As the number of primers increases, the possible sequence dependent interactions between primers of different primer pairs also increase, which results in the

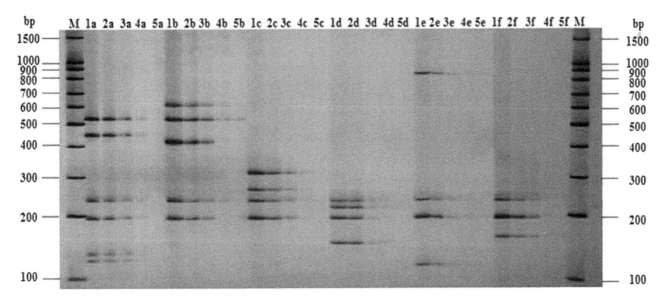

Figure 5. Sensitivity detection of single template by UP-M-PCR. a, MON 15985 of GM cotton; b, Huanong No. 1 of GM papaya; c, Bt11 of GM maize; d, GM rice with *bar* gene; e, RRS of GM canola with *hpt* gene; f, RRS of GM soybean. Lanes 1~5, template concentration from 50, 5, 0.5, 0.1 ng to 0.05 ng; lane M, 100 bp DNA Marker.

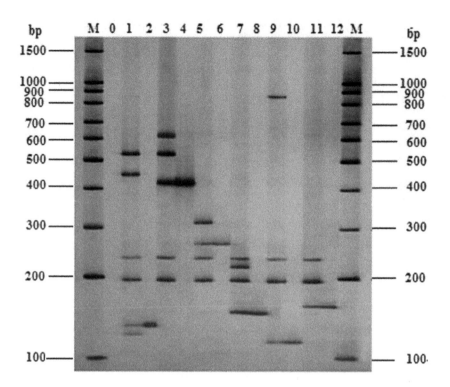

Figure 6. Application of UP-M-PCR. Lane 1,2, GM/non-GM cotton, with endogenous genes *sad1* (amplicon size 131 bp); Lane 3,4, GM/non-GM papaya, with endogenous genes *Pa* (amplicon size 403 bp); Lane 5,6, GM/non-GM maize, with endogenous genes *Ivr* (amplicon size 266 bp); Lane 7,8, GM/non-GM rice, with endogenous genes *sps* (amplicon size 150 bp); Lane 9,10, GM/non-GM canola, with endogenous genes *FatA* (amplicon size 116 bp); Lane 11,12, GM/non-GM soybean, with endogenous genes *Lec* (amplicon size 158 bp); Lane 0, negative control without template; lane M, 100 bp DNA Marker.

formation of primer-dimers. Small differences in amplification efficiencies for the different primer pairs might result in the preferential amplification of some of the PCR products, leaving other PCR products at subdetectable levels. Hence, primer design, PCR cycling conditions, and the concentration of each reaction component need to be cautiously optimized in order to avoid the formation of primer-dimers and to detect all DNA targets simultaneously without any primer.

In most cases, agarose gel electrophoresis is used for the separation of DNA fragments ranging from 10 kb~0.2 kb, but when separating particularly small pieces of DNA or small quantities of DNA, a vertical polyacrylamide gel is more appropriate, which has a higher resolution than that can be achieved with agarose gel electrophoresis. In this study, multiplex PCR products contained a series of DNA fragments in which most sizes were less than 500 bases and differed in size by only several bases (eg. *uidA* and *FatA*, only differed by 6 bases), and agarose gel electrophoresis cannot separated them efficiently. However, polyacrylamide gel has the advantage of exhibiting extraordinary resolution; pieces of DNA that differ in size by a single base pair can be separated in a polyacrylamide gel [20]. Thus sequencing gel electrophoresis (SGE) using polyacrylamide gel as separation medium, was chosen as the analysis tool for PCR amplicon fragments. What's more, the SGE system has longer lanes and higher separation voltage, which ensure the effective separation for DNA fragments of small difference. Primarily, SGE of DNA is only used for DNA sequencing. Nowadays its high resolution and large accommodation have made it widely used in many fields, such as RFLP, AFLP, SSR analysis and so on. The conditions of sequencing gel electrophoresis were carefully optimized in order to achieve best separating effect. The optimization of each factor was showed in

Figure S3, S4, S5, S6. According to the optimal DNA resolution of different acrylamide concentration [21], the final acrylamide concentration is determined at 5%. The specified voltage is 1–8 volts/cm and centimeters in this case specifies the length of the gel from top to bottom (i.e the direction the DNA will travel). To avoid differential heating in the center of the gel which could cause poor image development, polyacrylamide gels were run at optimal constant power 40~60 W (data not shown). The total electrophoresis time for complete separation of all amplicon fragments was about 90~120 min depending on the running power.

To realize the potential of SGE, a visualization method offering superior clarity and sensitivity is also required. The silver staining method has proven very effective in this regard. As a method, silver staining was originally developed to detect proteins separated by polyacrylamide gels [22]. It was further optimized and applied to visualize other biological molecules e.g. nucleic acids [23,24]. Silver staining of DNA has several advantages: (I) Image development and visualization is done under normal ambient light. Thus, the procedure can be performed entirely at the laboratory bench without the need for dark room or UV illumination facilities. (II) The image is resolved with the best possible sensitivity and detail, because silver is deposited directly on the molecules within the transparent gel matrix. Thus visualization is from the primary source and does not suffer any degradation or blurring that can accompany secondary imaging devices which involve fluorescence, autoradiography, focusing lenses, film development or digital image processing. (III) Silver staining offers similar sensitivity to autoradiography, but avoids radioactive handling, delays from development times and waste disposal issues [19]. Although this is an advantage in terms of scope, it nevertheless means that the protocol must be applied with due care; almost any other biological impurity such as stray human fingerprints incorporated into or onto the gel matrix on the gel surface will stain with perfect detail. It is thus important to use dust-free reagents of the best analytical grade, including the purest water available.

The UP-M-PCR method was originally developed by our lab and it firstly applied to the detection of stacked GM events Bt11×GA21 [19]. In this study we used the same PCR system and combined it with sequencing gel electrophoresis analysis to simultaneously detect 15 target genes in the GM crops successfully, but the sequence of UP is different from before, which just showed the flexibility of this novel PCR method. The key point of this approach is the idea of the reaction process, not a particular sequence of UP or some specific primers. Any sequence meeting the requirements of designing UP mentioned above can be applied in this PCR system. Otherwise, the careful optimization of the reaction system, including the choice of Tag Polymerase enzyme,

the concentration of each primer, etc. is very significant for the high-throughout multiplex PCR detection system, which is also the important result of our study. What's more, the perfect combination of UP-M-PCR and sequencing gel electrophoresis, as well as the optimization of the electrophoresis conditions, is another important innovative point in this study.

With the dramatic expansion of global area under cultivation of GM crops, there is an urgent need to step up the development of robust, efficient, and reliable methods for GM detection. The developed UP-M-PCR method with sequencing gel electrophoresis analysis used to simultaneously detect nine commonly used selectable marker and reporter genes and six endogenous genes in a single reaction can be a reliable tool for the screening of GM crops and for unintentional mixing of GM seeds with non-GM seeds. Besides, the novel UP-M-PCR can be used in all the fields where multiplex PCR is needed, which has promising application future and is worth being popularized

Table 1. Information of Compound Specific Primers Used in UP-M-PCR[a].

Primer name	Primer sequence	Length (bp)	Ref.
hpt-839-F	*TTTGGTCGTGGTGGTGGTTT*CGCCGATGGTTTCTACAA	879	13
hpt-839-R	*TTTGGTCGTGGTGGTGGTTT*GGCGTCGGTTTCCACTAT		
gus-565-F	*TTTGGTCGTGGTGGTGGTTT*AAATCGCCGCTTTGGACATA	605	this study
gus-565-R	*TTTGGTCGTGGTGGTGGTTT*TACTGGCTTTGGTCGTCATGA		
nptII-508-F	*TTTGGTCGTGGTGGTGGTTT*CCGACCTGTCCGGTGCCC	548	13
nptII-508-R	*TTTGGTCGTGGTGGTGGTTT*CCGCCACACCAGCCGGCC		
aadA-406-F	*TTTGGTCGTGGTGGTGGTTT*CCGCGCTGTAGAAGTCACCATTG	446	13
aadA-406-R	*TTTGGTCGTGGTGGTGGTTT*CCGGCAGGCGCTCCATTG		
Pa-363-F	*TTTGGTCGTGGTGGTGGTTT*GGCTCAATATGGTATTCACTACAGAAAT	403	this study
Pa-363-R	*TTTGGTCGTGGTGGTGGTTT*CATCGGTTTTGGCTGCATAA		
pat-262-F	*TTTGGTCGTGGTGGTGGTTT*GAAGGCTAGGAACGCTTACG	302	13
pat-262-R	*TTTGGTCGTGGTGGTGGTTT*GCCAAAAACCAACATCATGC		
lvr-226-F	*TTTGGTCGTGGTGGTGGTTT*CCAAACTGAATCCGGTCTGA	266	this study
lvr-226-R	*TTTGGTCGTGGTGGTGGTTT*GTGCGCTTCCTCTCGTTTTC		
35s-195-F	*TTTGGTCGTGGTGGTGGTTT*GCTCCTACAAATGCCATCATTGC	235	this study
35s-195-R	*TTTGGTCGTGGTGGTGGTTT*GATAGTGGGATTGTGCGTCATCCC		
bar-177-F	*TTTGGTCGTGGTGGTGGTTT*GCACAGGGCTTCAAGAGCGTGGTC	217	13
bar-177-R	*TTTGGTCGTGGTGGTGGTTT*GGGCGGTACCGGCAGGCTGAA		
nos-151-F	*TTTGGTCGTGGTGGTGGTTT*GTCTTGCGATGATTATCATATAATTTCTG	191	20
nos-151-R	*TTTGGTCGTGGTGGTGGTTT*CGCTATATTTTGTTTTCTATCGCGT		
Lec-118-F	*TTTGGTCGTGGTGGTGGTTT*GCCCTCTACTCCACCCCCA	158	21
Lec-118-R	*TTTGGTCGTGGTGGTGGTTT*GCCCATCTGCAAGCCTTTTT		
sps-110-F	*TTTGGTCGTGGTGGTGGTTT*GATCGCTTCCGCCATTAGCA	150	this study
sps-110-R	*TTTGGTCGTGGTGGTGGTTT*AACCGAGCGCGATCACTTGC		
sad1-91-F	*TTTGGTCGTGGTGGTGGTTT*CCACGAGACAGCCTATACCAAAA	131	22
sad1-91-R	*TTTGGTCGTGGTGGTGGTTT*CTTCTTCATCATGTCAGCAAATGC		
uidA-82-F	*TTTGGTCGTGGTGGTGGTTT*CACCACGGTGATATCGTCCAC	122	13
uidA-82-R	*TTTGGTCGTGGTGGTGGTTT* TTTCTTTAACTATGCCGGAATCCATC		
FatA-76-F	*TTTGGTCGTGGTGGTGGTTT*GGTCTCTCAGCAAGTGGGTGAT	116	this study
FatA-76-R	*TTTGGTCGTGGTGGTGGTTT*CGTCCCGAACTTCATCTGTAA		
UP	*TTTGGTCGTGGTGGTGGTTT*		this study

[a]The table shows the details of primer sequences, expected DNA fragment length and the source of primer used in UP-M-PCR. Each primer pair originates from the corresponding specific primer set (sequence in straight matter) and has a common sequence TTTGGTCGTGGTGGTGGTTT (20 bp) at its 5' - end in italics, which is also the sequence of the universal primer (UP) used in this developed new way.

Materials and Methods

Materials

The GM crops/events under study were all the main GM crops widely used nowadays, including GM maize event Bt11, GM cotton event MON 15985, GM rice with *bar* gene, GM papaya event Huanong No. 1, GM soybean event RRS and GM canola with *hpt* gene, along with their non-GM parent seeds of maize, cotton, rice, papaya, soybean and canola as controls. Before the extraction of DNA, they were ground respectively into powder with the size of 200 mesh in the fume hood in order to avoid cross-contamination.

Preparation of DNA Template

Extract and purify genomic DNA from the finely ground powder samples described above, using the Wizard® Genomic DNA Purification Kit, according to the manufacturer's instructions. For specificity and sensitivity test, DNA was diluted in ddH$_2$O to get a proper concentration. Evaluate the quality of the extracted DNA either on a 1% (wt/vol) agarose gel or by amplify a housekeeping gene by qualitative PCR, if the template DNA do not produce the expected amplicons, it suggests that the extracted DNA contains PCR inhibitors such as ethanol or xylene and these samples should not be used for further study.

Designing of Primers

Comparing genomic DNA sequences of the common GM crops cotton, papaya, maize, rice, canola and soybean with the universal adapter sequence using DNAMAN, the universal adapter sequence were not matched with any genomic DNA sequences. For the amplification of *hpt, npt II, aadA, pat, bar, nos, Lec, sps, uidA* and *FatA* genes, published primer pairs [13,25–27] were used. The universal primer (UP) and specific primer pairs for the amplification of *gus, Pa, Ivr, 35s* and *sad1* genes were designed using ABI PRISM Primer Express Version 2.0 software(Applied Biosystems company, FosterCity, CA) with an optimal melting temperature(Tm) of about 60°C. When designing the UP primer, the factors including having binding sites with most GM crops genome as little as possible, being rich in GC contents and having a melting temperature (*Tm*) of about 60°C etc. were particularly considered to insure the suitability of the UP in the novel mutiplex PCR. Compound specific primers used in UP-M-PCR for the specific detection of these genes originate from the corresponding specific primers. Each compound specific primer contains a common sequence TTTGGTCGTGGTGGTGGTTT at its 5′-end, which is just the sequence of UP in this new PCR system. Details of primer sequences are shown in Table 1.

Universal Primer-Multiplex PCR (UP-M-PCR)

Compared with traditional multiplex PCR system, UP-M-PCR contains more of a universal primer (UP) at a normal concentration (500 nmol/L), a little quantity (10~50 nmol/L) of 15 compound specific primers targeting for the 15 marker and endogenous genes. All the primers include a common sequence at its 5′-end, which is also the sequence of the UP. The amplification routine of UP-M-PCR is shown in Figure 7. At the initial stage of the reaction (about the former ten cycles), the compound specific primers take main action for amplification of target sequences due to their higher annealing temperature, while the universal primer almost has no amplification. With the compound specific primers used up and the amplified products incorporating the UP adaptor increasing, the UP begins to play a leading role to take the amplicons as templates and shows its ability to amplify the fragments of seven different targets.

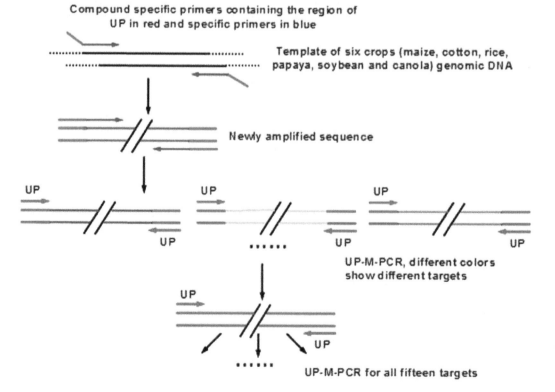

Figure 7. Amplification routine of UP-M-PCR. Each compound specific primer contained a universal sequence at the 5′-end (red) and the specific primer at the 3′-end (blue). The amplified fragments with the primer pairs of different targets are individually marked in different colors. The amplified fragments only by the universal primer are marked in red.

PCR Conditions

PCR analyses were carried out on Peltier Thermal Cycler Controller (MJ Research, BioRad Laboratories, Mass., U.S.A.). The efficiency of the primer pairs to amplify the target sequences was tested by simplex PCR using the corresponding genomic DNA with different target genes.

The concentrations of each primer were optimized in the single and multiplex PCR system, ranging from 500 to 5 nmol L^{-1} with series of dilutions, while the universal primer always had a normal concentration of 500 nmol L^{-1}. The final optimized compound specific primers mix contains 10 nM of 35s-195-F/R and nos-151-F/R (about 1/50 of UP), 16 nM of nptII-508-F/R and Pa-363-F/R (about 1/30 of UP), 50 nM of hpt-839-F/R, Ivr-262-F/R and Lec-110-F/R (about 1/10 of UP), and 25 nM of all other primers, including gus-565-F/R, aadA-406-F/R, pat-262-F/R, bar-177-F/R, sps-110-F/R, sad1-91-F/R, uidA-82-F/R and FatA-76-F/R (about 1/20 of UP).

To test the efficiency of Taq Polymerase to be employed in PCR assays, comparative tests were made with several Taq polymerases, such as Phire^{TM} Hot Start DNA polymerase, iProof^{TM} High-Fidelity DNA polymerase, and *TaKaRa Taq*^{TM}. All multiplex PCR assays were performed in a final volume of 80 μL with the following reagent concentrations: 250 ng template DNA mix, 5×Hot Start PCR buffer, 0.2 mM of dNTPs, 500 nmol/L universal primer, 1 μL DNA polymerase and the preoptimized compound specific primers mix.

To choose the best annealing temperature, a gradient PCR with annealing temperatures ranging from 56 to 64°C was performed. The extension temperature and the time for annealing and extension were also optimized in order to choose the best PCR conditions. The final amplification conditions were initial denaturation at 95°C for 10 min, 40 cycles consisting of denaturation at 95°C for 30 s, primer annealing at 60°C for 50 s, primer extension at 70°C for 50 s, and final extension at 70°C for 10 min. Each reaction was run in triplicate.

Analysis of PCR Products

PCR products were analyzed on sequencing gel electrophoresis (SGE). The gel was prepared with 5% polyacrylamide (acrylamide/bisacrylamide, 29/1; containing 3.5 M urea). Polyacrylamide gels are poured and run in 1× TBE at constant power 40~60 W to avoid differential heating in the center of the gel. After electrophoresis, silver staining method used to visualize DNA fragments, and images were recorded with a digital camera.

Supporting Information

Figure S1 Selection of universal primers. A, B, C: Reaction system adding UP1, UP2 and UP3 respectively. Lane 1a/1b, 2a/2b, 3a/3b: duplex PCR for amplifying hpt/pat, hpt/nptII, nptII/pat; lane 4a/4b: triplex PCR for amplifying hpt/nptII/pat; lane 1c/2c/3c/4c: NTC (no template control); lane M: DNA Marker DL 2000.

Figure S2 Optimization of primer concentration for UP-M-PCR. (A) UP-M-PCR for amplifying *hpt, gus, nptII, aadA, Pa, pat* and *Ivr* gene. Lane 1–4: after concentration adjustment; lane 5–8: before concentration adjustment. (B) UP-M-PCR for amplifying *35s, bar, nos* and *Lec* gene. Lane 1–3: after concentration adjustment; lane 4–6: before concentration adjustment. (C) UP-M-PCR for amplifying *sps, uidA, sad1* and *FatA* gene. Lane 1–3: after concentration adjustment; lane 4–6: before concentration adjustment; lane M: DNA Marker DL 2000.

Figure S3 Determination of electrophoretic condition. A: Constant voltage of 500 V; B: Constant Power of 60 W.

Figure S4 Determination of gel's concentration. A, B, C: Gel's concentration of 4%, 5%, 6%, respectively.

Figure S5 Determination of the urea concentration. A, B, C: Gel's concentration of 5%, with the urea concentration of 0, 7 mol/L, 3.5 mol/L, respectively.

Figure S6 Determination of sample treatment. A: Sample treated by denaturalization. B: Sample not treated by denaturalization.

Author Contributions

Conceived and designed the experiments: KH YL WX. Performed the experiments: WX ZZ NZ YY YS. Analyzed the data: WX ZZ KH YS. Contributed reagents/materials/analysis tools: WX NZ YY YS. Wrote the paper: WX ZZ KH YL.

References

1. Zhang YL, Zhang DB, Li WQ, Chen JQ, Peng YF, et al. (2003) A novel real-time quantitative PCR method using attached universal template probe. Nucleic Acids Res 31: e123.
2. Shrestha HK, Hwu K, Wang S, Liu L, Chang M (2008) Simultaneous detection of eight genetically modified maize lines using a combination of event- and construct-specific multiplex-PCR technique. J Agric Food Chem 56: 8962–8968.
3. Pinar A, Bozdemir N, Kocagoz T, Alacam R (2004) Rapid detection of bacterial atypical pneumonia agents by multiplex PCR. Cent Eur J Public Health 12: 3–5.
4. Ding C, Cantor CR (2003) A high-throughput gene expression analysis technique using competitive PCR and matrix-assisted laser desorption ionization time-of-flight. MSProc Natl Acad Sci 100: 3059–3064.
5. Hess CJ, Denkers F, Ossenkoppele GJ, Waisfisz Q, McElgunn CJ, et al. (2004) Gene expression profiling of minimal residual disease in acute myeloid leukaemia by novel multiplex-PCR-based method. Leukemia 18: 1981–1988.
6. Tettelin H, Radune D, Kasif S, Khouri H, Salzberg S (1999) Optimized multiplex PCR: efficiently closing a whole-genome shotgun sequencing project. Genomics 62: 500–507.
7. Inagaki S, Yamamoto Y, Doi Y, Takata T, Ishikawa T, et al. (2004) A new 39-plex analysis method for SNPs including 15 blood group loci. Forensic Sci Int 144: 45–57.
8. Elnifro E, Ashshi A, Cooper R, Klapper P (2000) Multiplex PCR: optimization and application in diagnostic virology. Clin Microbiol Rev 13: 559–570.
9. Shi M, Bleavins M, de la Iglesia F (1999) Technologies for detecting genetic polymorphisms in pharmacogenomics. Mol Diagn 4: 343–351.
10. Shi MM (2001) Enabling large-scale pharmacogenetic studies by high-throughput mutation detection and genotyping technologies. Clinical Chemistry 47: 164–72.
11. James D, Schmidt A, Wall E, Green M, Masri S (2003) Reliable detection and identification of genetically modified maize, soybean, and canola by multiplex PCR analysis. J Agric Food Chem 51: 5829–5834.
12. Zhang MH, Gao XJ, Yu YB, Qin J, Li QZ (2007) Detection of Roundup Ready soy in highly processed products by triplex nested PCR. Food Control 18(10): 1277–1281.
13. Randhawa GJ, Chhabra R, Singh M (2009) Multiplex PCR-Based Simultaneous Amplification of Selectable Marker and Reporter Genes for the Screening of Genetically Modified Crops. J Agric Food Chem 57: 5167–5172.
14. Randhawa GJ, Chhabra R, Singh M (2008) Molecular characterization of Bt cauliflower with Multiplex PCR and validation of endogenous reference gene in Brassicaceae family. Curr Sci 95(12): 1729–1731.
15. Randhawa GJ, Singh M, Chhabra R, Guleria S, Sharma R (2009) Molecular diagnosis of transgenic tomato with osmotin gene using multiplex polymerase chain reaction. Curr Sci 96(5): 689–694.
16. Randhawa GJ, Sharma R, Singh M (2009) Multiplex polymerase chain reaction for detection of genetically modified potato (Solanum tuberosum L.) with cry1Ab gene. Indian J Agric Sci 79(5): 368–371.

17. Bai WB, Xu WT, Huang KL, Yuan YF, Cao SS, et al. (2009) A novel common primer multiplex PCR (CP-M-PCR) method for the simultaneous detection of meat species. Food Control 20: 366–370.

18. Yuan YF, Xu WT, Zhai ZF, Shi H, Luo YB, et al. (2009) Universal Primer-Multiplex PCR approach for simultaneous detection of Escherichia coli, Listeria monocytogenes, and Salmonella spp. in food samples. Food Microbio Saf 74(8): 446–452.

19. Xu WT, Yuan YF, Luo YB, Bai WB, Zhang CJ, et al. (2009) Event-specific detection of stacked genetically modified maize Bt11×GA21 by UP-M-PCR and real-time PCR. J Agric Food Chem 57: 395–402.

20. Brant JB, Peter MG (2007) Silver staining DNA in polyacrylamide gels. Nat Protoc 2(11): 2649–2654.

21. Sambrook J, Russel DW (2001) Molecular Cloning: A Laboratory Manual 3rd Ed Cold Spring Harbor Laboratory Press, Cold Spring Harbor, NY.

22. Heukeshoven J, Dernick R (1985) Simplified method for staining of proteins in polyacrylamide gels and the mechanism of silver staining. Electrophoresis 6: 103–112.

23. Somerville LL, Wang K (1981) The ultrasensitive silver 'protein' stain also detects nanograms of nucleic acids. Biochem. Biophys Res Commun 102: 53–58.

24. Boulikas T, Hancock R (1981) A highly sensitive technique for staining DNA and RNA in polyacrylamide gels using silver. J Biochem Biophys Methods 4: 219–228.

25. Junichi M, Taichi O, Hiroshi A, Reiko T, Akihiro H, et al. (2009) Simultaneous detection of recombinant DNA segments introduced into genetically modified crops with multiplex ligase chain reaction coupled with Multiplex Polymerase Chain Reaction. J Agric Food Chem 57(7): 2640–2646.

26. Tomoaki Y, Hideo K, Takeshi M, Takashi K, Mayu I, et al. (2005) Applicability of quantification of genetically modified organisms to foods processed from maize and soy. J Agric Food Chem 53: 2052–2059.

27. Tigst D, Indira R (2008) Multiplex qualitative PCR assay for identification of genetically modified canola events and real-time event-specific PCR assay for quantification of the GT73 canola event. Food Control 19: 893–897.

5

Recent Weather Extremes and Impacts on Agricultural Production and Vector-Borne Disease Outbreak Patterns

Assaf Anyamba[1,4]*, Jennifer L. Small[1,5], Seth C. Britch[2], Compton J. Tucker[1], Edwin W. Pak[1,5], Curt A. Reynolds[1], James Crutchfield[3], Kenneth J. Linthicum[2]

1 National Aeronautics and Space Administration, Goddard Space Flight Center, Biospheric Sciences Laboratory, Greenbelt, Maryland, United States of America, 2 United States Department of Agriculture, Agricultural Research Service, Center for Medical, Agricultural, & Veterinary Entomology, Gainesville, Florida, United States of America, 3 United States Department of Agriculture, Foreign Agricultural Service, International Production & Assessment Division, Washington, District of Columbia, United States of America, 4 Universities Space Research Association, Columbia, Maryland, United States of America, 5 Science Systems and Applications Incorporated, Lanham, Maryland, United States of America

Abstract

We document significant worldwide weather anomalies that affected agriculture and vector-borne disease outbreaks during the 2010–2012 period. We utilized 2000–2012 vegetation index and land surface temperature data from NASA's satellite-based Moderate Resolution Imaging Spectroradiometer (MODIS) to map the magnitude and extent of these anomalies for diverse regions including the continental United States, Russia, East Africa, Southern Africa, and Australia. We demonstrate that shifts in temperature and/or precipitation have significant impacts on vegetation patterns with attendant consequences for agriculture and public health. Weather extremes resulted in excessive rainfall and flooding as well as severe drought, which caused ~10 to 80% variation in major agricultural commodity production (including wheat, corn, cotton, sorghum) and created exceptional conditions for extensive mosquito-borne disease outbreaks of dengue, Rift Valley fever, Murray Valley encephalitis, and West Nile virus disease. Analysis of MODIS data provided a standardized method for quantifying the extreme weather anomalies observed during this period. Assessments of land surface conditions from satellite-based systems such as MODIS can be a valuable tool in national, regional, and global weather impact determinations.

Editor: Tetsuro Ikegami, The University of Texas Medical Branch, United States of America

Funding: This work was made possible by funding from USDA Foreign Agricultural Service towards the Global Agricultural Monitoring project, DoD Armed Forces Health Surveillance Center's Global Emerging Infections Surveillance and Response System (AFHSC/GEIS) under the Human Febrile and Vector -Borne Illnesses (FVBI) Program and USDA Agricultural Research Service. The funders had no role in study design, data collection and analysis, decision to publish, or preparation of the manuscript.

Competing Interests: The authors have declared that no competing interests exist.

* E-mail: assaf.anyamba@nasa.gov

Introduction

Severe drought and flooding events occurred in the 2010–2012 period with pronounced impacts on agriculture and public health at regional and national scales. Such extreme weather events frequently result from factors driving interannual climate variability, such as the *El Niño*/Southern Oscillation (ENSO) phenomenon and related tropical and extra-tropical teleconnections [1–3]. Importantly, these extremes are increasingly being amplified by long-term shifts in climate and global warming [4,5]. Extreme weather patterns associated with ENSO have varying impacts on regional ecosystems, agriculture, and health [5–10], and these impacts were very evident during the 2010–2012 period. Anomalously wet or dry conditions can lead to ecological settings favouring emergence or re-emergence of disease vectors and pathogens of global public health relevance such as West Nile virus, malaria, dengue virus, cholera, Murray Valley encephalitis virus, and Rift Valley fever virus among others [11–16]. Extremes in temperature and rainfall can affect crop yields, crop pests, pasture productivity, and create hazardous fire conditions [17–21]. Heavy floods damage infrastructure and crops, and wash away productive topsoil. While socioeconomic impacts of such events

vary regionally, depending on existing resources for mitigation, in general they result in increased volatility in agricultural production and commodity prices, as well as higher costs for transport, infrastructure repair, and public health services.

Differences in assessment capability have made it difficult to quantify the global extent and impact of large-scale extreme weather events; however, the 2010–2012 period presents an opportunity to quantify impacts by analysis of satellite-based indicators of surface conditions during these years. We utilized 2000–2012 normalized difference vegetation index (NDVI) and land surface temperature (LST) measurements from the Moderate-Resolution Imaging Spectroradiometer (MODIS) instrument on-board NASA's Earth Observing System Terra and Aqua satellites [22] to illustrate the impacts of weather extremes on agriculture and vector-borne disease outbreak patterns for selected regional locations around the world shown in Figure 1. The anomalous conditions during 2010–2012 are the most extreme weather events in the Terra MODIS 12 year record, and the timing and unique intensity of these events is corroborated by analyses using longer term climate data sets [5,23–24]. We postulate that because both severe drought or flooding may create ecological conditions for the emergence of disease vectors and may substantially affect agricultural production, then under such extreme weather events as documented for 2010–2012 we should observe outbreaks of vector-borne diseases and sharp declines, in the cases of drought, and anomalous increases, in the cases of sufficient above-normal rainfall, in agricultural production. Our aim in this manuscript is to map the locations of documented 2010–2012 extremes at selected regional locations around the world using various satellite datasets, and describe and illustrate the impacts of these extremes by analyzing agricultural production data and vector-borne disease location data for the selected regions.

Materials and Methods

We analyzed two standard products from the MODIS instrument, the NDVI and LST, in combination with rainfall, crop production data, and vector-borne disease occurrence data extracted from an array of resources to investigate the impacts of weather extremes on agriculture and public health for selected regions of study shown in Figure S1 in File S1. The regions of study were selected based on a number of metrics and observations including assessment of areas of defined geographic anomalies in vegetation index data using the Global Agricultural Monitoring System at http://glam1.gsfc.nasa.gov/ and the Crop Explorer of major agricultural regions at http://www.pecad.fas.usda.gov/cropexplorer/. In addition, we monitored and evaluated vector-borne disease outbreak information from the Program for Monitoring Emerging Diseases (ProMED-mail), World Organization for Animal Health (OIE) (http://www.oie.int/) and US Centres for Disease Control and Prevention data records, available at http://diseasemaps.usgs.gov/wnv_us_human.html and information gathered through collaborations with the Global Emerging Infections Surveillance and Response System (GEIS) division of the DoD Armed Forces Health Surveillance Center (http://www.afhsc.mil/geisPartners). The selected regions and period of study experienced temporally persistent, spatially coherent anomalies and are unique in exemplifying clusters of agricultural production shortfalls and/or increase in vector-borne disease outbreaks associated with background weather anomalies taking place in each region (SI). Where any changes in agricultural production or a vector-borne disease outbreak were associated with extremes resulting from changes in the phase of ENSO, we referred to documentation from the NOAA Climate Prediction Center for information. Details of the data sets and analyses are given in the SI Materials and Methods (File S1), including Table S1 and Figures S1 to S4 in File S1. The NDVI quantifies the photosynthetic capacity of vegetation and how this varies in time [25,26], and LST provides a measure of the temperature of the land surface [27]. Both data sets can be used to monitor and assess agricultural growing conditions [28] and also infer ecological conditions that lead to the emergence and increase in abundance of disease vectors [16]. To gauge the impact of extreme weather events on regional ecosystems, we calculated monthly NDVI and LST anomalies by expressing differences between the monthly values and long-term mean 2000–2011 NDVI or LST values [29]. We also derived seasonal anomaly metrics by calculating cumulative three month anomalies, for example June, July, August (JJA) [SI]. The LST and NDVI monthly and seasonal anomalies were examined over the growing seasons for the selected six global agricultural regions worldwide (Figure 1 and Figure S1 in File S1) and were contrasted with agricultural production data extracted from the USDA Production, Supply, and Distribution Online (PSD) electronic database (http://www.fas.usda.gov/psdonline/) for the focal regions. Seasonal NDVI is particularly relevant to

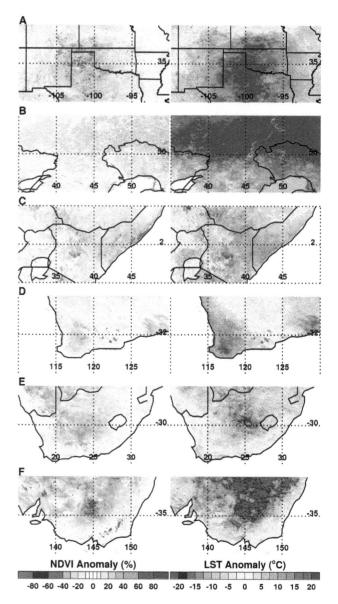

Figure 1. MODIS normalized difference vegetation index (NDVI) and land surface temperature (LST) anomalies for selected growing seasons and regions during 2010–2012. The NDVI quantifies photosynthetic potential of vegetation using satellite-measured reflected radiances in the infrared and red wavelengths [25,26]. We calculated monthly and seasonal NDVI and LST anomalies by expressing differences between current and mean 2000–2011 NDVI values. NDVI anomalies are expressed as percentage departures, and LST anomalies are absolute departures from long-term means. (A) Texas, US June, July, August (JJA) 2011, (B) SW Russia (Volga District) JJA 2010, (C) East Africa (Somalia/Kenya) DJF 2010/11, (D) SW Australia (Western Australia) SON 2010, (E) South Africa (Free State/North West) DJF 2010/11, (F) SE Australia (New South Wales) SON 2010.

agriculture, as cumulative NDVI is linearly related to gross primary production, and hence highly correlated to agricultural production [30]. Accordingly temperature (LST) during the growing season regulates crop growth [7,18,19] and in general vegetation, which in this case can be observed through NDVI time series. Anomaly calculations using MODIS data provided unbiased and spatially continuous representations of the impacts of extreme weather conditions on regional vegetation for the study period and were augmented with rainfall data extracted and derived for the focal regions from the Global Precipitation Climatology Project [31] (Table 1). Various publicly-available databases were surveyed to develop maps of outbreak locations of dengue, Rift Valley fever, Murray Valley encephalitis, and West Nile virus disease (SI), and these maps were juxtaposed with the focal anomalous regions identified in this study. We also compared the seasonal distribution of LST and NDVI values for the focal regions in the study period to gauge the direction of shifts of the variables from long-term means resulting from extreme conditions. Historical ENSO patterns and timing were determined from NOAA data portals (http://www.cpc.ncep.noaa.gov/products/ MD-index.shtml) to establish that extreme weather conditions associated with the cold *La Niña* phase of ENSO and associated teleconnections resulted either in severely depressed vegetation conditions/drought and above-average LST, or excessive rainfall, denser than average vegetation, and cooler than normal LST, with a heterogeneity of impacts on agricultural production and vector-borne disease transmission in the various focal regions around the world.

Ethics Statement

All disease case data analyzed were anonymized. We used GPS latitude-longitude coordinates to map approximate case locations, and we did not handle or deal with any human or animal specimens.

Results

Regions Affected by Drought: Below-Normal Rainfall and Above-Normal Temperatures

In 2010 and 2011, severe drought affected four regions: the southern United States, western Russia, Western Australia, and East Africa (Figure 1A–D); in 2012, the entire continental US was affected (Figure S2 in File S1). In all cases, drought reduced crop yields and led to increased wild fires. The southern half of the contiguous US experienced extreme and persistent drought from mid-2010 through September 2011, with the greatest impact in Texas (Figure 1A). The Texas drought epicenter had historic rainfall lows of up to ~66% below normal (Table 1), coupled with one of the hottest summers on record in 2011 [5,24,32], with monthly LSTs up to 8°C above normal (Figure 2A), and seasonal LST as high as 20°C above normal (Figure 1A). As a result of high temperatures, vegetation photosynthetic capacity was reduced by 40–60% below average (Figures 1A and 2A). Drought and scorching temperatures led to declines in crop conditions and abandonment of fields. In particular, this was because the Texas drought of 2011 was of historical proportions and classified as the

most extreme rainfall deficit year in over 100 years [24,32]. The severity of the drought overwhelmed the agricultural system with high evapo-transpiration rates, rationing of ground water resources for irrigation and other uses. Although producers switched irrigation from corn to more drought-tolerant cotton in some areas, cotton production still declined by 50% to 771 kilotons below the 2000–2011 average (Figure 3A). Extreme water deficits had reduced the capacity of the cotton crop to carry fruit as a result of lower rates of leaf photosynthesis and severely reduced the development of floral buds thus reducing yields [33]. Direct losses from the drought approached $10 billion in the state of Texas alone in 2011 [34,35]. Hot, dry conditions also resulted in a very active fire season across the southern US costing more than $1 billion [36]. Unprecedented 100-year climate conditions of extreme high temperatures (Figure 4A,E) and lack of rainfall were linked to the highest period of West Nile virus activity on record in Texas and the rest of the continental US: the 2012 epidemic of West Nile virus disease across the continental US (Figure 5) was the largest such outbreak since the introduction of West Nile virus into the country in 1999, and the spike in human West Nile virus disease cases in 2012 can in part be associated with extreme drought [12] and anomalously high positive shift in summer mean temperatures from ~30°C to 33°C (Figure 4A). Elevated temperatures increase the efficiency of transmission of West Nile virus by both *Culex pipiens* and *Cx. tarsalis* mosquitoes, and have positive effects on mosquito population development and survival, biting rates, and viral replication within these mosquito species [37–39].

Across Eurasia, the summer drought of June-August 2010 was centered in western Russia (Figure 1B) with the drought area extending to Belarus, Poland, Germany, Ukraine, and Kazakhstan [23,40]. Cumulative seasonal LSTs reached as high as 20°C above normal with declines in NDVI of up to 40% below normal (Figures 1B and 2B). Severe and persistent drought sharply reduced agricultural production of major grains in Russia's Central, Volga, and Ural Districts. Estimated wheat production for 2010 was just over 41,500 kilotons, approximately 14% below the 2000–2011 average (Figure 3B), and barley production for 2010 was estimated at 8,350 kilotons, which was 51% below the 2000–2011 average and the lowest in that time span. The combination of severe drought and high temperatures led to the abandonment of approximately 13.3 million hectares of cropland, including half the grain area in the Volga District. The extreme heat resulted in more than 15,000 deaths in Russia during the 2010 summer [23,40]. As in the southern US, the drought caused extreme fire conditions. From July-September 2010 more than 1.25 million hectares burned, causing an estimated $630 million in damages [41].

Table 1. Total seasonal rainfall, long-term means, and anomalies for 2010–2011 extracted from the Global Precipitation Climatology Project [31] for regions presented in Figure 1.

Region	Season	Total (mm)	Mean (mm)	Anomaly (%)
US (Texas)	June–August 2011	59.40	174.11	−65.88
SW Russia (Volga District)	June–August 2010	35.48	133.18	−73.36
East Africa (Somalia/Kenya)	December 2010–February 2011	7.68	51.86	−85.19
SW Australia (Western Australia)	September–November 2010	25.84	90.08	−71.31
South Africa (Free State/North West)	December 2010–February 2011	363.92	253.69	43.45
SE Australia (New South Wales)	September–November 2010	255.27	102.93	148.00

Anomalies are calculated based on the 1979–2011 climatology period.

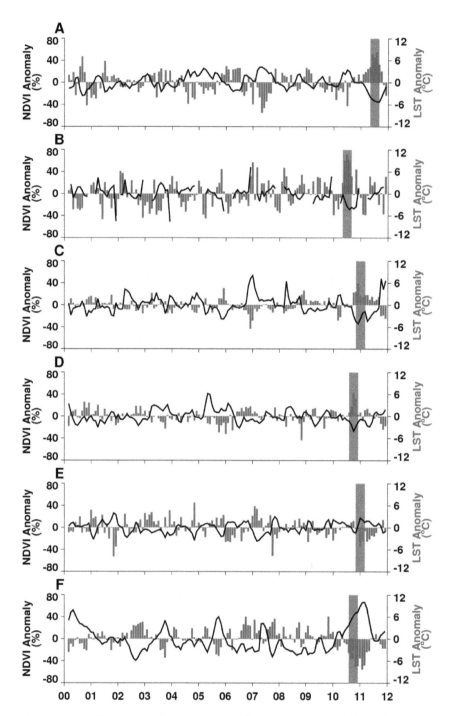

Figure 2. Monthly time series of 2000–2012 MODIS normalized difference vegetation index (NDVI; black line) and land surface temperature (LST; vertical red bars) anomalies for agricultural regions shown in Figure 1 and Figure S1 in File S1. Grey bars indicate selected periods of focused impact studies for each region.

East Africa experienced below-normal rainfall from early 2010 through mid-2011. Rainfall for December 2010 – February 2011 was ~85% below normal (Table 1) and seasonal LSTs were between 5–20°C above normal. The stress of these conditions on vegetation and pastures is illustrated by NDVI departures of 60% below normal (Figure 1C). The peak of the East Africa drought was during the 2010–2011 *La Niña* period (Figure 2C) and led to sharp declines in agricultural production, in particular sorghum, a staple food in this semi-arid region. For Somalia, the total sorghum production for 2011 was 25 kilotons, more than 80% below normal and the lowest for the last decade (Figure 3C). The drought diminished the productivity of pastures and caused widespread famine and high mortality in livestock throughout the region. Although it is difficult to gauge the total human impact of the East African famine, the United Nations estimated more than 13 million people required humanitarian assistance there by December 2011 [42]. Persistent above-normal temperatures were associated with the first known large-scale outbreak of dengue in

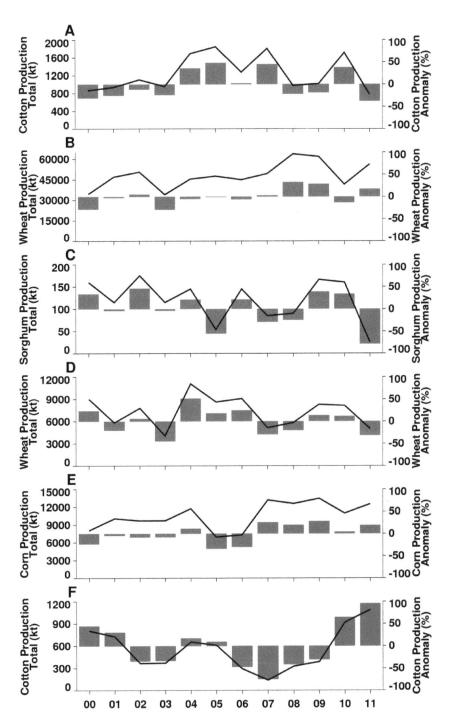

Figure 3. Crop production time series for selected regions in 2000–2011 expressed as annual totals (represented by the black line) and as annual anomalies, i.e., departures from the decadal mean (represented by the red bar chart).

East Africa. Dengue is a mosquito-borne disease associated with infrequent replenishment of stored water supply around households during hot, dry climatic conditions in densely populated areas [9,15,43], which has been shown to increase populations of *Aedes aegypti*, the primary mosquito vector of dengue [44]. Severe drought coupled with higher temperatures increased the abundance of container-breeding dengue virus vector mosquitoes in urban settings leading to the dengue outbreak that persists up to the present [45]. Outbreaks were centered in Mogadishu, Somalia,

and Mandera, Kenya (Figure 5), compounding the effects of famine on already vulnerable populations.

Australia experienced both extremes: drought over the western quarter and extreme wet conditions over the eastern half of the continent. Drought and floods in Australia are linked with ENSO cycles, as seen in 1916, 1917, 1950, 1954–1956, 1973–1975, and 2010–2011 [46]. In western Australia, drought prevailed during the 2010 growing season: seasonal LSTs were up to 20°C above normal, vegetation conditions were 30–60% below normal

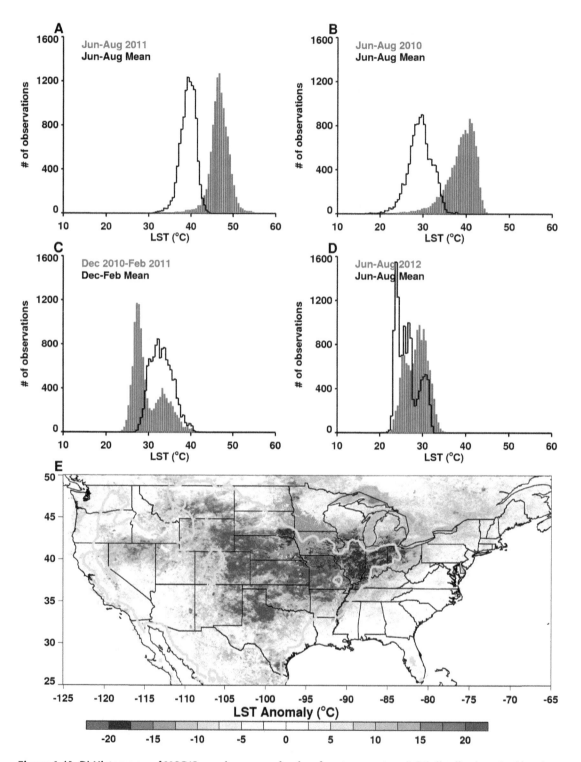

Figure 4. (A–D) Histograms of MODIS growing season land surface temperature (LST) distributions (red bars) compared against the long-term means (black lines) for selected regions of study. (E) Spatial pattern of LST anomalies for the US during June, July, August (JJA) 2012. The area shaded in olive green shows the dominant corn and soy growing region; the solid neon green line delineates the JJA 2012 seasonal 30°C isocline which encompasses the majority of the US agricultural region; and the dotted neon green line shows the JJA 2000–2011 long-term mean 30°C isocline. Persistent temperatures above 30°C destroy most crops and reduce yields [6,17,57].

(Figures 1D and 2D), and total rainfall for September – November 2010 was only ~26 mm, or ~71% below normal (Table 1). Regional wheat production was more than 30% below normal at just under 5,000 kilotons (Figure 3D). The persistent drought, low soil moisture, and above-normal temperatures centered on Western Australia's central agricultural region resulted in the worst overall agricultural production in 40 years [47].

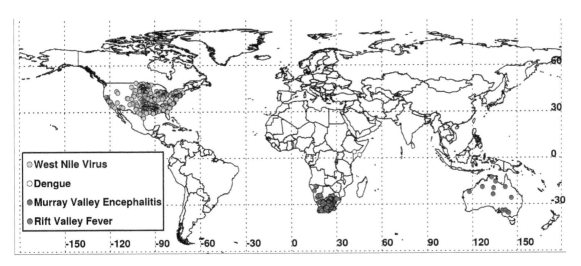

Figure 5. Global distribution of epidemics/epizootics of vector-borne disease outbreaks during 2010–2012 associated with weather extremes, showing the outbreak locations of West Nile virus disease (US, 2012), dengue (East Africa, 2011), Rift Valley fever (Southern Africa, 2011), and Murray Valley encephalitis (Australia, 2011).

Regions Affected by Excessive Rainfall and Below-Normal Temperatures

Three regions experienced excessive rain, floods, and below-normal temperatures during the September 2010 to May 2011 *La Niña* event [48]: the northern US and southern Canada; southeastern Australia; and southern Africa. In North America, heavy spring rains saturated fields, causing flooding and hampering field work in portions of the Corn Belt, Ohio Valley, Tennessee Valley, Missouri Valley, Pacific Northwest, and the northern Great Plains extending into southern Canada. The northern and central Great Plains and much of the Northeast and Atlantic Coast States accumulated rainfall totaling 200% or more above normal and experienced prolonged cool temperatures during this period (Figure S3 in File S1). Above-average precipitation also fell in the Pacific Northwest and rains triggered widespread flooding in the Mississippi and Missouri basins, putting thousands of acres of farmland under water and causing more than $5 billion in damage. Agricultural losses in southern Canada amounted to $1 billion, but impacts on US agricultural production were mixed: although states like Iowa reported bumper harvests, most agricultural states suffered reduced harvests from delayed planting of crops in spring 2011 because waterlogged fields were inaccessible [48,49].

The main corn producing region of South Africa, known as "The Maize Triangle" received heavy rainfall during the 2010–2011 *La Niña* [48], with December 2010-February 2011 total rainfall more than 360 mm, or ~43% higher than normal (Table 1). Abundant rainfall combined with cooler temperatures (Figures 1E and 2E) created favorable growing conditions throughout the country and resulted in the fifth successive year of 25% above-average maize production (Figure 3E). Cool, wet weather led to rapid green-up of vegetation (Figure 1E) and flooding of low-lying areas, or *dambos/pans*, creating ideal ecological conditions for hatching *Aedes* mosquito eggs infected with the Rift Valley fever virus. These conditions in association with the downward shifts in December-February 2010/2011 mean seasonal temperatures from ~40°C to 30°C in South Africa (Figure S4C in File S1) were conducive not only to increased mosquito populations, but increased virus infection in mosquitoes [50,51] and subsequent Rift Valley fever virus transmission. This resulted in the most extensive and widespread outbreak of Rift Valley fever in the region since the 1970s (Figure 5), negatively impacting the livestock industry and human health in southern Africa [51,52]. Overall, MODIS time series NDVI and LST anomalies during the 2010/11 *La Niña* were the most persistent and extreme anomalies for the southern Africa region for the 12 year record.

Southeastern Australia likewise experienced extreme wet conditions, with September-November 2010 rainfall total of ~255 mm, or 148% above normal (Table 1). High rainfall and cooler temperatures (Figure 1F) resulted in increased agricultural yields, putting Australia on track for one of the largest cotton crops in the last decade. The USDA estimated 2011 total cotton production in the region at 1,088 kilotons, approximately 95% higher than the 2000–2011 average (Figure 3F). Most cotton growing regions in New South Wales and Queensland received above-normal winter and spring rainfall, creating excellent planting conditions and sufficient irrigation resources. Overall, 2011 area planted and production estimates were at decadal highs and represent a significant recovery from severe droughts experienced during 2006–2009 (Figure 2F). However, the wet conditions also had negative impacts. The large accumulation of biomass in 2011 created heavy fuel loads that resulted in an intense fire season in early 2012, with large brush fires occurring in Victoria, south Australia, and New South Wales [53,54]. Rains also caused moderate to severe flooding in parts of eastern Australia, especially central-southern Queensland and northern New South Wales. At the peak of the *La Niña* event from November 2010 to January 2011, floods damaged crops in low-lying areas. The combined persistent and heavy rainfall, cooler temperatures, and vegetation growth created ideal conditions for an increase in populations of mosquito vectors of Murray Valley encephalitis virus, leading to outbreaks over northern and eastern Australia [55] (Figure 5). Murray Valley encephalitis mosquito vectors, primarily *Culex annulirostris*, favor cooler temperatures associated with heavy rainfall periods in the tropics and sub-tropics [56]. The downward shift from ~40°C to 30°C in December-January 2010/2011 mean seasonal temperatures compared to the long-term mean distribution for eastern Australia (Figure S4E in File S1) during this epidemic period shows that a cooler environment was conducive to increased mosquito populations and virus amplification, and increased virus transmission to humans.

Discussion

The above regional snapshots exemplify the contrasting patterns and impacts of extreme weather events and demonstrate that changing weather will vary geographically in both magnitude and extent. For example, Figure 4 shows seasonal LSTs against long-term mean seasonal LST distributions for selected regions during the study period. June-August LST distributions for both Texas, US (2011, Figure 4A) and Volga District, Russia (2010, Figure 4B) show significantly warmer and drier conditions compared to long term June-August means, with temperatures as high as ~45°C. Importantly, seasonal minimum temperatures for both regions were also ~10°C higher than normal, indicating an overall increase in seasonal LST, not simply increased LST variability. Temperatures above 30°C increase stress on vegetation and damage pastures, a variety of crops [6,7,17–20,57], and here led to sharp declines in agricultural production, such as a 50% decline in cotton production in Texas (Figure 3A) and a 25% decline in Volga District wheat (Figure 3B). In contrast, Figure 4C shows seasonal LSTs for December 2010-February 2011 over South Africa's Maize Triangle. Compared to Texas and Volga District, the weather in South Africa was cooler and wetter, with average temperatures in the range of 19–28°C that have been associated with increased corn yields [7,18,19,57]. Indeed, above-average corn production was observed over South Africa during the last five years (Figure 3E).

The 2011 drought effects in Texas were similar to the Iowa corn-growing area a year later (Figure 4D), where the 2012 seasonal minimum, mean, and maximum temperatures as high as 35°C, were all above normal, with higher monthly and long-term mean LSTs. The 2011–2012 US droughts were associated with negative public health impacts in addition to large scale agricultural impacts such as reduction in production of major commodities including corn and cotton. The spatial pattern in Figure 4E shows that in 2012 the drought epicenter had shifted to the central US unlike in 2011 when the drought was centered in Texas (Figure 1A). This shows the tendency for drought to persist, expand, and propagate over large areas once initiated. Extreme 2012 summer temperatures affected ~70% of the US corn growing area and led to extensive impoverishment of pasture and rangeland, plus record reductions in corn production. The widespread and persistent nature of drought and high temperatures also contributed to the large-scale epidemic of West Nile virus disease cases across the continental US [38] (Figure 5; SI).

Droughts of the magnitude we observed in the US, Russia, and southeastern Australia have produced recent volatility in agricultural commodity prices, increased costs for feed and water in the livestock industry, decreased supply of water resources, increased fire risk, and contributed to other detrimental societal impacts [17]. The 2012 growing season marked the third consecutive year of weather extremes that affected the northern hemisphere and resulted in depleted commodity stocks. The major 2010–2012 droughts were the most persistent and extreme for these regions in the 2000–2012 MODIS record. The 2011 drought in eastern

Africa was also the most extreme event in the MODIS record for the region. Conversely, *La Niña* rains in Australia and South Africa had largely positive impacts on agriculture, but associated vector-borne disease outbreaks negatively impacted public health. Our findings show that extreme seasonal shifts in weather conditions, regardless of direction, favor different vector-borne disease systems in different ways and may lead to increased risk of vector-borne disease outbreaks. Transmission of different vector-borne pathogens such as West Nile virus or Rift Valley fever virus may be enhanced by opposite weather extremes because they are different viruses with completely different ecologies of transmission, different hosts, different mosquito vectors, and different optimal habitats. For instance, anomalously hot and dry conditions can lead to increased storage of water around households and consequent increases in populations of container-breeding mosquitoes like *Aedes aegypti* that may transmit dengue virus. On the other hand, anomalously prolonged wet conditions and flooding, aside from triggering the large-scale emergence of disease-vector mosquitoes, can enhance production and sustainment of vegetation which directly favors the survival of mosquitoes and thereby their vectorial capacity, that is, their capacity to become infected by viruses, such as Rift Valley fever virus, and then transmit them. Such environmentally enhanced outbreaks will vary globally depending on the virus and its transmission ecology and the geographic location and baseline condition of disease endemism and seasonality [58], and could favor the globalization of such pathogens. Our analysis of temperature and vegetation conditions provides a method for quantifying weather extremes consistently from region to region, and demonstrates the value of satellite data in monitoring and mapping the magnitude and extent of such events. As extreme weather events become more common under a more variable climate, nations will face costly adaptation. Systematically-collected satellite data such as those we describe here can be a valuable contribution to national, regional, and global weather impact determinations.

Supporting Information

File S1 Supplementary Information. This SI section includes: Detailed Materials and Methods Text, Figures S1 to S4, Table S1, References, Figure Legends.

Acknowledgments

Holly Riebeek and Amy Houghton provided useful suggestions for improvement of the manuscript.

Author Contributions

Conceived and designed the experiments: AA KJL. Performed the experiments: AA JLS SCB. Analyzed the data: JLS AA SCB EWP JC. Contributed reagents/materials/analysis tools: AA JLS SCB CAR JC. Wrote the paper: AA SCB KJL JLS CJT JC CR EWP. Contributed knowledge on ecology and temperature effects on mosquito vectors: KJL SCB. Contributed knowledge on weather extreme impacts on vegetation and ecosystems: AA CJT. Contributed knowledge on extreme weather on agricultural production: CAR CR. Wrote data analysis software: JLS EWP.

References

1. Glantz MH (1991) Introduction. In: Glantz MH, Katz RW, Nicholls N, editors. Teleconnections Linking Worldwide Climate Anomalies. New York: Cambridge University Press. pp. 1–2.

2. Lyon B, Barnston AG (2005) ENSO and the spatial extent of interannual precipitation extremes in tropical land areas. J Clim 18: 5095–5109.

3. Hoerling MP, Kumar A (2000) Understanding and predicting extratropical teleconnections related to ENSO. In: Diaz HF, Markgraf V, editors. *El Niño* and the Southern Oscillation: Multiscale Variability and Global and Regional Impacts. New York: Cambridge University Press. pp. 57–88.

4. Hansen J, Sato M, Ruedy R (2012) Perception of climate change. Proc Natl Acad Sci U S A 109: 14726–14727, E2415–E2423.

5. Blunden J, Arndt DS (2013) State of the climate in 2012. Bull Am Meteorol Soc 94: S1–S258.

6. Battisti DS, Naylor RL (2009) Historical warnings of future food insecurity with unprecedented seasonal heat. Science 323: 240–244.

7. Legler DM, Bryant KJ, O'Brien JJ (1999) Impact of ENSO-related climate anomalies on crop yields in the US. Clim Change 42: 351–375.

8. Caviedes CN, Fik TJ (1992) The Peru-Chile eastern Pacific fisheries and climatic oscillation. In: Glantz MH, editor. Climate Variability, Climate Change, and Fisheries. Cambridge: Cambridge Univ Press. pp. 355–375.

9. Epstein PR (2005) Climate change and human health. N Engl J Med 353: 1433–1436.

10. Bouma MJ, Dye C (1997) Cycles of malaria associated with El Niño in Venezuela. J Am Med Assoc 278: 1772–1774.

11. Gubler DJ (2002) The global emergence/resurgence of arboviral diseases as public health problems. Arch Med Res 33: 330–342.

12. Epstein PR, Defilippo C (2001) West Nile virus and drought. Global Change Human Health 2: 105–107.

13. Pascual M, Rodó X, Ellner SP, Colwell R, Bouma MJ (2000) Cholera dynamics and El Niño-Southern Oscillation. Science 289: 1766–1769.

14. Nicholls NA (1986) A method for predicting Murray Valley encephalitis in southeast Australia using the Southern Oscillation. Aust J Exp Biol Med Sci 64: 587–594.

15. Chretien J-P, Anyamba A, Bedno SA, Breiman RF, Sang R, et al. (2007) Drought-associated chikungunya emergence along coastal East Africa. Am J Trop Med Hyg 76: 405–407.

16. Linthicum KJ, Anyamba A, Tucker CJ, Kelley PW, Myers MF, et al. (1999) Climate and satellite indicators to forecast Rift Valley fever epidemics in Kenya. Science 285: 397–400.

17. Rosenzweig CE, Iglesias A, Yang XB, Epstein PR, Chivian E (2001) Climate change and extreme weather events: Implications for food production, plant diseases, and pests. Global Change Human Health 2: 90–104.

18. Semenov MA, Porter JR (1995) Climatic variability and the modeling of crop yields. Agric Forest Meteorol 73: 265–283.

19. Schlenker W, Roberts MJ (2009) Nonlinear temperature effects indicate severe damages to US crop yields under climate change. Proc Natl Acad Sci U S A 106: 15594–15598.

20. Tubiello FN, Soussana JF, Howden SM (2007) Crop and pasture response to climate change. Proc Natl Acad Sci U S A 104: 19686–19690.

21. Mueller-Dombois D, Goldammer JG (1990) Fire in tropical ecosystems and global environmental change. In: Goldammer JG, editor. Fire in the Tropical Biota: Ecosystem Processes and Global Challenges. Heidelberg: Springer-Verlag. pp. 1–10.

22. Justice CO, Townshend JRG, Vermote EF, Masuoka E, Wolfe RE, et al. (2002) An overview of MODIS Land data processing and product status. Rem Sens Env 83: 3–15.

23. Trenberth KE, Fasullo JT (2012) Climate extremes and climate change: The Russian heat wave and other climate extremes of 2010. J Geophys Res 117: D17103.

24. Hoerling M, Kumar A, Dole R, Nielse-Gammon JW, Eischeid J, et al. (2013) Anatomy of an extreme event. J Climate 26: 2811–2832. doi: http://dx.doi.org/10.1175/JCLI-D-12-00270.1

25. Tucker CJ (1979) Red and photographic infrared linear combinations for monitoring vegetation. Remote Sens Environ 8: 127–150.

26. Myneni RB, Hall FG, Sellers PJ, Marshak AL (1995) The interpretation of spectral vegetation indexes. IEEE Trans Geosci Rem Sens 33: 481–486.

27. Wan Z (2008) New refinements and validation of the MODIS land-surface temperature/emissivity products. Rem Sens Env 112: 59–74.

28. Becker-Reshef I, Justice C, Sullivan M, Vermote E, Tucker CJ, et al. (2010) Monitoring global croplands with coarse resolution earth observations: The Global Agriculture Monitoring (GLAM) Project. Remote Sens 2: 1589–1609.

29. Anyamba A, Tucker CJ (2005) Analysis of Sahelian vegetation dynamics using NOAA-AVHRR NDVI data from 1981–2003. J Arid Environ 63: 596–614.

30. Prince SD, Tucker CJ (1986) Satellite remote sensing of rangelands in Botswana II: NOAA AVHRR and herbaceous vegetation. Int J Rem Sens 7: 1555–1570.

31. Adler RF, Huffman GJ, Chang A, Ferraro R, Xie P-P, et al. (2003) The version-2 Global Precipitation Climatology Project (GPCP) monthly precipitation analysis (1979-present). J Hydrometeorol 4: 1147–1167.

32. Winters KE (2013) A Historical Perspective on Precipitation, Drought Severity, and Streamflow in Texas during 1951–56 and 2011. Scientific Investigations Report 2013–5113. US Department of the Interior, US Geological Survey. Available: http://pubs.usgs.gov/sir/2013/5113/pdf/sir20135113.pdf. Accessed 2014 January 28.

33. Turner N, Hearn AB, Begg JE, Constable GA (1986) Cotton (Gossypium hirsutum L.): physiological and morphological responses to water deficits and their relationship to yield. Field Crops Res 14: 153–170.

34. Anderson DP, Welch JM, Robinson J (2012) Agricultural impacts of Texas's driest year on record. Choices 27: 1–3.

35. Nielsen-Gammon JW (2012) The 2011 Texas drought. Texas Water J 3: 59–95.

36. NOAA National Climatic Data Center (2011) State of the Climate: Wildfires for Annual 2011. Available: http://www.ncdc.noaa.gov/sotc/fire/2011/13. Accessed 2012 November 27.

37. Kilpatrick AM, Meola MA, Moudy RM, Kramer LD (2008) Temperature, viral genetics, and the transmission of West Nile virus by Culex pipiens mosquitoes. PLoS Pathog 4: e1000092.

38. Johnson BJ, Sukhdeo MVK (2013) Drought-induced amplification of local and regional West Nile virus infection rates in New Jersey. J Med Entomol 50: 195–204.

39. Moudy RM, Meola MA, Morin LL, Ebel GD, Kramer LD (2007) A newly emergent genotype of West Nile virus is transmitted earlier and more efficiently by Culex mosquitoes. Am J Trop Med Hyg 77: 365–370.

40. Matsueda M (2011) Predictability of Euro-Russian blocking in summer of 2010. Geophys Res Lett 38: L06801.

41. Munich RE (2010) Geo Risks Research Heat wave, drought, wildfires in Russia (Summer 2010). MR Touch Natural Hazards – Event report. Available: https://www.munichre.com/app_pages/www/@res/pdf/NatCatService/catastrophe_portraits/Event_report_HW_DR_WF_Russia_touch_en.pdf. Accessed 2013 April 10.

42. Congressional Research Service (2012) Horn of Africa: The Humanitarian Crisis and International Response. Congressional Research Service Report for Congress, R42046.

43. Padmanabha H, Soto E, Mosquera M, Lord CC, Lounibos LP (2010) Ecological links between water storage behaviors and Aedes aegypti production: implications for dengue vector control in variable climates. Ecohealth 7: 78–90.

44. Subra R (1983) The regulation of preimaginal populations of Aedes aegypti L. (Diptera: Culicidae) on the Kenya coast. I. Preimaginal population dynamics and the role of human behavior. Ann Trop Med Parasitol 77: 195–201.

45. IRIN (2011) Kenya: Medics overwhelmed as dengue fever spreads. IRIN Humanitarian News and Analysis. Available: http://www.irinnews.org/Report/93848/KENYA-Medics-overwhelmed-as-dengue-fever-spreads. Accessed 2013 April 10.

46. Nicholls N, Lavery B, Frederiksen C, Drosdowsky W, Torok S (1996) Recent apparent changes in relationships between the El Niño-Southern Oscillation and Australian rainfall and temperature. Geophys Res Letts 23: 3357–3360.

47. USDA Foreign Agricultural Service (2011) Australia: Grain and Feed Update February 2011. Global Agricultural Information Network (GAIN) Report Number AS1101. Available: http://gain.fas.usda.gov/Recent%20GAIN%20Publications/Grain%20and%20Feed%20Update_Canberra_Australia_1-20-2011.pdf. Accessed 2013 April 19.

48. NOAA/NWS (2011) Climate Diagnostics Bulletin (2010, 2011). Available: http://www.cpc.ncep.noaa.gov/products/CDB/CDB_Archive_pdf/pdf_CDB_archive.shtml. Accessed 2013 March 30.

49. USDA Foreign Agricultural Service (2010) Canada: Grain and Feed Update November Quarterly 2010. USDA Foreign Agricultural Service Global Agricultural Information Network (GAIN) Report Number CA0043. Available: http://gain.fas.usda.gov/Recent%20GAIN%20Publications/Grain%20and%20Feed%20Update_Ottawa_Canada_11-01-2010.pdf. Accessed 2013 April 19.

50. Turell MJ (1993) Effects of environmental temperature on the vector competence of Aedes taeniorhynchus for Rift Valley fever and Venezuelan equine encephalitis viruses. Am J Trop Med Hyg 49: 672–670.

51. Grobbelaar AA, Weyer J, Leman PA, Kemp A, Paweska JT, et al. (2011) Molecular epidemiology of Rift Valley fever virus. Emerg Infect Dis 17: 2270–2276.

52. Métras RT, Porphyre DU, Pfeiffer A, Kemp PN, Thompson, et al. (2012) Exploratory space-time analyses of Rift Valley fever in South Africa in 2008–2011. PLoS Negl Trop Dis 6: e1808.

53. Australian Government Bureau of Meteorology Queensland Climate Services Centre. (2010) Monthly Weather Review: Queensland December 2010. Brisbane: Commonwealth of Australia. Available: http://bit.ly/jcdZLt. Accessed 2014 January 15.

54. Chanson H (2011) The 2010–2011 Floods in Queensland (Australia): Photographic Observations, Comments and Personal Experience. Hydraulic Model Report No. CH82/11, School of Civil Engineering, The University of Queensland, Brisbane, Australia. Available: http://espace.library.uq.edu.au/view/UQ:239732. Accessed 2014 January 15.

55. Knox J, Cowan RU, Doyle JS, Ligtermoet MK, Archer JS, et al. (2012) Murray Valley encephalitis: a review of clinical features, diagnosis and treatment. Med J Aust 196: 322–326.

56. Van Den Hurk AF, Craig SB, Tulsiani SM, Jansen CC (2010) Emerging tropical diseases in Australia. Part 4. Mosquito-borne diseases. Ann Trop Med Parasit 8: 623–640.

57. Porter JR, Semenov MA (2005) Crop responses to climatic variation. Phil Trans R Soc B 360: 2021–2035.

58. Anyamba A, Linthicum KJ, Small J, Collins K, Tucker CJ, et al. (2012) Climate teleconnections and recent patterns of human and animal disease outbreaks. PLoS Negl Trop Dis 6: e1465. doi:10.1371/journal.pntd.0001465.

6

Impact of Single and Stacked Insect-Resistant Bt-Cotton on the Honey Bee and Silkworm

Lin Niu[1,4], Yan Ma[2], Amani Mannakkara[1,3], Yao Zhao[1], Weihua Ma[1], Chaoliang Lei[1]*, Lizhen Chen[1,4]*

1 Hubei Insect Resources Utilization and Sustainable Pest Management Key Laboratory, Huazhong Agricultural University, Wuhan, Hubei, China, **2** Institute of Cotton Research, Chinese Academy of Agricultural Sciences, Anyang, Henan, China, **3** Department of Agricultural Biology, Faculty of Agriculture, University of Ruhuna, Kamburupitiya, Sri Lanka, **4** College of Plant Science and Technology, Huazhong Agricultural University, Wuhan, Hubei, China

Abstract

Transgenic insect-resistant cotton (Bt cotton) has been extensively planted in China, but its effects on non-targeted insect species such as the economically important honey bee (*Apis mellifera*) and silkworm (*Bombyx mori*) currently are unknown. In this study, pollen from two Bt cotton cultivars, one expressing Cry1Ac/EPSPS and the other expressing Cry1Ac/Cry2Ab, were used to evaluate the effects of Bt cotton on adult honey bees and silkworm larvae. Laboratory feeding studies showed no adverse effects on the survival, cumulative consumption, and total hemocyte count (THC) of *A. mellifera* fed with Bt pollen for 7 days. No effects on the survival or development of *B. mori* larvae were observed either. A marginally significant difference between Cry1Ac/Cry2Ab cotton and the conventional cotton on the THC of the 3rd day of 5th *B. mori* instar larvae was observed only at the two highest pollen densities (approximately 900 and 8000 grains/cm^2), which are much higher than the pollen deposition that occurs under normal field conditions. The results of this study show that pollen of the tested Bt cotton varieties carried no lethal or sublethal risk for *A. mellifera,* and the risk for *B. mori* was negligible.

Editor: Nicolas Desneux, French National Institute for Agricultural Research (INRA), France

Funding: This work was supported by the National Special Transgenic Project from the Chinese Ministry of Agriculture (grant No. 2013ZX08011-002). The funders had no role in study design, data collection and analysis, decision to publish, or preparation of the manuscript.

Competing Interests: The authors have declared that no competing interests exist.

* E-mail: lzchen@mail.hzau.edu.cn (LC); ioir@mail.hzau.edu.cn (CL)

Introduction

China is one of the countries taking the lead in planting genetically modified (GM) crops, ranking sixth in the world by 2012 [1]. The planting area of transgenic Bt (*Bacillus thuringiensis* toxin) cotton reached 4.0 million hectares in China in 2012 [1]. Planting of Bt cotton cultivars has proven beneficial because of lower insecticide use and less damage from *Helicoverpa armigera*, the major pest of cotton [2,3]. An important technique in plant biotechnology is the stacking of resistance to multiple insects or of insect and herbicide resistance traits within a single cultivar [4]. Two new cotton varieties, Cry1Ac/Cry2Ab (both Bt toxins) and Cry1Ac/EPSPS (Bt toxin and 5-enolpyruvyl-shikimate-3-phosphate synthase), were developed in recent years, and they will be commercially available in the foreseeable future in China [5,6]. The Bt toxins (Cry1Ac and Cry2Ab) target lepidopteran pests [7,8], and the EPSPS gene makes the plants tolerant to the herbicide glyphosate [9,10].

Despite the benefits offered by GM plants, they also may have a negative impact on biodiversity and non-target organisms [11]. Thus, laboratory and extended lab/semi-field and field studies are necessary to assess such risks before commercialization [12]. As the first step of assessment of Bt cotton, laboratory tests need to be conducted to evaluate the risks of new cotton varieties on non-target organisms [13].

More than one-third of crops are pollinated by insects and other animals, among which honey bees account for about 80% of the total pollinating insects [14]. A recent study estimated the economic value of honey bee pollination for Chinese agriculture

to be worth ¥304.2 billion per year [15]. The honey bee *Apis mellifera* is the most important pollinator species around the world [16], with populations present in all countries growing GM crops [4,17], including Bt cotton [18]. Pollen is the sole protein source of *A. mellifera* colonies [19], and pollen of many important crops, including cotton [20], is collected by foraging bees [21]. Adults and larvae of *A. mellifera* are directly exposed to transgenic material via pollen consumption of GM crops, which are planted in mass monocultures [13].

The culture of the silkworm *Bombyx mori* is an important export industry that provides considerable income for people in many temperate Asian countries [22]. However, *B. mori* is susceptible to Cry1Ac and Cry2Ab proteins. The larvae of *B. mori* are fed entirely on mulberry leaves, and mulberry plants are often planted near or around the edges of cotton fields. Thus, the larvae may be exposed to the Bt insecticidal proteins expressed in Bt cotton pollen if the pollen is deposited on mulberry leaves [23].

As two economically important insects in China, *A. mellifera* and *B. mori* are key species to be tested for the potential adverse impacts of Bt cotton [18,23]. To date, few studies have assessed the potential negative impacts of Bt cotton on *A. mellifera* [18,20,24,25] and *B. mori* [23]. Existing results show that Bt toxins have no lethal effect on the two insects. Few studies of the sublethal effects of Bt toxins on *A. mellifera* [18,20,26,27] and *B. mori* [22,23,28] have been conducted either. However, as well as the side effects of pesticides on beneficial arthropods [29–31], the sublethal effects of Bt toxins on these two economically important insects might negatively impact larval development and immune capacity and lead to colony population decrease [31]. Thus, it is important to evaluate the sublethal effects of transgenic crops on honey bees and silkworms [22,32,33].

In China, the flowering period of cotton usually lasts from June to late August, a season during which honey bees have few available floral sources other than cotton. This period is also the time when silkworm rearing occurs [34]. Han et al. demonstrated that another Bt cotton (CCRI41) pollen exhibited highly variable expression of Cry1Ac throughout the season [18,25]. Therefore, the main goals of this study were to quantify the expression levels of the Bt toxins in the pollen of two transgenic cotton cultivars throughout the entire season and to determine the lethal and sublethal effects of the pollen on *A. mellifera* and *B. mori*. Li et al. measured the distribution of cotton pollen deposition and predicted the highest average pollen density to be 61.67 grains/ cm^2 at a distance of 0 m and 95.67 grains/cm^2 at a distance of 1 m from the edge of the cotton field [35]. Based on the density of cotton pollen deposited naturally on leaves of mulberry plants and considering that silkworms can not survive independently in the field, we conducted a series of laboratory bioassays to determine the effects of Bt cotton pollen on *B. mori*.

Materials and Methods

Ethics Statement

All necessary permits were obtained for the described study, which complied with all relevant regulations. All Bt cotton cultivars were planted in the experimental field at Huazhong Agricultural University, Wuhan, China, and the University gave permission to conduct the study at this site. The field studies did not involve endangered or protected species.

Pollen Collection

The two transgenic Bt cotton cultivars ZMSJ (expressing Cry1Ac/Cry2Ab) and ZMKCKC (expressing Cry1Ac/EPSPS) used in this study were gifts from the Institute of Cotton Research, the Chinese Academy of Agricultural Science. The local cotton variety, Emian 24 (non-GM cotton), was a gift from the National Key Laboratory of Crop Genetic Improvement, Huazhong Agricultural University. The ZMSJ cotton expresses two Bt proteins for the control of lepidopteran pests, such as the cotton bollworm *Helicoverpa armigera*. The ZMKCKC cotton expresses one gene for insect resistance and one gene for herbicide tolerance.

All cultivars were cultivated under recommended agronomic practices at the experiment field at Huazhong Agricultural University in early May 2011 without exposure to any pesticide. Pollen samples of each cultivar were collected using the multi-point field sampling method [18] on June 20th, July 20th, and August 20th (early bloom, mid-stage bloom and late bloom respectively). The freshly collected cotton pollen samples were sieved (830 μm mesh size) and stored at –80°C until they were used for experiments or analyses.

ELISA Quantitative Detection of Bt Proteins in Pollen

The quantities of Cry1Ac and Cry2Ab in each pollen sample were estimated using Envirologix Qualiplate Kits (EnviroLogix Quantiplate Kit, Portland, ME, USA). The detection limits for the two proteins were 0.1 ng/g and 0.52 ng/g, respectively. Before analysis, the fresh pollen samples were homogenized in 4 ml of extraction buffer and then kept at 4°C overnight for extraction of insecticidal proteins. After being centrifuged for 15 min at 7000 g, the supernatants of the extraction were used for the analyses.

Experimental Insects and Treatment Applications

Worker bees of *A. mellifera* were obtained from the apiary of Huazhong Agriculture University. Bees were fed with sucrose solution daily and colonies were not treated with insecticides. Emerging honey bee adults (0 d) were collected from a colony during summer for bioassays.

The silkworm we used is a hybrid of *B. mori*, Qiufeng × Baiyu, which is the main variety used for commercial cocoon production in Southern China. Silkworm eggs were placed in an incubator set at $25\pm0.5°C$ and $75\pm5\%$ relative humidity for hatching, and newly hatched larvae (neonates) were used in the bioassays.

For honey bees, feeding behavior was evaluated following the protocols described by Han et al. [18]. Emerging honey bees were kept in cages (15 × 10 × 20 cm) [36] with the top face covered with a piece of mesh, and they were used for the experiments after a 1-day period of adaptation to rearing conditions. Conventional cotton pollen, ZMSJ cotton pollen, and ZMKCKC cotton pollen were used in the experiments as three different treatments; Bt cotton pollen samples collected in July were used because they contained the highest amount of Bt toxin (see Table 1). Three different diets were prepared by mixing water, honey, and pollen at a ratio of 1:2:7 (weight) with no additional sugar provided. Five replicates (cage) were used per treatment with 40 bees per replicate, and during the bioassay process the honey bee mortality and pollen consumption were recorded daily. Honey bees were considered dead when they remained completely immobile, and they were removed from the cages every day [26]. After being exposed to the three different dietary treatments for 7 days, the surviving bees were prepared for the total hemocyte count (THC) experiment, in which 30 bees per treatment (five cages × six bees) were used. THC (number/μl) was determined using a phase contrast microscope (40×) with a hemocytometer [37].

For silkworms, experiments were performed following the protocols described by Li et al. with slight modification [38]. Based on the density of cotton pollen deposited naturally on leaves of mulberry plants [35] and Hansen's study [39], two different densities of each type of Bt cotton pollen and conventional cotton pollen were obtained by suspending 10 and 100 mg of pollen in 1 ml distilled water. For Bt pollen, we used samples collected in 20th July. Based on the ELISA results in Table 1, the content of Bt toxins was calculated and the data were shown in Table 2. Fresh mulberry leaves were collected from the Mulberry Experiment Garden at Huazhong Agricultural University, which is situated far from the cotton fields. Leaf disks were cut using a 7.0 cm^2-hole puncher and then dipped in the pollen suspension; the discs and suspension were shaken to ensure uniform distribution of pollen grains on the leaf surface. Under the microscope, the mean number of pollen grains on the leaves treated with 10 and 100 mg/ml pollen suspensions was approximately 900 and 8000 pollen grains/cm^2, respectively. A non-pollen treatment, in which leaf disks were treated only with distilled water, served as the negative control. Each treatment was replicated six times with 10 neonates each time. Larvae were fed with treated leaves from birth to pupation. Both developmental phase and mortality were monitored for all individuals every day until pupation, and the weight of molting larvae (molters) for 1st to 4th instars was also measured. To evaluate the hemocyte concentration, another six replicates were used for each treatment with 10 neonates per replicate. Hemolymph was collected from 5th instar larvae on days 1 (V-1), 3 (V-3), 5 (V-5), and 7 (V-7), and THC (number/μl) was determined using a blood cell counter as described by Tu et al. [40]. We use six insects for each THC test.

Table 1. Cry1Ac and Cry2Ab protein content in cotton pollen from the transgenic ZMSJ cotton, ZMKCKC cotton and the non-Bt cotton as assayed by ELISA method.

Transgene proteins	Cotton variety	Contents of transgene proteins ± SD (ng/g fresh pollen)		
		Jun.20	Jul.20	Aug.20
Cry1Ac	ZMSJ	159.0±29.2b	572.5±28.1a	58.5±38.8c
	ZMKCKC	175.7±48.6b	544.5±22.5a	485.0±39.6a
	Non-Bt	0	0	0
Cry2Ab	ZMSJ	92.0±22.0a	92.4±23.1a	77.2±17.4a
	ZMKCKC	0	0	0
	Non-Bt	0	0	0

Values with the different letters are significantly different at the P<0.05 level (ANOVA followed by Tukey's post-hoc test).

Data Analysis

The Cry proteins (Cry1Ac and Cry2Ab) content in cotton pollen was compared among the treatments using one-way analysis of variance (ANOVA) followed by Tukey's post-hoc test. The data from honey bees were analyzed with mixed models and used replicate (cage) as a random factor. The survival dynamics of honey bees were analyzed with Cox proportional hazards regression models, and the cumulative pollen consumption and THC results for honey bees were fitted to a log-linear model.

The survival response of *B. mori* to different dietary treatments was analyzed using the Kaplan-Meier procedure and Logrank test. Nonparametric tests (K independent samples: Kruskal-Wallis H-tests; two independent samples: Mann-Whitney U tests) were performed on the developmental duration of *B. mori* larvae (from 1st instar to 5th instar), because the assumptions for parametric analyses were not fulfilled. The molter weight and THC results of *B. mori* larvae were compared using ANOVA, and means were compared by Tukey's post-hoc test. All statistical tests were conducted using SAS Version 8.0 (SAS Institute Inc., Cary, NC, USA).

Results

ELISA Results for Cry1Ac and Cry2Ab in Cotton Pollens

The quantities of Cry1Ac or Cry2Ab in pollens of the two Bt cotton varieties were measured during the anthesis period from early bloom to late bloom. As expected, no Cry1Ac or Cry2Ab

Table 2. Cry1Ac and Cry2Ab protein content in different food types of the silkworm *Bombyx mori*.

Pollen	Density (mg/ml)	Contents of transgene proteins ± SD (ng/ml)	
		Cry1Ac	Cry2Ab
ZMSJ	100	57.3±2.8	9.2±2.3
	10	5.7±0.3	0.9±0.2
ZMKCKC	100	54.5±2.3	0
	10	5.4±0.2	0
Non-Bt	100	0	0
	10	0	0
Control	0	0	0

was detected in the non-Bt cotton. The amount of Cry2Ab protein in ZMSJ pollen was statistically steady throughout the season (all P>0.05). The highest amount of Cry1Ac protein in ZMSJ pollen was detected at mid-flowering, and the quantity was significantly lower in the early and late flowering periods (both P<0.001) (Table 1). For the ZMKCKC pollen, higher amounts of Cry1Ac protein were detected during mid and late bloom (with no difference between them, P=0.13), whereas the quantity was significantly lower in the early bloom period (both P<0.001).

Effect of Bt Pollen on Honey Bees

After 7 days, more than 70% of honey bees had survived in the treatments with Cry proteins and the control treatment, and no significant differences were detected between survival in the Bt pollen treatments and the control groups ($\chi^2 = 0.71$, df=2, P=0.70) (Fig. 1).

Cumulative consumption of pollen values were 28.4±2.2 mg for ZMSJ, 28.0±1.2 mg for ZMKCKC, and 32.5±2.6 mg for control, and no difference between the non-Bt and Bt pollen treatments was found ('food type' factor: $\chi^2 = 2.11$, df=2, P=0.35) (Table 3). The 'replicate' factor and its interaction with the 'food type' factor were not significant (replicate factor: $\chi^2 = 0.47$, df=4, P=0.98; and food type × replicate: $\chi^2 = 1.21$, df=8, P=0.99), meaning that different replicates had consistent results within the same food type.

Table 4 shows the THC of bees after exposure to Bt pollen for 7 days. No significant difference was found among all the treatments ('food type' factor: $\chi^2 = 1.43$, df=2, P=0.49). The 'replicate' factor and its interaction with the 'food type' factor were not significant (replicate factor: $\chi^2 = 1.03$, df=4, P=0.90; and food type × replicate: $\chi^2 = 3.79$, df=8, P=0.88), which indicates that, for the same food type, THC results were consistent among the different replicates.

Effect of Bt Pollen on Silkworms

Survivorship decreased from 96.7 to 76.7% across the whole larval period (the 1st to the 5th instar) among all treatments, and no statistical difference was observed in the survivorship for young larvae exposed to the two Bt pollen types, the non-Bt pollen, and the control diets ($\chi^2 = 3.40$, df=6, P=0.76) (Fig. 2).

No significant difference in duration of developmental phase among treatments was found for the first two instars and fully matured larvae (Kruskal-Wallis H-test; 1st instar: $\chi^2 = 6.87$, df=6, P=0.33; 2nd instar: $\chi^2 = 9.44$, df=6, P=0.15; 5th instar: $\chi^2 = 6.78$, df=6, P=0.34). For the 3rd larval stage, larvae of *B. mori* fed the control diet had a significantly shorter developmental

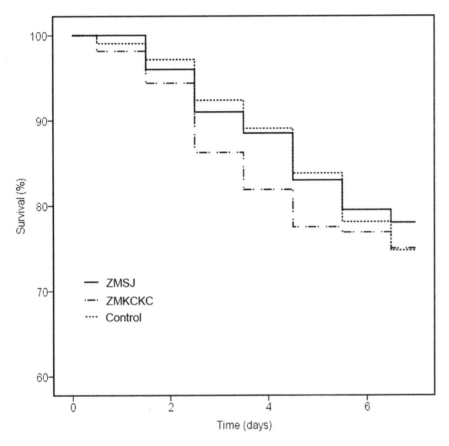

Figure 1. Survival analysis of honey bees from groups subjected to chronic exposure to ZMSJ pollen, ZMKCKC pollen and non-Bt pollen after 7 days. Data were analyzed with Cox proportional hazards regression models, and no significant differences were found among all the treatments at the P>0.05 level.

phase compared to those fed a high density of ZMSJ and ZMKCKC pollen (Mann-Whitney U test: P<0.001 and P = 0.04), or non-Bt pollen (P<0.001). In addition, 4th instar *B. mori* larvae fed the control diet also had a significantly shorter developmental phase compared to those fed a high density of Bt and non-Bt pollen (all P<0.001). However, no difference was found between Bt and non-Bt pollen treatments at different pollen densities for the 3rd and 4th larval stages (all P>0.05) (Fig. 3).

Total body weight of the molters was recorded and the weight distributions of the molters in the Bt pollen, non-Bt pollen, and control diet groups did not differ significantly after the first and final molting (1st instar: $F_{6,385} = 2.08$, P = 0.055; 4th instar: $F_{6,346} = 1.36$, P = 0.229). Just after the second and third moulting, larvae treated with high pollen density showed significant differences compared to the control (2nd instar: $F_{6,367} = 8.49$, P<0.001; 3rd instar: $F_{6,352} = 7.99$, P<0.001), but weights of Bt

Table 3. Cumulative consumption of pollen by honey bees from groups subjected to chronic exposure to ZMSJ pollen, ZMKCKC pollen and non-Bt pollen after 7 days.

Treatment			Cumulative consumption of pollen per bee ± SD (mg)
ZMSJ			28.4±2.2
ZMKCKC			28.0±1.2
Control			32.5±2.6
Source of variation	**df**	**χ^2**	**P value**
Food type	2	2.11	0.35
Replicate	4	0.47	0.98
Food type × replicate	8	1.21	0.99

Statistics from the log linear model used to analyze the cumulative consumption of honey bees at the end of the oral chronic exposure period among treatments (food type factor) and as function of replicate factor.

Table 4. Mean total hemocyte count of honey bees from groups subjected to chronic exposure to ZMSJ pollen, ZMKCKC pollen and non-Bt pollen after 7 days.

Treatment			Total hemocyte count ± SD (μl)
ZMSJ			600.0±63.5
ZMKCKC			571.7±24.0
Control			545.0±46.7
Source of variation	**df**	**χ^2**	**P value**
Food type	2	1.43	0.49
Replicate	4	1.03	0.90
Food type × replicate	8	3.79	0.88

Statistics from the log linear model used to analyze the total hemocyte count of honey bees at the end of the oral chronic exposure period among treatments (food type factor) and as function of replicate factor (five replicates per food type with six individual bees per replicate).

pollen fed larvae were almost identical to those of non-Bt pollen fed larvae (Fig. 4).

When we evaluated the THC of the 5th instar larvae, the results indicated that the hemocyte concentration increased with growth in the early and middle fifth instar phases and subsequently decreased during the prepupal stage (V-7). There were no significant differences in the THC of the V-1 and V-7 larvae among different treatments (V-1: $F_{6,35} = 1.62$, P = 0.172; V-7:

$F_{6,35} = 1.02$, P = 0.428). THC of the V-3 larvae reared on ZMSJ pollen at the two different densities was significantly higher than that of the control, but differences between other pollen treatments and the control were not found (V-3: $F_{6,35} = 17.76$, P<0.001). For the V-5 larvae, the hemocyte concentration was significantly higher than that of the control only in the high density of ZMSJ and non-Bt pollen treatments (V-5: $F_{6,35} = 18.04$, P<0.001), and no significant differences were observed between Bt and non-Bt treatments (Fig. 5).

Discussion

In order to minimize the environmental risks of cultivating GM crops, it is necessary to identify the possible adverse effects of transgenic cotton on non-target species during their development, especially for economically important insects in China. Our study is, to the best of our knowledge, the first to evaluate the effects of stacked Bt cotton on *B. mori*.

ELISA Results for Cry1Ac and Cry2Ab in Pollen

Knowing the concentration of toxic proteins expressed in pollen from transgenic cotton is very important for assessing its adverse impact on non-target insects [41]. It is crucial to identify a reliable expression level of insecticidal toxins in target GM crop tissues before conducting risk assessment because this value greatly impacts the effects on tested organisms [32]. The expression levels of transgenic proteins in the pollen of ZMSJ and ZMKCKC have not been reported previously.

In our study, the expression level of Cry1Ac in both Bt pollens varied greatly throughout the season, with the highest values in samples collected in July. This shows the importance of assessing

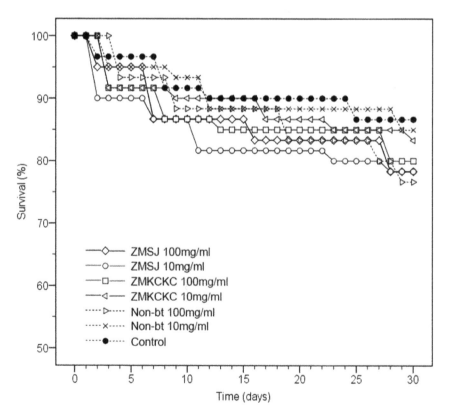

Figure 2. Survival analysis of silkworm larvae treated with different doses of Bt-pollen or non-Bt pollen. No significant differences in survival rates were found among all the treatments at the P>0.05 level (followed by Kaplan–Meier survival analysis).

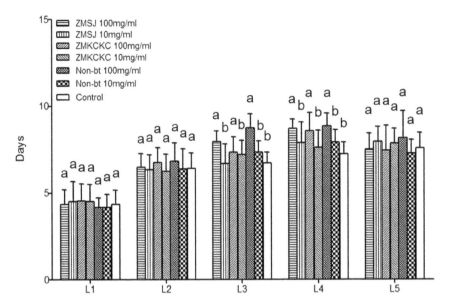

Figure 3. Duration of development of silkworm larvae treated with different doses of Bt pollen or non-Bt pollen. Values with the different letter are significantly different. Bars represent standard error. Levels of significance: P<0.05 (followed by nonparametric tests-K independent samples: Kruskal-Wallis H-tests; two independent samples: Mann-Whitney U tests).

toxin levels throughout the season. The level of the Cry2Ab toxin, however, was stable. Recent studies conducted on another GM cotton cultivar also demonstrated temporal variances in Cry1Ac protein expression [18,25,42].

Effects of Pollen from Single and Multiple Bt Cotton Varieties on Honey Bees

One future trend in plant biotechnology is the stacking of multiple resistance traits in a single cultivar [13]. Honey bees are exposed to mass flowering GM crops, which contain multiple toxins or resistance traits, but only a few studies have examined the effect of stacked Bt crop on bees [13,43]. Adverse effects of stacked transgenic cotton pollen on the survival, cumulative consumption, and THC of *A. mellifera* were not detected in this study. These findings suggest that the tested Bt cotton pollens have no deleterious effects on honey bees.

Neither larval nor adult honey bees have ever shown lethality when exposed to Bt proteins [44,45], and our data also suggest that synergistic effects of stacking Bt proteins at plant-produced levels are unlikely a be a risk to emerging adult bees. At a realistic exposure dose, the 7 day survivorship of Bt-pollen treated bees in our study was similar to that of bees exposed to the conventional cotton pollen (Fig. 1). The results are in line with recent tests on Cry1Ac/CpTI cotton pollen [18,41,46], stacked Bt maize pollen [43] or purified Bt proteins [44,45].

However, sublethal effects of the Bt pollen on larval development, feeding, learning performance, and foraging behavior might occur [13,27,37,47]. Honey bee larvae and young adults (less than 12 days old) mainly feed on pollen [48], and nurse bees consume 3.4 to 4.3 mg of pollen per day [49]. Therefore, the potential risks of GM crop pollen on feeding behavior of *A. mellifera* needed to be assessed. In our study, after 7 days of chronic exposure to two stacked Bt pollens, no feeding inhibition occurred. Similar results were reported for studies of single [27,47] or stacked Bt corn pollen [43]. Nevertheless, Han et al. reported an anti-feeding effect of Cry1Ac/CpTI cotton on honey bees [18]. However, that cultivar contained a different insect-resistant gene than the cultivar

used in our study. Comparing Cry1 with transgenic protease inhibitors in many studies, only the latter impacted the feeding behavior [50–53]. Better knowledge about the sublethal risks associated with ZMSJ and ZMKCKC pollen for honey bees may also be obtained studying the effects of pollen [13] or multiple Bt proteins [45] on larval development.

In pollinators, information about potential sublethal physiological effects is scarce [54]. However, such effects could impact important biological processes, notably immunity. Honey bees defend themselves from an especially diverse range of pathogens, including bacteria, fungi, viruses, nematodes, protozoa, mites, flies, and beetles [55,56]. Thus, it is important to determine if Bt toxins cause an immune reaction in honey bees. In this study, we assessed the risks of Bt cotton on the cellular immunity of honey bees. Higher THC is expected to be associated with higher resistance to disease [57]. Compared to the control, we found no negative effect of exposure to Bt pollen on THC in honey bees, which suggests that Bt pollens have no direct impact on honey bee health. This result is in line with a recent study that showed that most Cry proteins (>98%) in the bee gut were degraded, and had no harmful physiological effects on honey nurse bees [43].

Effects of Pollen from Single and Multiple Bt Cotton Varieties on Silkworm

We found that exposure of *B. mori* larvae to different densities of pollen from either Bt cotton cultivar expressing Cry1Ac and Cry2Ab proteins had no adverse effects on young larval survival or development. This finding is consistent with results for pollens from Bt corn [23,38] and rice [22]. Conversely, a transgenic Chinese cabbage pollen expressing Cry1Ac toxin adversely affected *B. mori* larvae when consumed [28]. Several factors may explain the differences in the response of *B. mori* to Bt pollens in different studies. For example, different pollens contain different levels of Cry protein levels. Furthermore, subspecies can vary in their susceptibility to Cry proteins [22].

Considering the importance of the insect hemocyte in the recognition and defense against microorganisms, we measured

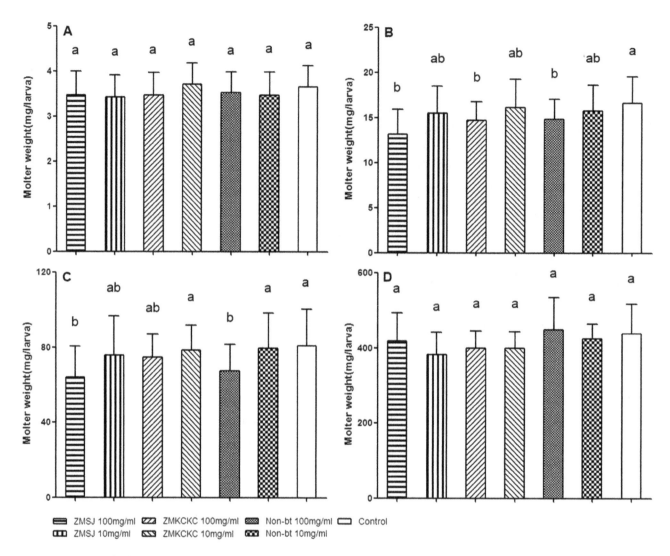

Figure 4. Weight of silkworm molters treated with different doses of Bt pollen or non-Bt pollen. A~D represents the molter weight of the 1st to 4th instar larvae. Values with the different letter are significantly different. Bars represent standard error. Levels of significance: P<0.05 (ANOVA followed by Tukey's post-hoc test).

THC levels in 5th instar *B. mori* larvae. At day 3 (V-3), the hemocyte concentration of larvae in the Cry1Ac/Cry2Ab cotton pollen treatment was increased relative to the control, indicating that the Cry1Ac/Cry2Ab cotton pollen caused an immune reaction. At day 5 (V-5), the high-density Cry1Ac/Cry2Ab cotton pollen treatment also had a significant influence on the immune system of *B. mori* larvae. However, in our experiments, the average density of cotton pollen deposited on mulberry leaves (approximately 900 and 8000 pollen grains/cm^2) is much higher than that occurs under normal field conditions (<200 grains/cm^2) [35], which indicates that the risk for *B. mori* was minimal. At day 7 (V-7, the day before larvae reached the pupal stage) there was no significant difference, showing that the pollen had no direct impact on the health of the preceding larval stages of *B. mori*.

Many factors affect the probability that *B. mori* larvae will be exposed to Bt cotton pollen. First, the cotton pollen density load on mulberry leaves is very important. The average density of cotton pollen naturally deposited on mulberry leaves [35] is lower than that of corn [38] or rice [22] pollen at the same distance from the edge of field. Under normal field conditions, the highest average

cotton pollen density is 61.67 grains/cm^2 at a distance of 0 m and 95.67 grains/cm^2 at a distance of 1 m from the edge of the cotton field [35]. In our experiments, the mean number of pollen grains on the leaves treated with 10 and 100 mg/ml pollen suspensions was approximately 900 and 8000 pollen grains/cm^2, which is substantially higher than the density that occurs under normal field conditions. Even in this worst-case feeding scenario, the Bt cotton pollen had little effect on the *B. mori* larvae.

Other important factors that can affect the impact of Bt pollen on non-target organisms are the degrees of hazard and exposure [51,52]. The hazard posed by Bt cotton pollen to *B. mori* primarily depends on the type of Cry gene present. The insecticidal crystal proteins (ICPs) that are encoded by the Cry1Ac and Cry2Ab genes have specific activity against certain lepidopterans larvae, and both have been used in many GM crops [8,53]. However, the mechanisms that underlie the specificity of these genes remain unclear. For example, Cry1Aa, Cry1Ab and Cry1Ac are all Cry1 genes, but Cry1Aa exhibits 400-fold greater toxicity against *B. mori* than Cry1Ac [22,58]. Thus, it is likely that the transgenic products

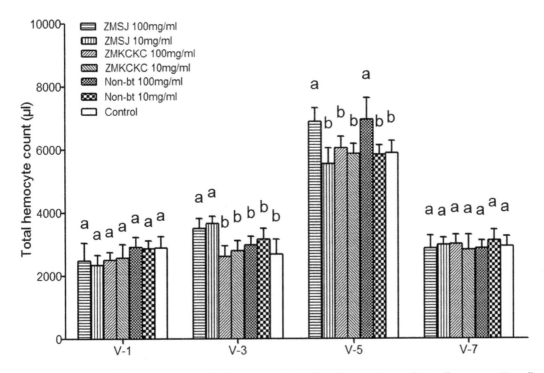

Figure 5. Mean total hemocyte count of silkworm treated with different doses of Bt pollen or non-Bt pollen. V-1, 3, 5, 7 represent the 1st, 3rd, 5th and 7th day of 5th instar larvae. Bars represent standard error. Values with the different letter are significantly different at the $P<0.05$ level (ANOVA followed by Tukey's post-hoc test).

of the Cry1Ac and Cry2Ab genes in our tested Bt cottons have low toxicity against *B. mori*.

The low concentration of the Bt proteins to which *B. mori* larvae were exposed may have led to minimal effect observed, even at a very high pollen density. Several factors affect the level of exposure, but it is mainly related to the expression level of the proteins. In our study, the expression of Cry2Ab was far lower than that of Cry1Ac (Table 1), which was mainly due to the promoter used in transformation. Therefore, in the development of transgenic crops, a suitable promoter should be selected to ensure that the gene is highly expressed in the tissues attacked by target pests and expressed at lower levels in the pollen [22]. The amount of Cry1Ac and Cry2Ab proteins released from the ingested pollens to the larval midgut may also be an important factor that affects exposure to the toxic proteins in Bt pollen. Yao et al. showed that the digestibility rate of pollen grains in the digestive tract of *B. mori* is very low (less than 30%) [22], which suggests that the pollen from Bt cotton poses little risk to silkworms even if it contains high levels of toxic Bt proteins. Considering all of these factors, the adverse effects of pollen from Bt cottons on the survival, development, and hemocyte concentration of the silkworm appear to be minimal.

In summary, our data indicate that consumption of Bt cotton pollens expressing Cry1Ac and Cry2Ab did not negatively affect young adult *A. mellifera* or *B. mori* larvae. However, further experiments are underway to determine whether sublethal effects could impact later silkworm generations or the honey bee queen. In addition, the susceptibility of silkworm larvae of other varieties to the pollen should be evaluated, as different varieties of silkworm may have different susceptibilities to Bt insecticidal proteins. Field studies of honey bees also need to be conducted, as such studies are the most direct way to assess the potential impacts of planting Bt cotton on a commercial scale on non-target organisms [27].

Acknowledgments

We thank Dr. Manqun Wang (Huazhong Agricultural University) for his valuable suggestions that improved the manuscript.

Author Contributions

Conceived and designed the experiments: LN CL LC. Performed the experiments: LN. Analyzed the data: LN YM AM YZ WM LC. Contributed reagents/materials/analysis tools: LN YM AM YZ WM CL LC. Wrote the paper: LN AM WM.

References

1. James C (2013) Global Status of Commercialized Biotech/GM Crops: 2012. ISAAA Brief No. 44. ISAAA: Ithaca, NY.
2. Wu KM, Lu YH, Feng HQ, Jiang YY, Zhao JZ (2008) Suppression of cotton bollworm in multiple crops in China in areas with Bt toxin–containing cotton. Science 321: 1676–1678.
3. Lu Y, Wu K, Jiang Y, Guo Y, Desneux N (2012) Widespread adoption of Bt cotton and insecticide decrease promotes biocontrol services. Nature 487: 362–365.
4. James C (2009) Global Status of Commercialized Biotech/GM Crops: 2009. ISAAA Brief No. 41. ISAAA: Ithaca, NY.
5. Luo JY, Cui JJ, Zhang S, Wang CY, Xin HJ (2012) Effects of transgenic Cry1Ac and Cry2Ab cotton on life table parameters and population dynamics of the cotton aphid. Chinese Journal of Applied Entomology 49: 906–910.
6. Zhang XH, Tian SR, Zhang TY, Li J, Yang ZG, et al. (2012) Study on differences in biological characteristics between transgenic cotton with insect-resistant and herbicide-resistant genes (Cry1Ac+EPSPS) and non-transgenic cotton. Acta Agriculturae Jiangxi 24: 12–16.
7. Höfte H, Whiteley H (1989) Insecticidal crystal proteins of *Bacillus thuringiensis*. Microbiological reviews 53: 242–255.
8. Dankocsik C, Donovan W, Jany C (2006) Activation of a cryptic crystal protein gene of *Bacillus thuringiensis* subspecies *kurstaki* by gene fusion and determination

of the crystal protein insecticidal specificity. Molecular microbiology 4: 2087–2094.

9. Padgette SR, Kolacz K, Delannay X, Re D, LaVallee B, et al. (1995) Development, identification, and characterization of a glyphosate-tolerant soybean line. Crop science 35: 1451–1461.

10. Liu XJ, Liu YH, Wang ZX, Wang XJ, Zhang YQ (2007) Generation of glyphosate-tolerant transgenic tobacco and cotton by transfor-mation with a 5-enolpyruvyl-shikimate-3-phosphate synthase EPSPS gene. Journal of Agricultural Biotechnology 15: 958–963.

11. Dale PJ, Clarke B, Fontes EMG (2002) Potential for the environmental impact of transgenic crops. Nature biotechnology 20: 567–574.

12. Romeis J, Hellmich RL, Candolfi MP, Carstens K, De Schrijver A, et al. (2011) Recommendations for the design of laboratory studies on non-target arthropods for risk assessment of genetically engineered plants. Transgenic Research 20: 1–22.

13. Hendriksma HP, Härtel S, Steffan-Dewenter I (2011) Testing pollen of single and stacked insect-resistant Bt-maize on in vitro reared honey bee larvae. PLoS One 6: e28174.

14. Klein AM, Vaissiere BE, Cane JH, Steffan-Dewenter I, Cunningham SA, et al. (2007) Importance of pollinators in changing landscapes for world crops. Proceedings of the Royal Society B: Biological Sciences 274: 303–313.

15. Liu PF, Wu J, Li HY, Lin SW (2011) Economic Values of Bee Pollination to China's Agriculture. Scientia Agricultura Sinica 44: 5117–5123.

16. Free JB (1993) Insect pollination of crops. 2nd ed Academic Press, London, UK.

17. Ruttner F (1988) Biogeography and taxonomy of honeybees: Springer-Verlag:Berlin.

18. Han P, Niu CY, Lei CL, Cui JJ, Desneux N (2010) Quantification of toxins in a Cry1Ac+CpTI cotton cultivar and its potential effects on the honey bee Apis mellifera L. Ecotoxicology 19: 1452–1459.

19. Decourtye A, Mader E, Desneux N (2010) Landscape enhancement of floral resources for honey bees in agro-ecosystems. Apidologie 41: 264–277.

20. Han P, Niu CY, Lei CL, Cui JJ, Desneux N (2010) Use of an innovative T-tube maze assay and the proboscis extension response assay to assess sublethal effects of GM products and pesticides on learning capacity of the honey bee Apis mellifera L. Ecotoxicology 19: 1612–1619.

21. Malone L, Burgess E (2009) Impact of genetically modified crops on pollinators. Environmental Impact of Genetically Modified Crops, CAB International: 199–222.

22. Yao H, Ye G, Jiang C, Fan L, Datta K, et al. (2006) Effect of the pollen of transgenic rice line, TT9-3 with a fused cry1Ab/cry1Ac gene from Bacillus thuringiensis Berliner on non-target domestic silkworm, Bombyx mori Linnaeus (Lepidoptera: Bombyxidae). Applied entomology and zoology 41: 339–348.

23. Li WD, Ye GY (2002) Evaluation of impact of pollen grains of Bt, Bt/CPTI transgenic cotton and Bt corn on the growth and development of the mulberry silkworm, bombyx mori linnaeus (lepidoptera: bombyxidae). Scientia Agricultura Sinica 35: 1543–1549.

24. Liu B, Shu C, Xue K, Zhou K, Li X, et al. (2009) The oral toxicity of the transgenic Bt+CpTI cotton pollen to honeybees (Apis mellifera). Ecotoxicology and Environmental Safety 72: 1163–1169.

25. Han P, Niu CY, Biondi A, Desneux N (2012) Does transgenic Cry1Ac+CpTI cotton pollen affect hypopharyngeal gland development and midgut proteolytic enzyme activity in the honey bee Apis mellifera L. (Hymenoptera, Apidae)? Ecotoxicology 21: 2214–2221.

26. Ramirez-Romero R, Desneux N, Decourtye A, Chaffiol A, Pham-Delègue M (2008) Does Cry1Ab protein affect learning performances of the honey bee Apis mellifera L. (Hymenoptera, Apidae)? Ecotoxicology and Environmental Safety 70: 327–333.

27. Dai PL, Zhou W, Zhang J, Cui HJ, Wang Q, et al. (2012) Field assessment of Bt cry1Ah corn pollen on the survival, development and behavior of Apis mellifera ligustica. Ecotoxicology and Environmental Safety 79: 232–237.

28. Kim YH, Kim H, Lee S, Lee SH (2008) Effects of Bt transgenic Chinese cabbage pollen expressing Bacillus thuringiensis Cry1Ac toxin on the non-target insect Bombyx mori (Lepidoptera: Bombyxidae) larvae. Journal of Asia-Pacific Entomology 11: 107–110.

29. Biondi A, Desneux N, Siscaro G, Zappalà L (2012) Using organic-certified rather than synthetic pesticides may not be safer for biological control agents: Selectivity and side effects of 14 pesticides on the predator Orius laevigatus. Chemosphere 87: 803–812.

30. He Y, Zhao J, Zheng Y, Desneux N, Wu K (2012) Lethal effect of imidacloprid on the coccinellid predator Serangium japonicum and sublethal effects on predator voracity and on functional response to the whitefly Bemisia tabaci. Ecotoxicology 21: 1291–1300.

31. Desneux N, Decourtye A, Delpuech JM (2007) The sublethal effects of pesticides on beneficial arthropods. Annu Rev Entomol 52: 81–106.

32. Romeis J, Bartsch D, Bigler F, Candolfi MP, Gielkens MM, et al. (2008) Assessment of risk of insect-resistant transgenic crops to nontarget arthropods. Nature biotechnology 26: 203–208.

33. Desneux N, Bernal JS (2010) Genetically modified crops deserve greater ecotoxicological scrutiny. Ecotoxicology 19: 1642–1644.

34. Shimizu M, Tajima Y (1972) Handbook of Silkworm Rearing. Agricultural Technique Manual 1. Fuji Publishing Co. LTD. Tokyo, Japan.

35. Li W, Wu K, Wang X, Guo Y (2003) Evaluation of impact of pollen grains of Cry1Ac and Cry1A+CpTI transgenic cotton on the growth and development of Chinese tussah silkworm (Antheraea pernyi). Journal of Agricultural Biotechnology 11: 488–493.

36. Pain J (1966) Nouveau modéle de cagettes expérimentales pour le maintien d'abeilles en captivité. Ann Abeille 9: 71–76.

37. Alaux C, Ducloz F, Crauser D, Le Conte Y (2010) Diet effects on honeybee immunocompetence. Biology Letters 6: 562–565.

38. Li W, Wu K, Wang X, Wang G, Guo Y (2005) Impact of pollen grains from Bt transgenic corn on the growth and development of Chinese tussah silkworm, Antheraea pernyi (Lepidoptera: Saturniidae). Environmental entomology 34: 922–928.

39. Hansen Jesse LC, Obrycki JJ (2000) Field deposition of Bt transgenic corn pollen: lethal effects on the monarch butterfly. Oecologia 125: 241–248.

40. Tu Z, Shirai K, Kanekatsu R, Kiguchi K, Kobayashi Y, et al. (1999) Effects of local heavy ion beam irradiation on the hemopoietic organs of the silkworm, Bombyx mori. J Sericult Sci Jpn 68: 491–500.

41. Liu B, Xu C (2003) Research progress on the impacts of transgenic resistant plants on pollinating bees. Acta Ecologica Sinica 23: 946.

42. Chen L, Cui J, Ma W, Niu C, Lei C (2011) Pollen from Cry1Ac/CpTI-transgenic cotton does not affect the pollinating beetle Haptoncus luteolus. Journal of Pest Science 84: 9–14.

43. Hendriksma HP, Küting M, Härtel S, Näther A, Dohrmann AB, et al. (2013) Effect of Stacked Insecticidal Cry Proteins from Maize Pollen on Nurse Bees (Apis mellifera carnica) and Their Gut Bacteria. PloS one 8: e59589.

44. Duan JJ, Marvier M, Huesing J, Dively G, Huang ZY (2008) A meta-analysis of effects of Bt crops on honey bees (Hymenoptera: Apidae). PLoS One 3: e1415.

45. Hendriksma HP, Härtel S, Babendreier D, von der Ohe W, Steffan-Dewenter I (2012) Effects of multiple Bt proteins and GNA lectin on in vitro-reared honey bee larvae. Apidologie 43: 549–560.

46. James C (2010) A global overview of biotech (GM) Crops: Adoption, impact and future prospects. GM crops 1: 8–12.

47. Rose R, Dively GP, Pettis J (2007) Effects of Bt corn pollen on honey bees: emphasis on protocol development. Apidologie 38: 368–377.

48. Haydak MH (1970) Honey bee nutrition. Annual Review of Entomology 15: 143–156.

49. Crailsheim K, Schneider L, Hrassnigg N, Bühlmann G, Brosch U, et al. (1992) Pollen consumption and utilization in worker honeybees (Apis mellifera carnica): Dependence on individual age and function. Journal of insect Physiology 38: 409–419.

50. Babendreier D, Kalberer NM, Romeis J, Fluri P, Mulligan E, et al. (2005) Influence of Bt-transgenic pollen, Bt-toxin and protease inhibitor (SBTI) ingestion on development of the hypopharyngeal glands in honeybees. Apidologie 36: 585–594.

51. Jesse LCH, Obrycki JJ (2003) Occurrence of Danaus plexippus L.(Lepidoptera: Danaidae) on milkweeds (Asclepias syriaca) in transgenic Bt corn agroecosystems. Agriculture, ecosystems & environment 97: 225–233.

52. Wolt JD, Peterson RKD, Bystrak P, Meade T (2003) A screening level approach for nontarget insect risk assessment: transgenic Bt corn pollen and the monarch butterfly (Lepidoptera: Danaidae). Environmental entomology 32: 237–246.

53. Mohan Babu R, Sajeena A, Seetharaman K, Reddy M (2003) Advances in genetically engineered (transgenic) plants in pest management–an over view. Crop Protection 22: 1071–1086.

54. Malone LA, Pham-Delègue MH (2001) Effects of transgene products on honey bees (Apis mellifera) and bumblebees (Bombus sp.). Apidologie 32: 287–304.

55. Schmid-Hempel R, Schmid-Hempel P (2002) Colony performance and immunocompetence of a social insect, Bombus terrestris, in poor and variable environments. Functional Ecology 12: 22–30.

56. Evans JD, Pettis JS (2005) Colony-Level impacts of immune responsiveness in honey bees, Apis mellifera. Evolution 59: 2270–2274.

57. Wilson-Rich N, Dres ST, Starks PT (2008) The ontogeny of immunity: development of innate immune strength in the honey bee (Apis mellifera). Journal of insect Physiology 54: 1392–1399.

58. Shinkawa A, Yaoi K, Kadotani T, Imamura M, Koizumi N, et al. (1999) Binding of phylogenetically distant Bacillus thuringiensis cry toxins to a Bombyx mori aminopeptidase N suggests importance of Cry toxin's conserved structure in receptor binding. Current microbiology 39: 14–20.

7

Genetically Modified Crops and Food Security

Matin Qaim[1]*, Shahzad Kouser[1,2]

1 Department of Agricultural Economics and Rural Development, Georg-August-University of Goettingen, Goettingen, Germany, **2** Institute of Agricultural and Resource Economics, University of Agriculture, Faisalabad, Pakistan

Abstract

The role of genetically modified (GM) crops for food security is the subject of public controversy. GM crops could contribute to food production increases and higher food availability. There may also be impacts on food quality and nutrient composition. Finally, growing GM crops may influence farmers' income and thus their economic access to food. Smallholder farmers make up a large proportion of the undernourished people worldwide. Our study focuses on this latter aspect and provides the first *ex post* analysis of food security impacts of GM crops at the micro level. We use comprehensive panel data collected over several years from farm households in India, where insect-resistant GM cotton has been widely adopted. Controlling for other factors, the adoption of GM cotton has significantly improved calorie consumption and dietary quality, resulting from increased family incomes. This technology has reduced food insecurity by 15–20% among cotton-producing households. GM crops alone will not solve the hunger problem, but they can be an important component in a broader food security strategy.

Editor: M. Lucrecia Alvarez, TGen, United States of America

Funding: This research was supported by the German Research Foundation (DFG). The funders had no role in study design, data collection and analysis, decision to publish, or preparation of the manuscript.

Competing Interests: The authors have declared that no competing interests exist.

* E-mail: mqaim@uni-goettingen.de

Introduction

Food security exists when all people have physical and economic access to sufficient, safe, and nutritious food. Unfortunately, food security does not exist for a significant proportion of the world population. Around 900 million people are undernourished, meaning that they are undersupplied with calories [1]. Many more suffer from specific nutritional deficiencies, often related to insufficient intake of micronutrients. Eradicating hunger is a central part of the United Nations' Millennium Development Goals [2]. But how to achieve this goal is debated controversially. Genetically modified (GM) crops are sometimes mentioned in this connection. Some see the development and use of GM crops as key to reduce hunger [3,4], while others consider this technology as a further risk to food security [5,6]. Solid empirical evidence to support either of these views is thin.

There are three possible pathways how GM crops could impact food security. First, GM crops could contribute to food production increases and thus improve the availability of food at global and local levels. Second, GM crops could affect food safety and food quality. Third, GM crops could influence the economic and social situation of farmers, thus improving or worsening their economic access to food. This latter aspect is of particular importance given that an estimated 50% of all undernourished people worldwide are small-scale farmers in developing countries [7].

In regard to the first pathway, GM technologies could make food crops higher yielding and more robust to biotic and abiotic stresses [8,9]. This could stabilize and increase food supplies, which is important against the background of increasing food demand, climate change, and land and water scarcity. In 2012, 170 million hectares (ha) – around 12% of the global arable land – were planted with GM crops, such as soybean, corn, cotton, and canola [10], but most of these crops were not grown primarily for direct food use. While agricultural commodity prices would be higher without the productivity gains from GM technology [11], impacts on food availability could be bigger if more GM food crops were commercialized. Lack of public acceptance is one of the main reasons why this has not yet happened more widely [12].

Concerning the second pathway, crops with new traits can be associated with food safety risks, which have to be assessed and managed case by case. But such risks are not specific to GM crops. Long-term research confirms that GM technology is not *per se* more risky than conventional plant breeding technologies [13]. On the other hand, GM technology can help to breed food crops with higher contents of micronutrients; a case in point is Golden Rice with provitamin A in the grain [14]. Such GM crops have not yet been commercialized. Projections show that they could reduce nutritional deficiencies among the poor, entailing sizeable positive health effects [15,16].

The third pathway relates to GM crop use by smallholder farmers in developing countries. Half of the global GM crop area is located in developing countries, but much of this refers to large farms in countries of South America. One notable exception is *Bacillus thuringiensis* (Bt) cotton, which is grown by around 15 million smallholders in India, China, Pakistan, and a few other developing countries [10]. Bt cotton provides resistance to important insect pests, especially cotton bollworms. Several studies have shown that Bt cotton adoption reduces chemical pesticide use and increases yields in farmers' fields [17–20]. There are also a few studies that have shown that these benefits are associated with increases in farm household income and living standard [21–23]. Higher incomes are generally expected to cause increases in food consumption in poor farm households. On the other hand, cotton is a non-food cash crop, so that the nutrition impact is uncertain.

Table 1. Number of farm households sampled in India in four survey rounds.

Farm households	2002	2004	2006	2008	Total
Adopters of Bt	131	246	333	375	1085
Non-adopters of Bt	210	117	14	5	346
Total	341	363	347	380	1431

Here we address this question and analyze the impact of Bt cotton adoption on calorie consumption and dietary quality in India. Bt cotton was first commercialized in India in 2002. In 2012, over 7 million farmers had adopted this technology on 10.8 million ha – equivalent to 93% of the country's total cotton area [10]. For the analysis, we carried out a household survey and collected comprehensive data over a period of several years. This is the first *ex post* study that analyzes food security effects of Bt cotton or any other GM crop with micro level data.

Materials and Methods

Ethics Statement

Our study builds on data from a socioeconomic survey of farm households in India. Details of this survey are explained further below. The institutional review board of the University of Goettingen only reviews clinical research; our study cannot be classified as clinical research. We consulted with the Head of the Research Department of the University of Goettingen, who confirmed that there is no institutional review board at our University that would require a review of such survey-based socioeconomic research.

Farm Household Survey

We carried out a panel survey of Indian cotton farm households in four rounds between 2002 and 2008. We used a multistage sampling procedure. Four states were purposively selected, namely Maharashtra, Karnataka, Andhra Pradesh, and Tamil Nadu. These four states cover a wide variety of different cotton-growing situations, and they produce 60% of all cotton in central and southern India [23]. In these four states, we randomly selected 10 cotton-growing districts and 58 villages, using a combination of census data and agricultural production statistics [18,19,23]. Within each village, we randomly selected farm households from complete lists of cotton producers. Sample households were visited individually, and the household head was taken through a face-to-face interview, for which we used a structured questionnaire. The questionnaire covered a wide array of agricultural and socioeconomic information, such as input-output details in cotton production, technology adoption, other income sources, and household living standards. The interviews were carried out in local languages by a small team of enumerators, who were trained and supervised by the researchers.

Prior to starting each interview, the study objective was explained. We also clarified that the data collected would be treated confidentially, analyzed anonymously, and be used for research purposes only. Based on this, the interviewees were asked for their verbal informed consent to participate. We decided not ask for written consent, because the interviews were not associated with any risk for participants. Furthermore, many of the sample farmers had relatively low educational backgrounds and were not used to formal paperwork. Very few households did not agree to participate; they were replaced with other randomly selected households in the same villages.

The first-round survey interviews took place in early 2003, shortly after the cotton harvest for the 2002 season was completed. The same survey was repeated at two-year intervals in early 2005 (referring to the 2004 cotton season), early 2007 (referring to the 2006 season), and early 2009 (referring to the 2008 season). In total, 533 households were interviewed during the 7-year period. Most of these households were visited in several rounds. The total sample consists of 1431 household observations (Table 1). In 2002, the proportion of Bt adopters was still relatively small, but it increased rapidly in the following years. By 2008, 99% of the sample households had adopted this technology. To our knowledge, this is the only longer-term panel survey of Bt cotton farm

Table 2. Descriptive statistics of farm households.

Variables	Adopters of Bt (N = 1085)	Non-adopters of Bt (N = 346)
Farm size (ha)	5.11 (5.85)	4.85 (5.51)
Cotton area cultivated (ha)	2.35 (2.35)	2.79 (19.67)
Area cultivated with Bt cotton (ha)	1.97*** (2.08)	0.00 (0.00)
Age of farmer (years)	45.58 (12.86)	45.94 (12.36)
Education of farmer (years)	7.58*** (4.94)	6.69 (5.03)
Per capita consumption expenditure (US$/year)	490.31*** (430.18)	311.72 (355.58)
Off-farm income (US$/year)	560.70 (1455.44)	504.27 (2289.87)
Calorie consumption per AE (kcal/day)	3329.41*** (719.38)	2829.88 (598.99)
Calories consumed from more nutritious foods per AE (kcal/day)[a]	703.89*** (374.90)	638.89 (345.41)
Household size (AE)	5.01 (2.42)	5.14 (2.24)
Food insecure households (%)[b]	7.93***	19.94

Mean values are shown with standard deviations in parentheses. N: Number of observations; AE: adult equivalent.
***Mean values between adopters and non-adopters of Bt are statistically significant at the 1% level.
[a]More nutritious foods include pulses, fruits, vegetables, and all animal products.
[b]Consumption of less than 2300 kcal per AE and day.

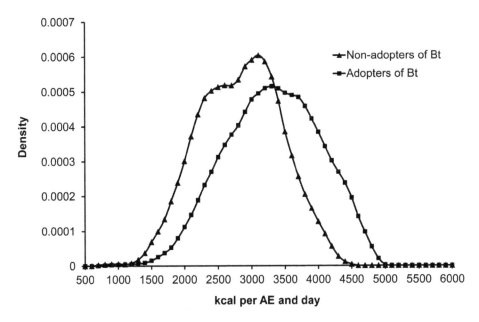

Figure 1. Density functions of household calorie consumption for adopters and non-adopters of Bt cotton. Functions were estimated non-parametrically using the Epanechnikov kernel with 1085 and 346 observations for adopting and non-adopting households, respectively. AE: adult equivalent.

households in a developing country (the data set with the variables used in this article is available as Data S1).

Calorie Consumption Data

The survey questionnaire included a detailed food consumption recall, which is a common tool to assess food security at the household level [24]. For a 30-day recall period, households were asked about the quantity consumed of different food items and the corresponding monetary value. The questions covered food consumed from own production, market purchases, gifts, and transfers.

The quantity data for the different food items were converted to calories consumed by using calorie conversion factors for India [25,26]. The total household calorie consumption from the 30-day recall was then divided by 30 to obtain a calorie value per day. Taking into account the age and gender structure of households, as well as physical activity levels of household members, the number of adult equivalents (AE) was calculated for each household. Male adults involved in farming count as 1.0 AE, female adults involved in farming as 0.8 AE. Male and female adults with lower physical activity levels count as 0.8 and 0.7, respectively. For children and adolescents, appropriate adjustments were made [25–27]. The daily household calorie consumption was divided by the number of AE in a household to obtain the calories consumed per AE and day.

Values for minimum dietary energy requirements found in the literature vary, which is due to several reasons [24]. Values stated per capita are lower than those stated per AE, because children have lower calorie requirements than adults. Moreover, not all studies take physical activity levels into account already in the AE calculations, as we do. The average daily calorie requirement for a moderately active AE in India is 2875 kcal/day [25]. According to the World Health Organization, a safe minimum daily intake should not fall below 80% of the calorie requirement, meaning 2300 kcal per AE. Minimum values around 2300 kcal per day for adult men are also found in other studies [28]. Based on this, we take 2300 kcal per AE as the threshold, that is, households with daily calorie consumption below 2300 kcal per AE are considered food insecure.

Most of the calories consumed in rural India are from cereals such as wheat, rice, millet, and sorghum that are rich in carbohydrates but less nutritious in terms of protein and micronutrient contents. Hence, in addition to total calories consumed we calculated the number of calories consumed from more nutritious foods to assess dietary quality. In the category "more nutritious foods", we include pulses, fruits, vegetables, and all animal products (i.e., milk, milk products, meat, fish, and eggs). Recent research suggests that the share of calories consumed from higher value, non-staple foods can also be used as an indicator of nutritional sufficiency [29]. The reason is that poor and undernourished households will largely choose foods that are the cheapest available sources of calories, namely cereals in the context of rural India. Only when they have surpassed subsistence, consumers will begin to substitute towards foods that are more expensive sources of calories [29].

It should be mentioned that food consumption data from household surveys may not provide very accurate data to measure nutritional status [24,30]. Sometimes, consumption data overestimate calorie intakes, because food losses, waste, and other uses within the household cannot be properly accounted for. However, this limitation applies to both adopters and non-adopters of Bt, so

Table 3. Bt cotton area among adopting households.

	2002	2004	2006	2008	Total
Mean Bt area (ha)	0.94	1.64	2.15	2.37	1.97
Standard deviation	1.32	1.87	2.14	2.22	2.08
Number of observations	131	246	333	375	1085

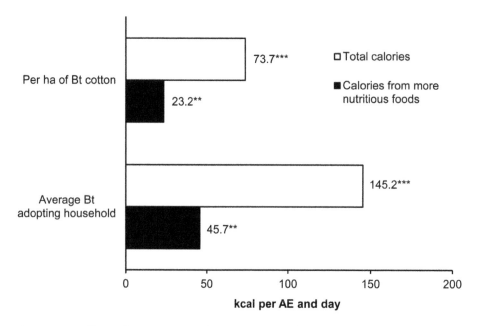

Figure 2. Net effects of Bt adoption on household calorie consumption. Results based on calorie consumption regression models estimated with panel data and household fixed effects (within estimator). Full model results are shown in Table 4. Calories from more nutritious foods include pulses, fruits, vegetables, and animal products. Effects for the average adopting household take into account the number of ha of Bt cotton actually grown. **Significant at the 5% level. ***Significant at the 1% level.

that the comparison between Bt and non-Bt, which is relevant for the impact assessment, is unaffected.

Regression Models

To estimate the impact of Bt cotton adoption on calorie consumption, we regress total daily calorie consumption per AE

Table 4. Calorie consumption models.

Variables	Model (1) Total calories (RE model)	Model (2) Total calories (FE model)	Model (3) Calories from more nutritious foods (FE model)
Bt area (ha)	79.08*** (18.85)	73.71*** (21.40)	23.17** (10.05)
Farm size (ha)	9.27** (4.22)	−0.69 (7.80)	1.97 (3.56)
Education of farmer (years)	9.41** (4.40)	–	–
Off-farm income (US$/year)	0.07*** (0.02)	0.05*** (0.02)	0.01* (0.007)
Household size (AE)	−62.48*** (10.71)	−89.46*** (14.43)	−29.33*** (6.89)
Karnataka (dummy)[a]	88.36 (57.97)	–	–
Andhra Pradesh (dummy)[a]	21.46 (58.00)	–	–
Tamil Nadu (dummy)[a]	212.86** (84.56)	–	–
2004 (dummy)[b]	−34.35 (48.97)	−5.98 (51.60)	−45.25* (25.33)
2006 (dummy)[b]	13.68 (54.48)	30.09 (61.12)	−112.87*** (29.41)
2008 (dummy)[b]	−92.92 (60.51)	−74.59 (69.51)	−72.70** (30.20)
Constant	3229.31*** (90.46)	3537.08*** (78.16)	843.23*** (41.42)
Number of observations	1431	1431	1431
R^2	0.13	0.09	0.10
Hausman test (chi-square statistic)	16.82**		

The dependent variable in models (1) and (2) is the total number of kcal consumed per AE and day. The dependent variable in model (3) is the number of kcal consumed from more nutritious foods (i.e., pulses, fruits, vegetables, and animal products) per AE and day. All coefficient estimates can be interpreted as marginal effects; robust standard errors are shown in parentheses. AE: adult equivalent; RE: random effects; FE: fixed effects.
*, **, ***Significant at the 10%, 5%, and 1% level, respectively.
[a]The reference state is Maharashtra.
[b]The reference year is 2002.

Table 5. Impact of Bt adoption on food security among cotton-producing households.

	Food insecure households (%)[a]	Change in food insecurity relative to status quo (%)
Non-adopters of Bt cotton (status quo)	19.94	
If non-adopters adopted Bt on their total cotton area	15.90	−20.26
If non-adopters adopted Bt on 85% of their cotton area	16.76	−15.95

The proportion of food insecure households in the status quo refers to the subsample of 346 non-adopters. For these households, changes in calorie consumption through Bt adoption were simulated, assuming full Bt adoption (on 100% of their cotton area) and partial Bt adoption (on 85% of their cotton area, as observed in the subsample of Bt adopters). For the simulations, the net effect of Bt on total calorie consumption per ha was used (Figure 2).
[a]Consumption of less than 2300 kcal per adult equivalent and day.

on Bt adoption, measured as the number of hectares of Bt cotton grown by a household in a particular year. Since Bt adoption increases farm profits and household incomes [23], we expect a positive and significant treatment effect. However, calorie consumption is also influenced by other factors that need to be controlled for. We control for education of the household head (measured in terms of the number of years of schooling); education plays an important role for both income generation and consumption behavior. We also include a variable for household size (measured in terms of AE). Moreover, we control for farm size in terms of area owned, which is a proxy for agricultural asset ownership more generally. Farm income is not included in the model, as this is directly influenced by Bt adoption. However, off-farm income, measured in US$ per year, is controlled for. We also include state dummies for Karnataka, Andhra Pradesh, and Tamil Nadu (Maharashtra is the reference state), capturing climatic and agroecological differences. Given the panel structure of the data with four survey rounds, we use year dummies for 2004, 2006, and 2008 (2002 is the reference year).

Panel data models are often estimated with a random effects estimator [31]. However, a random effects estimator can lead to

biased impact estimates when there is unobserved heterogeneity between Bt adopting and non-adopting households. Such bias resulting from endogeneity of the treatment variable is referred to as selection bias in the impact assessment literature [23,31]. Unobserved heterogeneity may potentially result from differences in household characteristics (e.g., Bt adopting farmers may have higher motivation, better management skills, or better access to information) or farm characteristics (e.g., differences in soil quality, or water access). Our panel data allow us to control for such unobserved heterogeneity. Since we surveyed the same households repeatedly over a 7-year period when Bt adoption increased, for many households we have observations with and without Bt adoption. Hence, we rely on a within household estimator, which is also called a fixed effects estimator. Differencing within households with the fixed effects estimator eliminates time-invariant unobserved factors, so that they can no longer bias the impact estimates [31]. A Hausman test is used to confirm the appropriateness of the fixed effects specification [19,31].

We estimate an additional model using calories from more nutritious foods (i.e., pulses, fruits, vegetables, and animal products) instead of total calorie consumption as dependent

Table 6. Robustness checks of Bt effects with different model specifications.

Variables	Model (1) Total calories	Model (2) Calories from more nutritious foods	Model (3) Total calories	Model (4) Total calories
Bt area 2002–04 (ha)	135.25*** (28.95)	17.94 (13.24)	–	–
Bt area 2006–08 (ha)	54.67** (23.33)	24.79** (10.46)	–	–
Cumulative Bt area (ha)	–	–	17.20 (12.20)	−28.08** (13.21)
Bt area (ha)	–	–	–	105.63*** (26.82)
Number of observations	1431	1431	1431	1431

Variables	Model (5) Total calories	Model (6) Total calories	Model (7) Total calories	Model (8) Total calories
Bt area (ha)	73.71*** (21.40)	76.19*** (27.62)	110.01*** (27.48)	53.40* (30.99)
Bt (dummy)	–	–	–	599.84*** (70.29)
Number of observations [a]	1431	1016	852	852

All models are estimated with household fixed effects. Other explanatory variables were included in estimation, as in Table 4, but are not shown here for brevity. The dependent variable in all models is calorie consumption measured in kcal per AE and day. Coefficient estimates can be interpreted as marginal effects; robust standard errors are shown in parentheses.
*, **, ***Significant at the 10%, 5%, and 1% level, respectively.
[a]In model (6), all observations of households that had adopted Bt in 2002 were dropped. In models (7) and (8), all observations of households that had adopted Bt in all survey rounds were dropped.

variable. This additional model helps to analyze impacts of Bt cotton adoption on dietary quality. A positive coefficient for the treatment variable would indicate that Bt adoption increases the consumption of more nutritious foods, thus not only contributing to more calories but also to better dietary quality.

Results and Discussion

Descriptive statistics are shown in Table 2. The average farm household owns 5 ha of land, without a significant difference between Bt adopters and non-adopters. Around half of this area is grown with cotton. Other crops cultivated include wheat, millet, sorghum, pulses, and in some locations rice, among others. Households are relatively poor; average annual per capita consumption expenditures range between 300 and 500 US$.

Bt adopting households consume significantly more calories than non-adopting households, and a smaller proportion of them is food insecure (Figure 1, Table 2). This suggests that the cash income gains through Bt adoption may have improved food security among cotton-producing households. Yet, this simple comparison does not yet prove a causal relationship.

Impact of Bt Cotton Adoption on Food Security

To further analyze the relationship between Bt adoption and calorie consumption, we use panel regression models, as explained above. The main explanatory variable of interest is the Bt cotton area of a farm household, for which descriptive statistics are shown in Table 3. The average Bt area among technology adopters in the sample is close to 2 ha, which is equivalent to 85% of the total cotton area of these farms. A breakdown by survey year shows that the average Bt area increased from less than 1.0 ha in 2002 to 2.4 ha in 2008. Hence, not only the number of Bt adopters but also the Bt area per adopting household increased considerably over time.

The regression results are shown in Table 4. Each ha of Bt cotton has increased total calorie consumption by 74 kcal per AE and day. For the average adopting household, the net effect is 145 kcal per AE (Figure 2), implying a 5% increase over mean calorie consumption in non-adopting households. Most of the calories consumed in rural India stem from cereals that are rich in carbohydrates but less nutritious in terms of protein and micronutrients. Yet the results show that Bt adoption has significantly increased the consumption of calories from more nutritious foods, thus also contributing to improved dietary quality.

We applied the total calorie consumption effect of Bt to the subsample of non-adopters to simulate the food security impact of adoption: if all non-adopters switched to Bt, the proportion of food insecure households would drop by 15–20% (Table 5). Most of these nutritional benefits have materialized already, as over 90% of all cotton farm households in India have adopted Bt technology by now.

Robustness Checks

We tested the robustness of the Bt effects by estimating calorie consumption models with alternative specifications. These additional estimates are shown in Table 6. We first look at possible changes in impact over time. In model (1), the Bt area variable is split into two periods, namely 2002–04 and 2006–08. In both periods, the Bt impact on calorie consumption was positive and significant, but the effect was bigger in 2002–04 than in 2006–08. The reason for this change is not that income effects of Bt adoption would shrink; recent research showed that the profit gains of Bt cotton in India were constant or even increased over

time [23]. The change in the calorie effect per ha of Bt is rather due to the fact that the Bt area per farm increased considerably in the later period, as was shown above. Measured per farm household, the calorie consumption effect of Bt was actually very similar in 2002–04 and 2006–08.

The smaller calorie consumption effect per ha of Bt with an increasing Bt area on a farm is consistent with Engel's law, which states that the proportion of the household budget spent on food decreases as income rises [32]. Unsurprisingly, the same trend is not observed when we focus on higher value, non-staple foods. The results of model (2) in Table 6 suggest that the Bt effect on calories from more nutritious foods has been increasing over time. Hence, Bt cotton adoption leads to a lower staple calorie share, implying higher nutritional sufficiency and better dietary quality [29].

In model (3) of Table 6, we analyze whether the Bt effect is cumulative, meaning that households that have adopted Bt earlier or on larger areas benefit over-proportionally. This might be the case when profit gains from Bt adoption are reinvested, possibly entailing larger consumption benefits in subsequent periods. To test for this option, we constructed a cumulative Bt area variable, adding up the Bt area on a farm in a particular year and Bt areas on the same farm in previous survey rounds. The coefficient of this variable is insignificant; cumulative effects do not seem to be important. If we include this variable together with the standard Bt area variable, the cumulative coefficient turns negative while the actual treatment effect increases (model 4). Again, this is consistent with Engel's law, implying that larger areas with Bt lead to lower proportions of the income gains being spent on calories.

In models (6) and (7), we analyze to what extent changes in the sample affect the estimation results. For easy comparison, results from the full-sample reference model, which were discussed above, are repeated in model (5). It is sometimes observed that early adopters of a new technology benefit more than late adopters. This may be due to cumulative effects, which we already tested for. In addition, general equilibrium adjustments may contribute to differential impacts between early and late adopters [33]. In model (6), we exclude all households that had adopted Bt already in the first survey round in 2002. The change in the Bt effect is very small, so we conclude that late adopters enjoy the same nutritional benefits per ha of Bt as early adopters.

This specification in model (6) with early adopters excluded is also an additional robustness check for possible issues of endogeneity and selection bias. The fixed effects panel estimator controls for time-invariant heterogeneity between adopters and non-adopters of Bt. But it cannot control for possible time-variant differences, which might play a role if early adopters are more innovative also with respect to other opportunities not captured in our data. The similarity of the results in models (5) and (6) substantiates that the estimated Bt impacts do not suffer from selection bias. In model (7), we exclude all observations of households that had adopted Bt in all survey rounds, so that the results are purely based on within household comparisons. The treatment effect remains highly significant. It even increases in magnitude, suggesting that the full-sample result is rather a cautious, lower-bound estimate. Finally, model (8) includes a dummy for Bt adoption in addition to the Bt area variable used before. The dummy produces a large coefficient, underlining the positive food security impact of Bt adoption. But the Bt area effect remains positive and significant, too, which confirms that using a continuous treatment variable is appropriate.

Overall, the additional results with alternative specifications strengthen the findings and show that the positive impacts of Bt cotton adoption on food security in India are very robust.

Conclusions

The results of this research confirm that the income gains through Bt cotton adoption among smallholder farm households in India have positive impacts on food security and dietary quality. GM crops are not a panacea for the problems of hunger and malnutrition. Complex problems require multi-pronged solutions. But the evidence suggests that GM crops can be an important component in a broader food security strategy. So far, food security impacts are still confined to only a few concrete examples. The nutritional benefits could further increase with more GM crops and traits becoming available in the future. Appropriate policy and regulatory frameworks are required to ensure that the needs of poor farmers and consumers are taken into account and that undesirable social consequences are avoided.

Acknowledgments

We thank Vijesh Krishna and two anonymous reviewers of this journal for very useful comments.

Author Contributions

Analyzed the data: MQ SK. Wrote the paper: MQ SK. Conceived and designed the survey: MQ.

References

1. FAO (2012) The State of Food Insecurity in the World (Food and Agriculture Organization of the United Nations, Rome).
2. United Nations (2012) The Millennium Development Goals Report 2012 (United Nations, New York).
3. Juma C (2011) Preventing hunger: biotechnology is key. Nature 479: 471–472.
4. Borlaug N (2007) Feeding a hungry world. Science 318: 359.
5. Shiva V, Barker D, Lockhart C (2011) The GMO Emperor has No Clothes (Navdanya International, New Delhi).
6. Friends of the Earth (2011) Who Benefits from GM Crops: An Industry Built on Myths (Friends of the Earth International, Amsterdam).
7. World Bank (2007) World Development Report 2008: Agriculture for Development (World Bank, Washington, DC).
8. Fedoroff NV, Battisti DS, Beachy RN, Cooper PJM, Fischhoff DA, et al. (2010) Radically rethinking agriculture for the 21st century. Science 327: 833–834.
9. Tester M, Langridge P (2010) Breeding technologies to increase crop production in a changing world. Science 327: 818–822.
10. James C (2012) Global Status of Commercialized Biotech/GM Crops: 2012, ISAAA Briefs No.44 (International Service for the Acquisition of Agri-biotech Applications, Ithaca, NY).
11. Sexton S, Zilberman D (2012) Land for food and fuel production: the role of agricultural biotechnology. In: The Intended and Unintended Effects of US Agricultural and Biotechnology Policies (eds. Zivin, G. & Perloff, J.M.), 269–288 (University of Chicago Press, Chicago).
12. Jayaraman K, Jia H (2012) GM phobia spreads in South Asia. Nature Biotechnology 30: 1017–1019.
13. European Commission (2010) A Decade of EU-Funded GMO Research 2001–2010 (European Commission, Brussels).
14. Paine JA, Shipton CA, Chaggar S, Howells RM, Kennedy MJ, et al. (2005) Improving the nutritional value of Golden Rice through increased pro-vitamin A content. Nature Biotechnology 23: 482–487.
15. Stein AJ, Sachdev HPS, Qaim M (2008) Genetic engineering for the poor: Golden Rice and public health in India. World Development 36: 144–158.
16. De Steur H, Gellynck X, Van Der Straeten D, Lambert W, Blancquaert D, et al. (2012) Potential impact and cost-effectiveness of multi-biofortified rice in China. New Biotechnology 29: 432–442.
17. Huang J, Mi J, Lin H, Wang Z, Chen R, et al. (2010) A decade of Bt cotton in Chinese fields: assessing the direct effects and indirect externalities of Bt cotton adoption in China. Science China Life Sciences 53: 981–991.
18. Krishna VV, Qaim M (2012) Bt cotton and sustainability of pesticide reductions in India. Agricultural Systems 107: 47–55.
19. Kouser S, Qaim M (2011) Impact of Bt cotton on pesticide poisoning in smallholder agriculture: A panel data analysis. Ecological Economics 70: 2105–2113.
20. Qaim M (2009) The economics of genetically modified crops. Annual Review of Resource Economics 1: 665–693.
21. Ali A, Abdulai A (2010) The adoption of genetically modified cotton and poverty reduction in Pakistan. Journal of Agricultural Economics 61: 175–192.
22. Subramanian A, Qaim M (2010) The impact of Bt cotton on poor households in rural India. Journal of Development Studies 46: 295–311.
23. Kathage J, Qaim M (2012) Economic impacts and impact dynamics of Bt (Bacillus thuringiensis) cotton in India. Proc. Natl. Academy of Sciences USA 109: 11652–11656.
24. de Haen H, Klasen S, Qaim M (2011) What do we really know? Metrics for food insecurity and undernutrition. Food Policy 36: 760–769.
25. Gopalan C, Rama Sastri BV, Balasubramanian SC (2004) Nutritive Value of Indian Foods (Indian Council of Medical Research, National Institute of Nutrition, Hyderabad).
26. National Sample Survey Organization (2007) Nutritional Intake in India 2004–05, Report No.513 (Government of India, New Delhi).
27. Rao CHH (2005) Agriculture, Food Security, Poverty and Environment – Essays on Post-Reform India (Oxford University Press, New Delhi).
28. FAO (2001) Human Energy Requirement, Food and Nutrition Technical Report 1 (Food and Agriculture Organization of the United Nations, Rome).
29. Jensen RT, Miller NH (2010) A Revealed Preference Approach to Measuring Hunger and Undernutrition. NBER Working Paper No.16555 (National Bureau of Economic Research, Cambridge, MA).
30. Bouis HE (1994) The effect of income on demand for food in poor countries: are our food consumption databases giving us reliable estimates? Journal of Development Economics 44: 199–226.
31. Cameron AC, Trivedi PK (2005) Microeconometrics: Methods and Applications (Cambridge University Press, Cambridge).
32. Leathers HD, Foster P (2004) The World Food Problem: Tackling the Causes of Undernutrition in the Third World (Lynne Rienner Publishers, Boulder, CO).
33. De Janvry A, Sadoulet E (2002) World poverty and the role of agricultural technology: direct and indirect effects. Journal of Development Studies 38(4): 1–26.

8

Bacterial Communities Associated with the Surfaces of Fresh Fruits and Vegetables

Jonathan W. Leff[1], Noah Fierer[1,2]*

1 Cooperative Institute for Research in Environmental Sciences, University of Colorado, Boulder, Colorado, United States of America, **2** Department of Ecology and Evolutionary Biology, University of Colorado, Boulder, Colorado, United States of America

Abstract

Fresh fruits and vegetables can harbor large and diverse populations of bacteria. However, most of the work on produce-associated bacteria has focused on a relatively small number of pathogenic bacteria and, as a result, we know far less about the overall diversity and composition of those bacterial communities found on produce and how the structure of these communities varies across produce types. Moreover, we lack a comprehensive view of the potential effects of differing farming practices on the bacterial communities to which consumers are exposed. We addressed these knowledge gaps by assessing bacterial community structure on conventional and organic analogs of eleven store-bought produce types using a culture-independent approach, 16 S rRNA gene pyrosequencing. Our results demonstrated that the fruits and vegetables harbored diverse bacterial communities, and the communities on each produce type were significantly distinct from one another. However, certain produce types (i.e., sprouts, spinach, lettuce, tomatoes, peppers, and strawberries) tended to share more similar communities as they all had high relative abundances of taxa belonging to the family Enterobacteriaceae when compared to the other produce types (i.e., apples, peaches, grapes, and mushrooms) which were dominated by taxa belonging to the Actinobacteria, Bacteroidetes, Firmicutes, and Proteobacteria phyla. Although potentially driven by factors other than farming practice, we also observed significant differences in community composition between conventional and organic analogs within produce types. These differences were often attributable to distinctions in the relative abundances of Enterobacteriaceae taxa, which were generally less abundant in organically-grown produce. Taken together, our results suggest that humans are exposed to substantially different bacteria depending on the types of fresh produce they consume with differences between conventionally and organically farmed varieties contributing to this variation.

Editor: Gabriele Berg, Graz University of Technology (TU Graz), Austria

Funding: This work was supported with funding from the Alfred P. Sloan Foundation's Microbiology of the Built Environment Program. The funders had no role in study design, data collection and analysis, decision to publish, or preparation of the manuscript.

Competing Interests: The authors have declared that no competing interests exist.

* E-mail: Noah.Fierer@colorado.edu

Introduction

Fresh produce, including apples, grapes, lettuce, peaches, peppers, spinach, sprouts, and tomatoes, are known to harbor large bacterial populations [1–7], but we are only just beginning to explore the diversity of these produce-associated communities. We do know that important human pathogens can be associated with produce (e.g., *L. monocytogenes*, *E. coli*, *Salmonella*), and since fresh produce is often consumed raw, such pathogens can cause widespread disease outbreaks [8–11]. In addition to directly causing disease, those microbes found in produce may have other, less direct, impacts on human health. Exposure to non-pathogenic microbes associated with plants may influence the development of allergies [12], and the consumption of raw produce may represent an important means by which new lineages of commensal bacteria are introduced into the human gastrointestinal system. More generally, produce-associated microbes can have important effects on the rates of food spoilage [13], and many of the microbes found on kitchen surfaces appear to come from produce sources [14].

Previous work investigating microbial communities on fresh produce has generally focused on culturable pathogenic bacteria and fungi (*sensu* [9]) with only a few recent studies having assessed the composition of produce-associated microbial communities

using culture-independent techniques. From this previous work, a few key patterns emerge: (1) Different produce types and cultivars can harbor different abundances of specific bacterial groups [9], (2) farming and storage conditions may influence the composition and abundances of microbial communities found on produce [3,5,15–18], and (3) non-pathogenic microbes may interact with and inhibit microbial pathogens found on produce surfaces [7,9,19–21]. Despite this body of work, we still have a limited understanding of the diversity of produce-associated microbial communities, the factors that influence the composition of these communities, and the distributions of individual taxa across produce types (particularly those taxa that are difficult to culture).

We expected the overall composition of microbial communities to vary across produce types for a variety of reasons. First, we know from previous work on tree leaf surfaces that different plant lineages are likely to harbor very distinct bacterial communities [22,23]. Moreover, we know that a range of environmental factors which can shape microbial community composition, including pH and moisture availability, can vary across produce types [6,13,24]. Likewise, differences in growing conditions, transport procedures, and storage conditions could influence the diversity and composition of produce-associated microbial communities. For example, we would expect produce grown closer to the ground to have higher relative abundances of soil microbial taxa and produce stored at cold temperatures for longer periods of time may harbor greater abundances of cold-tolerant bacteria [6,15,16,25].

Farming practices may also have an important, but under-studied, influence on the composition of produce-associated microbial communities. Consumers in developed nations are commonly exposed to differences in farming practices through their choice between organic and conventionally farmed produce items. Organic farming practices can differ from conventional farming practices in a variety of ways, including the types of fertilizer and pesticides that are used, and these differences have the potential to impact microbial community structure on produce surfaces [4,17,18,26]. However, we do not know if these potential effects of farming practices on produce-associated microbial communities are evident across a wide range of produce types and whether such effects persist up until the point that produce is purchased and consumed.

The objective of this study was to characterize the bacterial communities on the surfaces of multiple types of fruits and vegetables at the point of sale. We focused on those produce types that are frequently consumed raw, as we are likely exposed to far more live bacteria when we consume raw foods compared to cooked foods. Specifically, we addressed two fundamental questions: (1) How does bacterial community structure differ among produce types? and (2) Do differences in farming practices, such as those used on conventional and organic farms, have the potential to influence the composition of bacterial communities on the surfaces of produce items as experienced by end consumers? Because we know that culture-based techniques do not adequately capture a large portion of bacterial diversity on produce [27], we addressed these questions using high-throughput pyrosequencing analysis of the 16 S rRNA gene found in bacterial DNA extracted from the surfaces of the produce items.

Materials and Methods

Sample Collection

Fresh produce items were purchased from three differently-branded grocery stores in Boulder, CO, USA. These items consisted of eleven produce varieties, and for nine of these, both organic and conventional-labeled versions were obtained. Produce varieties and numbers of replicates are described in Table 1. In the USA, organically farmed produce differs from conventionally farmed produce in that synthetic pesticides and fertilizers, ionizing radiation, and sewage are generally not allowed in its production (http://www.usda.gov/). We acknowledge that differences observed between conventional and organic-labeled produce could be attributable to a number of factors that are not necessarily reflective of the differences in farming practices represented by the label. These include potential differences in farm location, transport, storage conditions, and storage time. However, these factors are difficult to control and thus our goal in this study was to assess the *potential* for broad-scale differences in farming practices to affect bacterial communities on produce items available to end consumers.

Lettuce and spinach samples were pre-rinsed and sold pre-packaged, other produce items were collected either in store packaging (grapes, lettuce, mushrooms, spinach, sprouts, and strawberries) or sterile plastic bags (apples, peaches, peppers, and tomatoes). Replicate samples were collected from discrete packages (of the same brand) at each store when sold pre-packaged and replicate samples of other produce types were collected from discrete fruits. Bacterial samples were collected

from each produce sample within a store on the same day, and each of the three stores were sampled within a single week. Bacterial samples were taken from produce samples using either sterile cotton swabs (following the procedure described in [14]) or by using a sterile water rinse to reduce the collection of chloroplasts (Table 1). For the rinsing procedure, water and produce samples were added to sterile plastic bags, and gently shaken for 5 min. Bacteria in the rinse water were collected onto 0.2 μm filters (Corning, Inc., Tewksbury, MA, USA) by vacuum filtration. Swabs and filters were stored at -20°C for less than 2 weeks prior to molecular analysis. DNA was extracted from swabs and filters using the PowerSoil-htp kit (Mo Bio Laboratories, Inc., Carlsbad, CA, USA) using modifications described previously [28].

Determination of Bacterial Community Composition and Diversity

16 S rRNA gene sequences were analyzed via barcoded pyrosequencing to quantify the diversity and community composition of the bacterial communities associated with each of the 215 produce samples collected. They were amplified and sequenced from the extracted genomic DNA using a procedure described in [23]. Briefly, sequences were PCR amplified in triplicate using a primer pair (799 f/1115 r) which does not amplify chloroplast DNA [23,29]. The reverse primer contained a 12-bp barcode sequence unique to each sample. The triplicate reactions were combined, DNA concentrations were measured, and equal quantities of DNA from each sample were combined together. The pooled DNA sample was cleaned using the UltraClean PCR Clean-Up Kit (Mo Bio Laboratories, Inc., Carlsbad, CA, USA) and sequenced at the Engencore facility at the University of South Carolina on the Roche 454 sequencing platform.

The 16 S rRNA gene sequences were processed using the QIIME v. 1.4.0 pipeline [30] to determine the diversity and composition of the produce-associated bacterial communities. Default parameters were used except that only sequences between 240 and 400 bp with both primers removed were retained for downstream analyses, and taxonomic identities were assigned to operational taxonomic units (OTUs) using the RDP classifier [31] trained on the Greengenes microbial 16 S rRNA gene sequence dataset (February 4, 2011 revision; greengenes.lbl.gov), clustered at a 97% similarity threshold. Because we obtained a variable number of sequences per sample (from only a few sequences to >4,000), the sequence data were rarefied at 200 sequences per sample to account for this variation. The rarefaction resulted in some samples being lost prior to further analysis, and information on the numbers of samples included in downstream analyses is provided in Table 1. At 200 sequences per sample, we were not able to survey the full extent of bacterial diversity in each sample, but previous work demonstrates that this depth of sampling is sufficient for accurate assessments of alpha and beta diversity patterns on both leaf surfaces [23] and in other microbial habitats [32]. Amplicon sequences were deposited in the public EMBL-EBI database (http://www.ebi.ac.uk/) and may be accessed using the accession number, ERP002018.

Statistical Analyses

To assess differences in microbial community composition across the produce items (beta diversity), we calculated both phylogenetic metrics (weighted and unweighted UniFrac distances, [33,34]) and a taxonomic metric (Bray-curtis dissimilarities calculated from log-transformed OTU abundances). Differences in overall bacterial community composition among the produce types and between farming practice type (organic versus conven-

Table 1. Produce varieties and sample numbers.

Produce variety	Bacteria sampling method	Sample replicates (purchased)		Sample replicates (after rarefaction)	
		Conventional	Organic	Conventional	Organic
Apple (*Malus domestica* "Granny Smith")	swab	12	12	9	8
Grapes (*Vitis vinifera*)	rinse	12	8	10	8
Lettuce (*Lactuca sativa* var. *longifolia*)	rinse	12	8	10	7
Mushrooms (*Agaricus bisporus*)	swab	12	4	12	4
Peach (*Prunus persica*)	swab	12	8	12	6
Pepper (*Capsicum annuum* "bell")	swab	12	12	10	12
Spinach (*Spinacia oleracea*)	rinse	11	12	8	12
Strawberries (*Fragaria* × *ananassa*)	swab	12	12	12	12
Tomato (*Solanum lycopersicum*)	swab	12	12	12	11
Alfalfa sprouts (*Medicago sativa*)	rinse	12		10	
Mung bean sprouts (*Vigna radiate*)	rinse	8		7	

tional) were assessed using a permutational multivariate ANOVA test (PERMANOVA) with produce type and farming practice as fixed factors and the grocery store brand as a random factor. PERMANOVA tests were also used to test for the effects of farming practice on bacterial community composition within individual produce types. Significant differences in taxonomic richness were assessed across produce types using the non-parametric Kruskal-Wallis test and between conventional and organic labeled produce items using a t-test. Significant differences in the relative abundances of individual bacterial taxa across produce types or factor levels were determined using ANOVA and the false discovery rate (FDR) correction. T-tests were used when comparing the relative abundances of individual taxa between conventional and organic analogs. All multivariate analyses were performed using PRIMER 6 [35], and univariate analyses were performed using R [36].

Results

Differences in Bacterial Community Diversity and Composition Across Produce Types

Although variable, taxonomic richness levels differed among the eleven produce types ($P<0.001$) with richness being highest on peaches, alfalfa sprouts, apples, peppers, and mushrooms and lowest on bean sprouts and strawberries (Fig. 1). Bacterial communities were highly diverse regardless of the produce type with between 17 and 161 families being represented on the surfaces of each produce type. However, the majority of these families were rare; on average, only 3 to 13 families were represented by at least two sequences per produce type. In some cases, OTUs assigned to a single bacterial family were dominant. For example, 88, 58, and 53% of OTUs on bean sprouts, spinach, and strawberries were assigned to the family Enterobacteriaceae, respectively. In contrast, the communities on apples were relatively even with no single family representing more than 8% of the sequences (Fig. 2).

Across the produce types, bacterial communities also differed with respect to their taxonomic structure, and produce type had a far larger influence on the observed variation in bacterial community composition than farming practice or store brand (Table 2). Furthermore, pairwise tests revealed that the community composition on the surface of each produce type differed significantly from one another ($P=0.001$ in all cases; Fig. S1). Still, certain produce types shared more similar community structure than others. On average, tree fruits (apples and peaches) tended to share communities that were more similar in composition than they were to those on other produce types, and produce typically grown closer to the soil surface (spinach, lettuce, tomatoes, and peppers) shared communities relatively similar in composition. Surface bacterial communities on grapes and mushrooms were each strongly dissimilar from the other produce types studied (Fig. 2).

Across all samples, the most abundant bacterial families were Enterobacteriaceae [30% (mean)], Bacillaceae (4.6%), and Oxalobacteraceae (4.0%). However, some families had high relative abundances on individual produce types (Fig. 2). Nearly all of the abundant bacterial families (representing ≥3% of sequences in any produce type) differed in their relative abundance among produce types. Among these families, only 2 of 19 bacterial families did not significantly differ in relative abundances across the produce types (Fig. 2). Enterobacteriaceae, for example, was the most abundant family on bean sprouts, spinach, lettuce, tomatoes, peppers, alfalfa sprouts and strawberries (at least 20%) but had substantially lower relative abundances on apples, peaches, grapes and mushrooms (Fig. 3). As previously mentioned, Enterobacteriaceae is one major group responsible for the clustering patterns described above and in Fig. 2 as those communities with high relative abundances of Enterobacteriaceae tended to cluster apart from those with lower relative abundances. Apples and peaches tended to have greater relative abundances of Microbacteriaceae and Sphingomonadaceae than other produce types. Grape surface communities displayed relatively strong contributions from the families Bacillaceae and Acetobacteraceae, and mushrooms, which showed the strongest differences from other produce types, had large relative abundances of Micrococcaceae, Sphingobacteriaceae, and Pseudomonadaceae (Fig. 2). Patterns in community composition differences at the family level were also reflected by differences in the dominant genera across the produce types. *Pantoea sp.* had a high relative abundance in most of the produce types that also had a high relative abundance of Enterobacteriaceae (those with >20% reported above). However, other produce types were

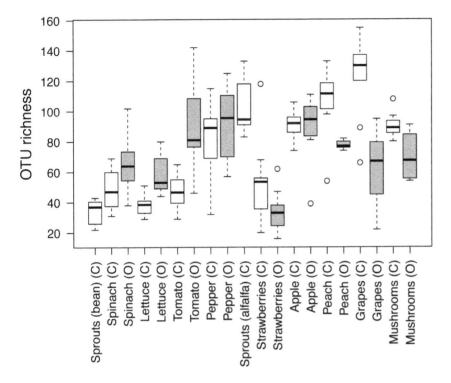

Figure 1. Boxplot of taxon richness for each produce type and conventional (C) and organic (O) equivalents. Samples were rarefied at 200 sequences per sample. Circles represent outliers.

generally characterized by dominant genera specific to that produce type (Table 3).

Potential for Farming Practice to Impact Bacterial Communities

Differences in taxonomic richness on the surfaces of conventional and organic-labeled analogs depended on the produce type (Fig. 1). Organic-labeled produce had significantly greater OTU richness compared to conventional-labeled produce on spinach, lettuce, and tomatoes, and significantly lower OTU richness on peaches and grapes ($P<0.05$ for all cases, Fig. 1).

Bacterial community composition also differed significantly between conventional and organic-labeled produce samples when taking into account variation due to produce type and store brand ($P=0.001$), with variation in farming practice more strongly related to variation in community composition than store brand (Table 2). Furthermore, community structure differed significantly between conventional and organic-labeled produce samples within each produce type ($P<0.05$ in all cases; Fig. 4). Although the taxa driving the observed differences between conventional and organic-labeled produce were not consistent across the produce types (Table 4), conventional-labeled varieties had a greater relative abundance of Enterobacteriaceae taxa across several produce types, including spinach, lettuce, tomatoes, and peaches (Table 4). On average, enterobacteria were 64% more abundant on the surfaces of conventional labeled spinach, lettuce, tomatoes, and peaches when compared with their organic labeled equivalent ($P<0.05$ in all cases), but these differences were not evident on the surfaces of other produce types ($P>0.05$; Fig. 3). Differences among organic and conventional labeled individuals of other produce types were generally associated with families that were specific to that produce type (Table 4). For example, the

communities on grapes were distinguished by a greater relative abundance of Bacillaceae on the organic-labeled grapes (Table 4).

Discussion

Our results generally demonstrated high bacterial diversity across the eleven fruits and vegetables we analyzed. Six phylogenetically diverse phyla were well represented by the sequences in at least one produce type: Actinobacteria, Bacteroidetes, Firmicutes, Proteobacteria, and TM7 (Fig. 2). The bacterial taxa we observed were consistent with findings from other studies that have used culture-independent techniques to describe taxon abundances. We found the surface bacterial communities of spinach, lettuce, and tomatoes to be numerically dominated by Gammaproteobacteria, a pattern which has also been noted in previous studies [5,15,16,37,38]. Similarly, Ottesen et al. [18] observed that Alphaproteobacteria was the most abundant bacterial class on apples, and we found the family Sphingomonadaceae within the class Alphaproteobacteria was the most abundant family present on apples. It is more difficult to directly compare our results with the large body of research on produce-associated bacteria that has been conducted using culture-based techniques as such techniques do not typically quantify proportions of bacteria belonging to specific taxonomic groups, rather binning them into operationally-defined groups determined by the culturing media used. Furthermore, culture-based studies detect a different fraction of the bacterial community assessed using culture-independent techniques, and, in most cases, a small fraction of the total bacterial diversity [27].

We observed distinct bacterial communities and substantial variation in bacterial richness across the produce types we analyzed. The family Enterobacteriaceae, which was relatively abundant in many of the samples, contributed strongly to this variation. Enterobacteriaceae taxa dominated the community

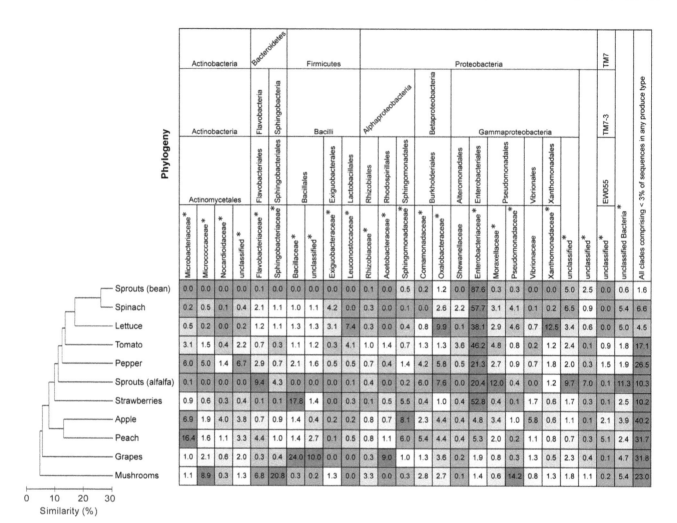

Figure 2. Relationships between bacterial communities on each produce type and relative abundances of bacterial families. The dendrogram is based on mean Bray-Curtis dissimilarities and shows differences among produce types in the overall composition of the bacterial communities. The heatmap shows mean relative abundances (%) of bacterial families on produce types. Only families and unclassified groupings representing at least three percent on any produce type are represented.

Table 2. PERMANOVA results of main factors.

Factor (type)	Diversity metric*	Pseudo-F	P	Component of variation
Produce type (Fixed)				
	Bray-Curtis	5.91	0.001	931
	Unweighted UniFrac	4.19	0.001	5.4×10^{-2}
	Weighted UniFrac	19.0	0.001	1.4×10^{-2}
Farming practice label (Fixed)				
	Bray-Curtis	2.96	0.001	78
	Unweighted UniFrac	2.16	0.001	4.1×10^{-3}
	Weighted UniFrac	7.32	0.001	1.0×10^{-3}
Store (Random)				
	Bray-Curtis	1.70	0.001	54
	Unweighted UniFrac	1.53	0.001	3.7×10^{-3}
	Weighted UniFrac	2.48	0.001	4.6×10^{-4}

*Bray-curtis dissimilarities were log transformed.

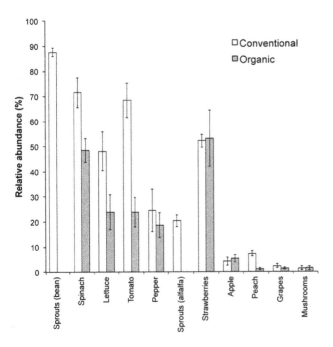

Figure 3. Mean relative abundances (±1 S.E.M) of bacteria belonging to the family Enterobacteriaceae. Each produce type and conventional and organic-labeled equivalents are shown. No organic-labeled equivalents were sampled for either type of sprouts.

Table 3. Bacterial OTUs representing large proportions (>5%) of their bacterial community on a given produce type.

Produce type	OTU classification[a]	Relative abundance (%)[b]
Sprouts (bean)		
	Pantoea sp.	57.5
	Klebsiella/Raoultella sp.	14.4
Spinach		
	Pantoea sp.	32.4
	Klebsiella/Raoultella sp.	9.0
Lettuce		
	Xanthomonas sp.	10.0
	Pantoea sp.	8.9
	Pectobacterium sp.	8.0
	Leuconostoc sp.	6.9
	Janthinobacterium sp.	5.7
Tomato		
	Klebsiella/Raoultella	26.9
	Pectobacterium sp.	9.8
Pepper		
	Pantoea sp.	11.1
Sprouts (alfalfa)		
	Acinetobacter sp.	9.3
Strawberries		
	Buchnera aphidicola	23.6
	Bacillus sp. 1	17.1
	Pantoea sp.	10.4
Apple		
	Photobacterium sp.	5.6
Peach		
	Microbacterium sp.	6.2
	Undetermined microbacteriaceae	6.1
Grapes		
	Bacillus sp. 1	18.2
	Gluconacetobacter sp	6.0
	Bacillus sp. 2	5.0
Mushrooms		
	Pseudomonas sp.	11.3
	Pedobacter sp.	5.5

[a]Classifications determined using BLAST with the NCBI nucleotide database.
[b]Values represent means.

composition in the majority of produce types, but several produce types (apples, peaches, grapes, and mushrooms) harbored a very low proportion of bacteria from this family (Fig. 3). This pattern also generally coincided with patterns in richness–produce types with greater proportions of taxa belonging to Enterobacteriaceae generally had a lower taxonomic richness (Fig. 1). Other bacterial families rarely had high relative abundances on more than two produce types (Fig. 2). Taken together, these results highlight that there is minimal overlap in the dominant bacterial taxa among produce types and that there is no 'typical' produce-associated community. Nonetheless, one Enterobacteriaceae taxon, putatively classified as *Pantoea* sp., was particularly abundant on many of the produce types harboring large proportions of Enterobacteriaceae (Table 3). This taxon might play an important role in the ecology of their hosts as certain *Pantoea* spp. are plant pathogens [39,40], but others may protect their hosts from disease or promote growth [19,41]. Overall, it is not surprising there were high relative abundances of Enterobacteriaceae across many of the produce types as members of this family are known to colonize certain fruits and vegetables [42,43]. What remains to be determined is why this family was dominant on certain produce types and relatively rare on others.

Likewise, it is difficult to unequivocally determine the specific factors responsible for driving the divergence between the bacterial communities on different produce types, but it is likely that several factors contribute to the patterns observed. Phyllosphere bacterial communities are known to strongly differ across plant species [23] likely due to variations in metabolites, physical characteristics, and symbiotic interactions with the host plant and other microbial inhabitants [37,44]. These characteristics may similarly select for specific microbial taxa on fruits and vegetables [13,37]. Additionally, the produce-growing medium could serve as a reservoir of bacteria that inoculate fruits and vegetables prior to harvest. However, our data do not provide evidence that this is an

important mechanism for driving the relative abundances of the dominant taxa. For example, bean sprouts and spinach harbored very similar communities but the sprouts were grown hydroponically while the spinach was grown in soil (Fig. 2). Differences in handling, transport, and storage could also play a role in structuring the microbial communities [15,16,25]. Only the lettuce and spinach samples, for example, were rinsed prior to packaging, and storage times likely differed among the produce items. Furthermore, differences in storage temperatures among produce items due to refrigeration could influence the relative abundance of cold-tolerant bacteria [15,16]. Additional research

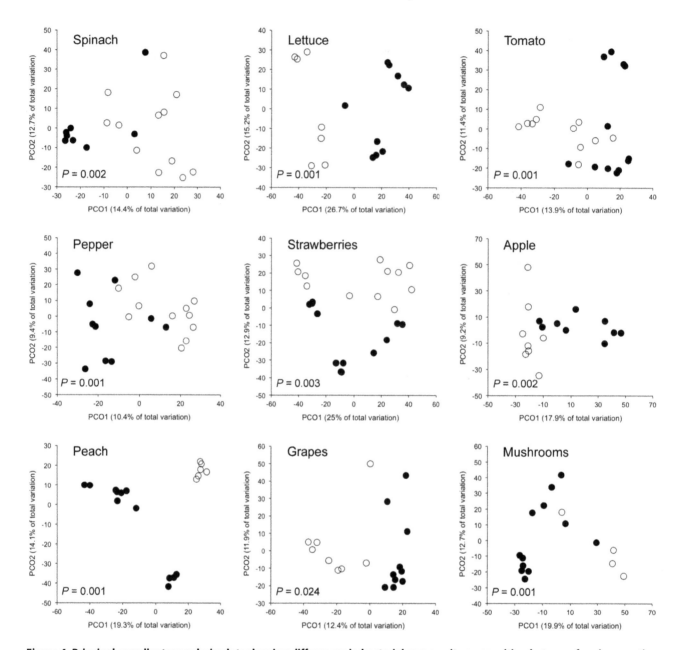

Figure 4. Principal coordinates analysis plots showing differences in bacterial community composition between farming practices. Plots are based on Bray-Curtis dissimilarities comparing surface bacterial communities of conventional-labeled (filled circles) and organic-labeled (open circles) produce items within each produce type. P-values were calculated using PERMANOVA.

needs to be conducted to disentangle the contribution of these factors in structuring produce-associated bacterial communities.

In addition to variation among produce types, we also found a somewhat weaker, but significant effect, of organic versus conventional label on the produce-associated communities (Fig. 4). This effect could be attributable to a number of factors including: growing location, fertilizer use, pesticide use, other agricultural practices, and shipping and handling procedures. Likewise, some of these differences could have been due to the direct application of bacterial agents used in organic pesticides (e.g., *Bacillus* spp.) or other bacteria found in the organic manures. Nevertheless, our results suggest that differences in farming practices could be influencing the relative abundance of specific taxa on the surfaces of fresh produce available at grocery stores. Overall, Enterobac-

teriaceae showed consistently greater relative abundances on conventional-labeled spinach, lettuce, tomatoes, and peaches when compared to organic-labeled varieties (Table 4). Differences between the microbiota on conventional and organically farmed produce items have been reported in other studies [4,17,18,26], but the differences in specific taxa may not always be consistent. For example, Oliveira et al. [4] observed a greater abundance of Enterobacteriaceae on organically farmed lettuce than its conventionally farmed equivalent via culturing techniques. Nonetheless, our data do suggest that shifts in community composition can persist for extended periods of time from the field to the grocery store and presumably, into the home of the consumer. This highlights the potential for differences in the microbiota between conventionally and organically farmed produce items to impact

Table 4. Bacterial families that differ in their relative abundances between conventional and organic-labeled equivalents within produce types.

Produce type	Taxonomy*				*P*	Relative abundance (%)	
	Phylum	Class	Order	Family		Conventional	Organic
Spinach							
	Proteobacteria	Gammaproteobacteria	Enterobacteriales	Enterobacteriaceae	0.008	71.5	48.5
	Firmicutes	Bacilli	Bacillales	Paenibacillaceae	0.028	0.3	3.6
	Firmicutes	Bacilli	Exiguobacterales	Exiguobacteraceae	0.048	2.7	5.2
Lettuce							
	Firmicutes	Bacilli	Lactobacillales	Leuconostocaceae	0.029	12.5	0
	Proteobacteria	Gammaproteobacteria	Enterobacteriales	Enterobacteriaceae	0.035	48.2	23.8
	Proteobacteria	Gammaproteobacteria	Pseudomonadales	Moraxellaceae	0.039	0.4	6.4
Tomato							
	Proteobacteria	Gammaproteobacteria	Enterobacteriales	Enterobacteriaceae	<0.001	66.8	23.8
	Actinobacteria	Actinobacteria	Actinomycetales	unclassified	0.010	0.6	3.8
	Proteobacteria	Gammaproteobacteria	Pseudomonadales	Moraxellaceae	0.015	2.9	6.8
	Proteobacteria	Alphaproteobacteria	Rhodobacterales	Rhodobacteraceae	0.015	1.1	3.1
Pepper							
	Actinobacteria	Actinobacteria	Actinomycetales	Microbacteriaceae	0.026	9.4	3.2
Apple							
	Bacteroidetes	Sphingobacteria	Sphingobacteriales	Flexibacteraceae	0.010	5.0	0.6
	Actinobacteria	Actinobacteria	Actinomycetales	Nocardioidaceae	0.036	7.1	0.6
	Proteobacteria	Alphaproteobacteria	Rhizobiales	Methylobacteriaceae	0.037	4.5	0.8
Peach							
	Actinobacteria	Actinobacteria	Actinomycetales	Microbacteriaceae	<0.001	7.4	34.5
	Proteobacteria	Gammaproteobacteria	Enterobacteriales	Enterobacteriaceae	<0.001	7.3	1.2
	Proteobacteria	Alphaproteobacteria	Rhodobacterales	Rhodobacteraceae	<0.001	1.7	4.5
	Bacteroidetes	Flavobacteria	Flavobacteriales	Flavobacteriaceae	0.012	3.4	6.4
	Proteobacteria	Betaproteobacteria	Burkholderiales	Oxalobacteraceae	0.026	6.4	0.4
	Firmicutes	Bacilli	Bacillales	Bacillaceae	0.034	0.5	3.3
Grapes							
	Firmicutes	Bacilli	Bacillales	Bacillaceae	0.007	9.6	42.0
	Firmicutes	Clostridia	Clostridiales	Clostridiaceae	0.043	3.9	1.4
	Actinobacteria	Actinobacteria	Actinomycetales	Micrococcaceae	0.049	3.0	0.9
Mushrooms							
	Proteobacteria	Alphaproteobacteria	Rhizobiales	Rhizobiaceae	0.001	4.3	0.3
	Actinobacteria	Actinobacteria	Actinomycetales	Micrococcaceae	0.003	11.7	0.4
	Bacteroidetes	Sphingobacteria	Sphingobacteriales	Sphingobacteriaceae	0.015	16.3	34
	Proteobacteria	Betaproteobacteria	Burkholderiales	Comamonadaceae	0.024	3.5	1.0

*Only families greater than or equal to 3% in one group and differed with a p-value less than 0.05 (t-test) are shown. No families met these criteria on the surfaces of strawberries.

human health. However, as it was not our objective to differentiate between closely related taxa that may have pathogenic and non-pathogenic representatives, future research is required to assess whether the bacterial community changes associated with organic and conventional-labeled produce may impact human exposures to potential pathogens.

Our results demonstrate differences among produce types in the diversity and composition of the produce-associated bacterial communities and the potential for farming practice to affect the types of bacteria that may be consumed. Moreover, they help to establish a basis on which to pose several further questions. For example: Do the differences in communities among produce types and farming practices influence microbial degradation of produce? Do these differences infer variation in the abundance of human pathogens or human health? Do they influence taste/quality of the produce being sold? It will be important to initiate controlled experiments to determine which factors are driving the differences in bacterial communities among the different produce types and conventional and organic-labeled varieties. In particular, focused studies examining how pesticide and fertilizer use impact produce-associated microbial communities would be useful as these factors are critical in differentiating conventional and organic farming practices. There is a substantial body of literature focused on the potential effects of farming practices on food chemistry and quality with many studies finding inconsistent results [45]; this work demonstrates that the effects of different farming practices on produce-associated microbial communities can be significant and are clearly worthy of further investigation.

Supporting Information

Figure S1 Principal coordinate analysis plot showing bacterial community composition by produce type. This plot is based on Bray-Curtis dissimilarities of samples rarefied at 200 sequences per sample.

Acknowledgments

We thank Gilbert Flores, Scott Bates, and Chris Lauber for their input during the project and Jessica Henley for her assistance with sample processing.

Author Contributions

Conceived and designed the experiments: JWL NF. Performed the experiments: JWL. Analyzed the data: JWL NF. Contributed reagents/materials/analysis tools: NF. Wrote the paper: JWL NF.

References

1. King AD, Magnuson JA, Török T, Goodman N (1991) Microbial flora and storage quality of partially processed lettuce. J Food Sci 56: 459–461.
2. Badosa E, Trias R, Parés D, Pla M, Montesinos E, et al. (2008) Microbiological quality of fresh fruit and vegetable products in Catalonia (Spain) using normalised plate-counting methods and real time polymerase chain reaction (QPCR). J Sci Food Agric 88: 605–611.
3. Ponce AG, Agüero MV, Roura SI, del Valle CE, Moreira MR (2008) Dynamics of Indigenous Microbial Populations of Butter Head Lettuce Grown in Mulch and on Bare Soil. J Food Sci 73: M257–M263.
4. Oliveira M, Usall J, Viñas I, Anguera M, Gatius F, et al. (2010) Microbiological quality of fresh lettuce from organic and conventional production. Food Microbiol 27: 679–684.
5. Rastogi G, Sbodio A, Tech JJ, Suslow TV, Coaker GL, et al. (2012) Leaf microbiota in an agroecosystem: spatiotemporal variation in bacterial community composition on field-grown lettuce. ISME J: 1–11.
6. Nguyen-the C, Carlin F (1994) The microbiology of minimally processed fresh fruits and vegetables. Crit Rev Food Sci Nutr 34: 371–401.
7. Liao CH, Fett WF (2001) Analysis of native microflora and selection of strains antagonistic to human pathogens on fresh produce. J Food Prot 64: 1110–1115.
8. Beuchat LR (1996) Pathogenic microorganisms associated with fresh produce. J Food Prot 59: 204–216.
9. Critzer FJ, Doyle MP (2010) Microbial ecology of foodborne pathogens associated with produce. Curr Opin Biotechnol 21: 125–130.
10. Harris LJ, Farber JN, Beuchat LR, Parish ME, Suslow TV, et al. (2003) Outbreaks associated with fresh produce: Incidence, growth, and survival of pathogens in fresh and fresh-cut produce. Compr Rev Food Sci F 2: 78–141.
11. Fatica MK, Schneider KR (2011) Salmonella and produce: Survival in the plant environment and implications in food safety. Virulence 2: 573–579.
12. Hanski I, von Hertzen L, Fyhrquist N, Koskinen K, Torppa K, et al. (2012) Environmental biodiversity, human microbiota, and allergy are interrelated. Proc Natl Acad Sci U S A 109: 8334–8339.
13. Gram L, Ravn L, Rasch M, Bruhn JB, Christensen AB, et al. (2002) Food spoilage–interactions between food spoilage bacteria. Int J Food Microbiol 78: 79–97.
14. Flores GE, Bates ST, Caporaso JG, Lauber CL, Leff JW, et al. (2012) Diversity, distribution and sources of bacteria in residential kitchens. Environ Microbiol In press.
15. Rudi K, Flateland SL, Hanssen JF, Bengtsson G, Nissen H (2002) Development and evaluation of a 16 S ribosomal dna array-based approach for describing complex microbial communities in ready-to-eat vegetable salads packed in a modified atmosphere. Appl Environ Microbiol 68: 1146–1156.
16. Lopez-Velasco G, Welbaum GE, Boyer RR, Mane SP, Ponder MA (2011) Changes in spinach phylloepiphytic bacteria communities following minimal processing and refrigerated storage described using pyrosequencing of 16 S rRNA amplicons. J Appl Microbiol 110: 1203–1214.
17. Granado J, Thürig B, Kieffer E, Petrini L, Fliessbach A, et al. (2008) Culturable fungi of stored "golden delicious" apple fruits: a one-season comparison study of organic and integrated production systems in Switzerland. Microb Ecol 56: 720–732.
18. Ottesen AR, White JR, Skaltsas DN, Newell MJ, Walsh CS (2009) Impact of organic and conventional management on the phyllosphere microbial ecology of an apple crop. J Food Prot 72: 2321–2325.
19. Enya J, Shinohara H, Yoshida S, Tsukiboshi T, Negishi H, et al. (2007) Culturable leaf-associated bacteria on tomato plants and their potential as biological control agents. Microb Ecol 53: 524–536.
20. Shi X, Wu Z, Namvar a, Kostrzynska M, Dunfield K, et al. (2009) Microbial population profiles of the microflora associated with pre- and postharvest tomatoes contaminated with Salmonella typhimurium or Salmonella montevideo. J Appl Microbiol 107: 329–338.
21. Teplitski M, Warriner K, Bartz J, Schneider KR (2011) Untangling metabolic and communication networks: interactions of enterics with phytobacteria and their implications in produce safety. Trends Microbiol 19: 121–127.
22. Kim M, Singh D, Lai-Hoe A, Go R, Abdul Rahim R, et al. (2012) Distinctive phyllosphere bacterial communities in tropical trees. Microb Ecol 63: 674–681.
23. Redford AJ, Bowers RM, Knight R, Linhart Y, Fierer N (2010) The ecology of the phyllosphere: geographic and phylogenetic variability in the distribution of bacteria. Environ Microbiol 12: 2885–2893.
24. Kroupitski Y, Pinto R, Belausov E, Sela S (2011) Distribution of Salmonella typhimurium in romaine lettuce leaves. Food Microbiol 28: 990–997.
25. Zagory D (1999) Effects of post-processing handling and packaging on microbial populations. Postharvest Biol Technol 15: 313–321.
26. Schmid F, Moser G, Müller H, Berg G (2011) Functional and structural microbial diversity in organic and conventional viticulture: organic farming benefits natural biocontrol agents. Appl Environ Microbiol 77: 2188–2191.
27. Yashiro E, Spear RN, McManus PS (2011) Culture-dependent and culture-independent assessment of bacteria in the apple phyllosphere. J Appl Microbiol 110: 1284–1296.
28. Fierer N, Hamady M, Lauber CL, Knight R (2008) The influence of sex, handedness, and washing on the diversity of hand surface bacteria. Proc Natl Acad Sci U S A 105: 17994–17999.
29. Chelius MK, Triplett EW (2001) The diversity of Archaea and Bacteria in association with the roots of Zea mays L. Microb Ecol 41: 252–263.
30. Caporaso J, Kuczynski J, Stombaugh J (2010) QIIME allows analysis of high-throughput community sequencing data. Nat Methods 7: 335–336.
31. Wang Q, Garrity GM, Tiedje JM, Cole JR (2007) Naive Bayesian classifier for rapid assignment of rRNA sequences into the new bacterial taxonomy. Appl Environ Microbiol 73: 5261–5267.
32. Kuczynski J, Liu Z, Lozupone C, McDonald D (2010) Microbial community resemblance methods differ in their ability to detect biologically relevant patterns. Nat Methods 7: 813–819.
33. Lozupone C, Knight R (2005) UniFrac: a new phylogenetic method for comparing microbial communities. Appl Environ Microbiol 71: 8228–8235.
34. Lozupone C, Lladser ME, Knights D, Stombaugh J, Knight R (2011) UniFrac: an effective distance metric for microbial community comparison. ISME J 5: 169–172.
35. Clarke K, Gorley R (2006) PRIMER v6: User Manual/Tutorial.
36. R Development Core Team (2012) R: A Language and Environment for Statistical Computing.
37. Hunter PJ, Hand P, Pink D, Whipps JM, Bending GD (2010) Both leaf properties and microbe-microbe interactions influence within-species variation in bacterial population diversity and structure in the lettuce (Lactuca Species) phyllosphere. Appl Environ Microbiol 76: 8117–8125.
38. Telias A, White JR, Pahl DM, Ottesen AR, Walsh CS (2011) Bacterial community diversity and variation in spray water sources and the tomato fruit surface. BMC Microbiol 11: 81.
39. Gitaitis RD, Gay JD (1997) First report of a leaf blight, seed stalk rot, and bulb decay of onion by Pantoea ananas in georgia. Plant disease 81: 1096.
40. Coutinho TA, Venter SN (2009) Pantoea ananatis: an unconventional plant pathogen. Mol Plant Pathol 10: 325–335.
41. Dastager SG, Deepa CK, Puneet SC, Nautiyal CS, Pandey A (2009) Isolation and characterization of plant growth-promoting strain Pantoea NII-186. From Western Ghat Forest soil, India. Lett Appl Microbiol 49: 20–25.
42. Wright C, Kominos SD, Yee RB (1976) Enterobacteriaceae and Pseudomonas aeruginosa recovered from vegetable salads. Appl Environ Microbiol 31: 453.
43. Abadias M, Usall J, Anguera M, Solsona C, Viñas I (2008) Microbiological quality of fresh, minimally-processed fruit and vegetables, and sprouts from retail establishments. Int J Food Microbiol 123: 121–129.
44. Lindow SE, Brandl MT (2003) Microbiology of the Phyllosphere. Appl Environ Microbiol 69: 1875–1883.
45. Smith-Spangler C, Brandeau ML, Hunter GE, Bavinger JC, Pearson M, et al. (2012) Are organic foods safer or healthier than conventional alternatives? A systematic review. Ann Intern Med 157: 348–366.

9

Root Interactions in a Maize/Soybean Intercropping System Control Soybean Soil-Borne Disease, Red Crown Rot

Xiang Gao[1], Man Wu[1], Ruineng Xu[1], Xiurong Wang[1], Ruqian Pan[1], Hye-Ji Kim[2], Hong Liao[1]*

1 State Key Laboratory for Conservation and Utilization of Subtropical Agro-bioresources, Root Biology Center, South China Agricultural University, Guangzhou, China, **2** Department of Tropical Plants and Soil Sciences, College of Tropical Agriculture and Human Resources, University of Hawaii at Manoa, Honolulu, Hawaii, United States of America

Abstract

Background: Within-field multiple crop species intercropping is well documented and used for disease control, but the underlying mechanisms are still unclear. As roots are the primary organ for perceiving signals in the soil from neighboring plants, root behavior may play an important role in soil-borne disease control.

Principal Findings: In two years of field experiments, maize/soybean intercropping suppressed the occurrence of soybean red crown rot, a severe soil-borne disease caused by *Cylindrocladium parasiticum* (*C. parasiticum*). The suppressive effects decreased with increasing distance between intercropped plants under both low P and high P supply, suggesting that root interactions play a significant role independent of nutrient status. Further detailed quantitative studies revealed that the diversity and intensity of root interactions altered the expression of important soybean *PR* genes, as well as, the activity of corresponding enzymes in both P treatments. Furthermore, 5 phenolic acids were detected in root exudates of maize/ soybean intercropped plants. Among these phenolic acids, cinnamic acid was released in significantly greater concentrations when intercropped maize with soybean compared to either crop grown in monoculture, and this spike in cinnamic acid was found dramatically constrain *C. parasiticum* growth *in vitro*.

Conclusions: To the best of our knowledge, this study is the first report to demonstrate that intercropping with maize can promote resistance in soybean to red crown rot in a root-dependent manner. This supports the point that intercropping may be an efficient ecological strategy to control soil-borne plant disease and should be incorporated in sustainable agricultural management practices.

Editor: Ching-Hong Yang, University of Wisconsin-Milwaukee, United States of America

Funding: This work was supported by the National Natural Science Foundation of China (grant no. 31025022) and National Key Basic Research Special Funds of China (grant no. 2011CB100301). The funders had no role in study design, data collection and analysis, decision to publish, or preparation of the manuscript.

Competing Interests: The authors have declared that no competing interests exist.

* E-mail: hliao@scau.edu.cn

Introduction

Reducing disease severity through increasing the within-field multiple crop species, intercropping has been widely reported in the literature and commercially implemented [1]. For example, disease-susceptible rice varieties have been grown in combination with resistant varieties to achieve higher yield and greatly suppress blast severity compared to growth in monoculture [2]. Also, growing mixtures of resistant oats and susceptible barley results in decreased stem rust severity, and thereby increases yield [3]. However, the mechanisms underlying these heterogeneous genotype and crop species effects on disease control remain unclear.

Soil-borne diseases are caused by phytopathogenic microbes in the soil, which may damage plants upon penetration of the root or basal stem [3]. With a potential to seriously harm the agroecosystem, and with broad geographic ranges, soil-borne diseases can significantly impact global food production [4]. Furthermore, eradication of the pathogen is difficult as long as soil-borne disease pathogens accumulate in cultivated soils. To date,

effective methods to control soil-borne diseases remain lacking. Therefore, the simple and economic alternative of agricultural management is often a preferred method for preventing crop losses caused by soil-borne diseases. It has long been recognized that crop heterogeneity suppresses soil-borne disease infection and appears to be a more promising solution than petrochemical-based pesticide application. Among the successful examples of this strategy are inhibition of tomato bacterial wilt caused by *Pseudomonas solanacearum* in a Chinese chive-tomato intercropping system [5], suppression of Fusarium wilt in rice with watermelon intercropping [6,7], and alleviation of tomato early blight disease by intercropping with marigold [8].

One potential target for control through crop heterogeneity is red crown rot, which is one of the most severe soil-borne diseases for plants. It is caused by the fungus *Cylindrocladium parasiticum* (*C. parasiticum*, teleomorph *Calonectria ilicicola*) and is named after the reddish brown basal stalk tissue of infected plants [9–11]. This pathogen has a wide host range, and red crown rot has been found in many countries with severe crop losses possible throughout its distribution. Soybean (*Glycine max* L. Merr.), an economically important host of red crown rot, is a major cash crop and has been planted world wide. In South China, estimates for soybean yield loss due to red crown rot range to as high as 50% [11]. At present, there are no effective pesticides to control red crown rot.

Intercropping has been widely used in Asia, Latin America and Africa to enhance crop productivity through improved nutrient uptake, enhanced land equivalent ratio (LER), increased yield, and better disease control [12,13]. Legume/cereal intercropping is the most common intercropping system, which not only allows crops to take up more nutrients than in monoculture, but also reduces disease occurrence [14,15]. However, little attention has been paid to the belowground environment of the plant root systems. As the main organ to contact the soil and perceive signals from neighboring plants [16], root function and behavior should be particularly important in soil-borne disease control. In one case, Fang et al. [17] reported that root interactions in the maize/soybean intercropping can integrate responses to P status and root behaviors among neighboring plants, and thereby impact maize P nutrition. A previous study by the current authors found that co-inoculation with rhizobia and mycorrhizal fungi can inhibit soybean red crown rot [18]. However, uncertainty remains about what happens when only mycorrhizal fungi, which favor maize over soybean, are introduced. In this study, effects of maize intercropping with soybean on red crown rot in soybean were studied in two years of field experiments. Further in-door experiments were then carried out to evaluate the underlying mechanisms for inhibition of soybean red crown rot by intercropping with maize.

Methods and Materials

1. The Boluo (E114.28°, N23.18°) field site is the experimental base of the South China Agricultural University Root Biology Center. Therefore, the authority that issued the permit for this location is South China Agricultural University.

2. No specific permissions were required for this location. I am here to confirm that the field studies did not involve endangered or protected species.

Field trials

Field experiments were carried out on acidic lateritic red soils at the Boluo field site (E114.28°, N23.18°) in the Guangdong Province of China from July to October in 2009 and 2010 with the average day/night temperature around 25–33°C and monthly rainfall around 150 mm. Basic soil chemical characteristics were as follows: pH, 5.37; organic matter, 17.63 g kg^{-1}; available P (Bray I method), 15.68 mg P kg^{-1}; available N, 86.64 mg N kg^{-1}; available K, 75.28 mg K kg^{-1}. The field site had a history of up to 10 years of continuous soybean cultivation, in which severe soybean red crown rot (RCR) damage occurred [11]. Two years of field trials were conducted in a split-block design with two P supplies (high P and low P), and four cultivation modes (soybean monoculture, maize monoculture, two maize/soybean intercropping systems) for a total of 8 treatments (Figure A in File S1). High P (HP) was implemented by adding calcium superphosphate (SSP) at the rate of 80 kg P$_2$O$_5$ ha^{-1}, and low P (LP) plots had no P fertilizer added. The two maize/soybean intercropping systems differed in planting distances between soybean and maize plants, which were 20 cm for ISC1 and 5 cm for ISC2. Urea and KCl were added to all plots as 80 kg N ha^{-1} and 60 kg K ha^{-1} for supplementing N and K supplies. Each treatment had four

replicates with a total of 32 plots (i.e. 2 P × 4 mode × 4 replicates). The planting area of each plot was 54 m^2. Soybean (*Glycine max*) variety HN89 and maize (*Zea mays*) variety ZD958 were employed in the field experiment.

Seventy days after planting, 100 plants were selected from each plot to investigate the severity of RCR. Disease incidence and severity caused by *C. parasiticum* were determined according to Gao et al. [18]. Disease incidence was defined as the percentage of infected subterranean stems. Disease severity was recorded on a 0–5 scale: i.e., 0, no visible symptoms; 1, small necrotic lesions on the subterranean stem; 2, necrotic lesions extending around the subterranean stem, 3, necrotic lesions extending to the soil surface; 4, severe necrosis on the subterranean stem and roots, as well as severe chlorosis appearing in leaves; 5, plant death. The disease index was summarized within each plot as $\{[(n_1 \times 1)+(n_2 \times 2)+(n_3 \times 3)+...+(n_N \times N)]/[N \times (n_1+n_2+n_3...+n_N)]\} \times 100$, where $n_1...n_N$ was the number of subterranean stems in each of the respective disease category, and N was the highest disease severity score [2,18].

Seventy days after planting, two representative healthy soybean plants from each plot were harvested for measuring root length and symbiotic traits. Total root length in the 20-cm upper soil layer of the 5-cm strip between intercropped plants was measured to represent the intensity of interacting roots between the two crop species in the field as described by Fang et al. [16]. In short, trenches were dug by shoveling between rows of intercropped plants. The walls were carefully scraped with a screwdriver to reveal the tips of the roots. Plastic transparent sheets (25×30 cm) were positioned adjacent to the exposed soil wall. The roots in the 20 cm upper soil layer were marked on the sheets, and the root length in the 5 cm section between the two plants was quantitatively measured with root image analysis software (WinRhizo Pro, Régent Instruments). Arbuscular mycorrhizal fungi (AMF) colonization rates and nodule numbers on soybean roots were measured according to Wang et al. [19].

At maturity, soybean biomass and yield were average for 20 healthy and 20 infected plants within each soybean monoculture plot. Growth reduction was calculated as: reduction (%) = (yield/biomass of healthy plants - yield/biomass of infected plants) yield/biomass of healthy plants × 100. This value was determined for each of the four replicates separately.

Root barrier sand culture

Root barrier sand culture experiments were conducted in the greenhouse of Root Biology Center of South China Agricultural University in 2011 using soybean variety HN89 and maize variety ZD958. *Cylindrocladium parasiticum* (*C. parasiticum*, GenBank Accession No. GU073284) was isolated from infected soybean roots at the Boluo field site and used as the pathogen inoculant [11]. The sand culture experiment consisted of 3 factors (P level, root barrier and pathogen inoculation), including two P levels, three root barrier arrangements and two pathogen inoculations, for a total of 12 treatments. There were 6 replicates for each treatment. The two P levels were HP (500 µM P) and LP (15 µM P). Plastic pots with two equal compartments were used to provide three types of root barriers as illustrated in Figure A in File S1 and described as follows: solid barrier to eliminate root interaction and exudates movement, a nylon mesh barrier (30 µm pores) to prevent interspecies root intermingling while permitting root exudates exchange, and no barrier to allow roots and exudates to completely interact. Soybean plants were inoculated with *C. parasiticum* infected wheat seeds or sterilized media as a control before sowing. Wheat seeds of infected *C. parasiticum* were obtained from 14-day-old media inoculated with *C. parasiticum*. The spore concentration was determined using a hemocytometer and adjusted to 1×10^5 spores per mL [18]. Before sowing, infected wheat seeds were placed at depth of 5 cm in sand culture pots. Two independent biological experiments were conducted, with very similar results between experiments, so only the results from one experiment were presented.

Plants were irrigated daily using modified 1/2 strength Hoagland nutrient solution with either of the two P additions listed above [20]. Sixty days after planting, incidence and severity of red crown rot was measured as described above. In order to determine the root infection by *C. parasiticum*, root colony forming units (CFUs) were measured from sand culture plants. Randomly selected soybean roots were cut into 1 cm segments and washed thoroughly and surface-sterilized with 0.5% (v/v) NaClO for 3 min, rinsed three times in sterilized water, and then blotted on sterilized filter paper. About 1 g root segments were ground in 50 mL of sterile deionized water using a blender at high speed for 1 min. The homogenized root suspension was diluted 10 fold with sterile deionized water. 100 μL aliquots of diluted solutions were evenly spread on 4 plates containing Rose Bengal Medium using a sterile glass rod. There were 4 replicates and thus 16 plates for each treatment. All plates were incubated at 28°C in the dark for 4 days. Colonies of *C. parasiticum* were identified and counted on each plate to determine the CFUs per gram of roots for each treatment [18].

Another sand culture experiment was carried out for pathogen-related (*PR*) gene expression analysis. Except for the inoculation process, all the treatments were the same as above. Plants were grown in sand for thirty days without inoculation, and then inoculated with *C. parasiticum*. Spore suspensions of *C. parasiticum* were obtained from 14-day-old V8-juice media which were collected by adding 10 mL of sterile water to each Petri dish and rubbing the surface with a sterile L-shaped spreader. The suspension was subsequently filtered through 3-layers of cheese-cloth. The spore concentration was determined using a hemacytometer and adjusted to 1×10^5 spores per mL [18]. Thirty days after planting, plant stem base was infected with 20 mL *C. parasiticum* spore suspensions each pot. Three soybean roots from each pot were randomly harvested at 1 and 5 days after inoculation (DAI) to detect the expression pattern of eight related *PR* genes in response to inoculation using quantitative real time PCR (RT-PCR) described by Gao et al. [18]. Specific primer sequences and putative functions of tested plant *PR* genes were listed in the Table S1 in File S2. Simultaneously, the activities of polyphenol oxidase (PPO) and phenylalanine ammonia-lyase (PAL) were determined based on methods reported in Song et al. [21].

Root exudates composition and *in vitro C. parasiticum* assays

A hydroponic experiment was carried out to collect root exudates in 2012. Soybean and/or maize seedlings were grown in half strength Hoagland nutrient solution with either of the two P additions (LP and HP) as described above. The cultivation modes included monoculture maize (MC) or soybean (MS), and intercropping of soybean and maize (ISC). There were six treatments (three cultivation modes by 2 P levels) with four replicates. Thirty days after planting, root exudates from each treatment were collected and processed, and a bioassay for colony diameter growth and sporulation was conducted as described in Gao et al. [18]. Roots were gently removed from pots and washed with deionized water. Cleaned roots were submerged in a plastic cup containing 500 mL of 0.5 μM CaCl$_2$ to collect exudates for 6 hours. Root exudates were filtered through a 0.45 μm Millipore

membrane and stored at -20 °C. During collection, each cup containing 3 plants was covered by a black plastic lid to avoid contamination and light [7,22]. Pathogen growth diameter and sporulation was quantified by adding 2 mL root exudates to V8-juice medium before it solidified to yield a total volume of 20 mL per Petri dish. Plates were incubated at 28 °C in the dark. Colony diameter and sporulation of *C. parasiticum* were determined 5 and 14 days after incubation using ruler and hemocytometer [18]. Phenolic compounds in the root exudates were identified using an HPLC system (SPD-20A, Shimadzu, Tokyo) with gallic acid, p-coumaric acid, phthalic acid, vanillic acid, syringic acid, ferulic acid, salicylic acid and cinnamic acid included as standard phenolic compounds [22]. According to the above HPLC analyses results, the dominant phenolic acids, including ferulic, gallic, p-coumaric, cinnamic and salicylic acids (Sigma, USA), were subsequently applied exogenously at five concentrations, 0, 10, 20, 30 and 40 mg/L, in V8-juice media to test for allelopathic effects on growth and sporulation of *C. parasiticum* as described above.

Data analysis

Data were statistically analyzed by Two-way ANOVA using Excel 2003 (Microsoft Corporation, 1985–2003) and SAS 9.1 (SAS Inc., Cary, NC, USA) for multiple comparisons.

Results

Reduction of soybean growth and yield caused by red crown rot in the field

Red crown rot significantly inhibited soybean growth, and subsequently reduced soybean yield in comparison to uninfected plants in field experiments (Figure B in File S1). Compared to healthy plants, infected soybean biomass in the monoculture treatment was reduced in 2009 by 23% and 27%, and in 2010 by 28% and 43% at LP and HP, respectively. Soybean yield was reduced in 2009 by 21% and 33%, and in 2010 by 17% and 44% at LP and HP, respectively. These observations illustrate the severe damage potential of red crown rot on soybean, particularly with high P supply.

Disease severity of soybean red crown rot in the field

Intercropping significantly reduced disease incidence and severity index of soybean red crown rot, and high P slightly enhanced red crown rot in both monoculture and intercropping (Table 1). The planting distance of soybean and maize plants also significantly affected the disease severity as indicated by the lowest disease incidence and index occurring when soybean was grown in close proximity with maize (ISC2, 5 cm apart). Compared to monoculture soybean (MS), the disease incidence in ISC2 was reduced by 53% and 59% in 2009, and by 43% and 47% in 2010 at LP and HP, respectively. The disease index in ISC2 was decreased by 49% and 46% in 2009, and by 57% and 50% in 2010, at LP and HP, respectively (Table 1). This suggests that the occurrence and development of red crown rot on soybean growth can be alleviated by intercropping with maize, with particular attention paid to planting distance.

Mycorrhization, nodulation and root interactions in the field

Co-inoculation with AMF and rhizobia has been shown to alleviate soybean red crown rot [18]. However, in this study, the mycorrhization rate and nodule numbers were not affected by intercropping or planting distance (Table S2 in File S2), which

Table 1. Disease severity of soybean caused by *C. parasiticum* as affected by cultivation mode, P level and planting distance in the two years of field experiments on acid soils.

		Incidence (%)		Disease index	
		LP	HP	LP	HP
2009	MS	43±3.4Ab	62±5Aa	33±3Aa	40±3Aa
	ISC1	32±2Ba	39±4Ba	26±1Ab	32±2Ba
	ISC2	20±2Ca	26±2Ca	17±1Ba	22±3Ca
2010	MS	45±3Ab	58±4Aa	36±3Ab	48±2Aa
	ISC1	33±1Bb	41±2Ba	26±2Bb	36±2Ba
	ISC2	26±2Ca	31±1Ca	16±1Cb	24±2Ca

Note: Disease incidence and index were measured as described in Materials and Methods. HP: 80 kg P_2O_5 ha^{-1} added as calcium superphosphate, LP: no P fertilizer added. MS: soybean monoculture, ISC1: maize/soybean intercropping with 20 cm spacing; ISC2: maize/soybean intercropping with 5 cm spacing. All the data are the mean of four replicates ± SE. The same upper-case letter after numbers in the same column for the same trait in the same year indicates no significant difference among cultivation modes at 0.05 ($P<0.05$); The same lower-case letter after numbers in the same row for the same trait in the same year indicates no significant difference between two P levels at 0.05 ($P<0.05$).

indicates that neither mycorrhization nor nodulation are major contributors to the variation of soybean red crown rot observed in the field. In contrast, intermingled roots between the two different crop species were significantly affected by planting distance (Fig. 1). The total root length in the upper 20 cm soil layer of the 5 cm section centrally located between plants in ISC2 was greater in 2009 by 179% and 234% compared to ISC1 (20 cm apart), and in 2010 by 183% and 181% at LP and HP, respectively. Moreover, low P supply also affected root interactions. The root length in the upper, middle area at low P compared to high P increased by 33% and 24% in ISC1, and by 11% and 25% in ISC2 in 2009 and 2010, respectively (Fig. 1). This suggests that the intensity of intermingled roots between soybean and maize plants might be the major player in reducing the disease severity of soybean red crown rot by maize/soybean intercropping.

Disease severity of soybean red crown rot in sand culture

The disease severity of soybean red crown rot was dramatically decreased with increasing root interactions in sand culture. Compared to the solid barrier treatment, both mesh and no barrier treatments had significantly lower disease incidences and indices, and fewer pathogen CFUs at both P levels (Fig. 2, Figure C in File S1). When soybean and maize roots completely interacted with each other (no barrier), the lowest disease incidences and indices and fewest pathogen CFUs were observed. An intermediate level of disease incidence was observed in plants grown with the mesh barrier treatment, and the highest level of disease was seen in plants separated by solid barriers. This is consistent with field results that an increase in root interactions reduced the occurrence of soybean red crown rot.

Expressions of important *PR* genes and activities of corresponding enzymes in sand culture

The results displayed in Figure 3 showed that transcript abundances of all tested *PR* genes in soybean roots were induced to a high level by inoculation with *C. parasiticum*, and this induction was enhanced by increasing root interactions. Except *PR4* on day 5 and *PR3, PR10, PR12* and *PAL* on day 1, the relative expression values of most tested *PR* genes were highest in the no barrier treatment, followed by the mesh barrier treatment. For example, *PPO* transcription abundance increased by 7.3- and 7.4- fold, and *PAL* by 12.9- and 4.1- fold in LP and HP, respectively, at 5 DAI in a comparison of the no barrier treatment with the solid barrier

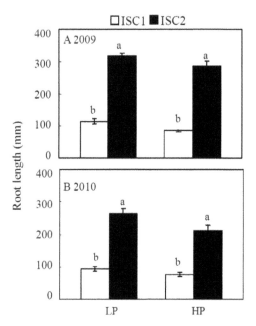

Figure 1. Root interactions between soybean and maize plants in the field. Root interactions were measured as the total root length in the upper 20 cm soil layer of the 5 cm section centrally located between the maize and soybean plants. HP: 80 kg P_2O_5 ha^{-1} added as calcium superphosphate, LP: no P fertilizer added. ISC1: maize/soybean intercropping with 20 cm spacing; ISC2: maize/soybean intercropping with 5 cm spacing. All the data are the mean of four replicates ± SE. F value from Two-way ANOVA: 8.35 for P treatment ($P<0.05$), 159.08 for cultivation mode ($P<0.001$), 1.99 for interaction (not significant). Bars with different letter(s) vary significantly among treatments as determined by Duncan's multiple range test ($P<0.05$).

treatment (Fig. 3G and H). Furthermore, the enzyme activities of PPO and PAL in soybean roots were significantly enhanced by root interactions with maize (Figure D in File S1). PPO activity increased in the no barrier treatment compared to the solid barrier treatment at 1 DAI by 114% and 99%, and in the mesh treatment by 66% and 43% in LP and HP, respectively. Also, PAL activity increased in the no barrier treatment compared to the solid barrier

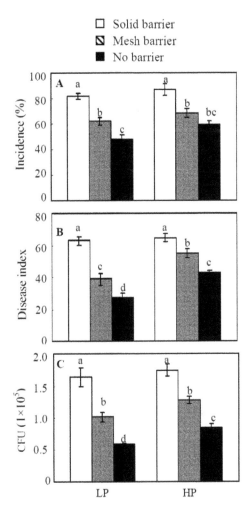

☐ Solid barrier
◨ Mesh barrier
■ No barrier

Figure 2. Disease severity of soybean red crown rot in sand culture. A, disease incidence, B, disease index, C, CFU. LP, 15 μM P; HP, 500 μM P. All of the data are the means of four replicates ±SE. Bars with different letter(s) vary significantly among the different inoculation treatments as determined by Duncan's multiple range test ($P<0.05$).

treatment at 1 DAI by 65% and 227%, and in the mesh treatment by 39% and 105% in LP and HP, respectively.

In vitro C. parasiticum assays using root exudates

In comparison with soybean monoculture (MS), the addition of root exudates from the maize/soybean intercropping (ISC) or maize monoculture (MC) dramatically inhibited pathogen growth (Fig. 4). At LP, colony diameter of *C. parasiticum* decreased by 54% and 38%, and sporulation declined by 37% and 27% when growth in MS exudates was compared to that in ISC or MC exudates, respectively. Furthermore, P supply also affected pathogen growth as indicated by the observed 34%, 31% and 49% increases in colony diameter in HP compared to LP when adding the root exudates from MS, MC and ISC, respectively. This suggests that maize roots can be triggered to secrete antagonistic substances against soybean pathogen growth in the intercropping system, which might be enhanced under low P supply.

HPLC analysis of root exudates

We investigated the composition of root exudates using HPLC analysis, and found that the composition of root exudates varied

significantly among the two crop species and the different cultivation modes (i.e., monoculture of soybean or maize, intercropping with soybean and maize). Roots of maize plants released greater amounts of phenolic acids than those of soybean (Figure E in File S1). We further analyzed the presence and levels of 8 common phenolic acids in the root exudates using HPLC with chemical standards. This analysis identified 5 phenolic acids, including ferulic, gallic, *p*-coumaric, cinnamic and salicylic acids, in the root exudates from MC or ISC, but only gallic acid was detected in soybean monoculture root exudates (Table 2). The concentration of the phenolic acids other than gallic acid in root exudates increased when maize was intercropped with soybean. For example, cinnamic acid concentration in exudates from the ISC treatment compared to the MC treatment increased by 20% and 43% at LP and HP, respectively. Phosphorus supply also influenced the concentrations of most phenolic acids. Ferulic, *p*-coumaric and cinnamic acids in root exudates of the ISC grown plants exhibited 68%, 58% and 22% increases at LP compared to HP.

Effects of phenolic acids on *C. parasiticum* growth

To further investigate the effects of the above five phenolic acids on *C. parasiticum* growth and sporulation, we conducted an *in vitro* *C. parasiticum* assay using individual phenolic acids. The results showed that when the concentration reached 20 mg/L, all of the tested phenolic acids inhibited the growth and sporulation of *C. parasiticum* (Fig. 5). At lower concentrations, only cinnamic and ferulic acids inhibited *C. parasiticum* growth, while, lower concentration gallic and salicylic acids slightly stimulated *C. parasiticum* growth. Furthermore, the inhibitory effects of cinnamic acid on *C. parasiticum* growth increased with increasing concentration, and even completely stopped *C. parasiticum* growth when the concentration reached 40 mg/L. These results suggest that cinnamic acid is the most effective phenolic acid in suppressing *C. parasiticum* growth and development.

Discussion

It has been well documented that increasing crop species diversity can yield benefits in plant disease control [2,3]. However, most studies have focused on above-ground plant functions and performances. Few plant disease studies pay attention to the root, the main organ of plants to sense environmental attacks and related signals in the soil. To the best of our knowledge, this is the first report on interspecies signaling between root systems that results in improved control of soil-borne disease, with evidence coming both from field trials and greenhouse-laboratory experiments at both physiological and molecular levels.

Intercropping is the intermingled growth of at least two crops grown together in the same field, which is the most popular cultivation model for increasing crop genetic diversity [23]. It has been widely noted that intercropping systems can reduce disease damage [4,15,24], probably due to fewer attacks by disease organisms compared to when crops are grown in monoculture [2,24]. In the current study, based on two years of field experiments where soybean red crown rot was severe in monoculture, we found that intercropping soybean with maize significantly reduced disease severity of soybean red crown rot (Table 1). This was further confirmed in sand culture with different root barrier treatments. Most significantly, when root interactions were completely blocked, severe red crown rot still occurred in soybean plants, even when intercropped with maize (Fig. 2). This indicates that above-ground interactions in the maize/soybean intercropping system do not contribute to control of soybean red

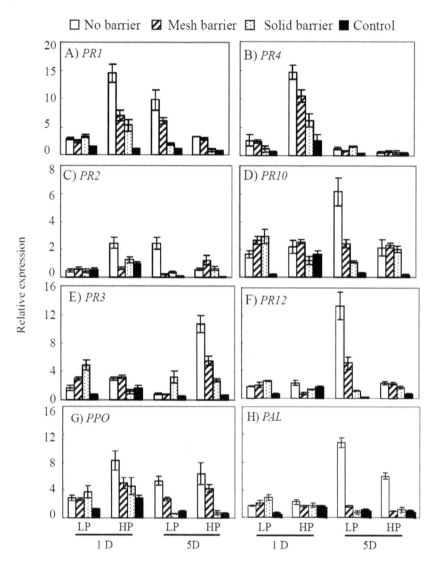

□ No barrier   ▨ Mesh barrier   ▥ Solid barrier   ■ Control

Figure 3. Expression of eight defense-related (*PR*) genes in soybean roots. LP: 15 μM P; HP: 500 μM P. Soybean roots were inoculated with *C. parasiticum* (see Materials and Methods for details). Solid barrier: eliminate root interactions and exudates movement between the roots of the two plant species; mesh barrier: prevent root intermingling of two species while permitting root exudates exchange; no barrier: allow roots and exudates to completely interact. Each bar represents the mean of three replicates ± SE.

crown rot. In contrast, the least severe red crown rot was recorded when the roots freely interacted with each other in the no barrier treatment, indicating that interactions among roots in intercropping is important in control of this soil-borne disease. Moreover, we also found that in the field, the reduction of soybean red crown rot was obviously enhanced with decreasing planting distance between soybean and maize plants, which increased interaction intensity between the two different crop species (Table 1 and Fig. 1). This finding suggests that besides genetic diversity, the intensity of root interaction is also important for control of soybean red crown rot in the maize/soybean intercropping system. This point is also supported by reports that in order to effectively control disease spread, roots of mixed crops need to be sufficiently intermingled [24,25].

Root exudates have been recognized to play vital roles in the rhizosphere [26–28] in large part due to continuous production and secretion of allelochemical compounds into the rhizosphere [29,30]. There are many reports attributing defense against phytopathogens to root exudates released by neighboring non-host

plants [6–7,31]. In the present sand culture experiment, the mesh barrier treatment had less effects on disease control than the no barrier treatment, yet still significantly reducing the severity of soybean red crown rot (Fig. 2), implying that root exudates alone can inhibit soybean red crown rot. This was supported by *in vitro* assays where the significant growth inhibition of *C. parasiticum* was observed when root exudates were applied. Root exudates collected from intercropped soybean and maize inhibited *C. parasiticum* growth more effectively than from maize alone (Fig. 4), suggesting that intercropped maize releases more allelopathic compounds than in monoculture.

Phenolic acids have been well established as major rhizosphere allelochemicals that suppress phytopathogens [22,32], and they are detectable in root exudates by HPLC [7,29]. In order to identify the specific compounds inhibitory to *C. parasiticum* growth from the maize/soybean intercropping system, the 8 most common phenolic acids in root exudates were initially tested, with 4 phenolic acids specifically released by maize roots (Figure E in File S1). Among these 4 compounds, release of cinnamic acid

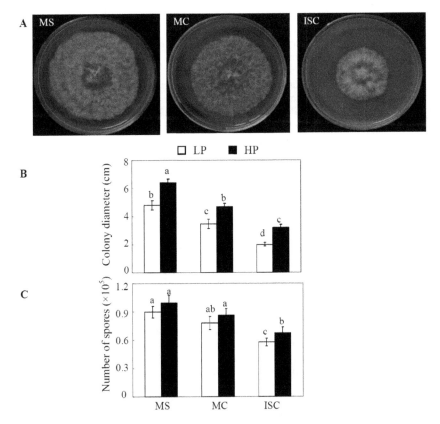

Figure 4. *C. parasiticum* growth as affected by root exudates. A, growth performance, B, colony diameter, C, sporulation. LP, 15 μM P; HP, 500 μM P. MS: soybean monoculture; MC: maize monoculture; ISC: maize/soybean intercropping. F value from Two-way ANOVA: colony diameter, 685.62 for P treatment ($P<0.001$), 1119.75 for cultivation mode ($P<0.001$), 12.73 for interaction ($P<0.001$); sporulation, 38.33 for P treatment ($P<0.001$), 116.49 for cultivation mode ($P<0.001$), 1.02 for interaction (not significant). Each bar represents the mean of four replicates ± SE. Bars with different letter(s) vary significantly among treatments as determined by Duncan's multiple range test ($P<0.05$).

was significantly enhanced by maize/soybean intercropping (Table 2), and this compound dramatically constrained *C. parasiticum* growth *in vitro* (Fig. 5). Cinnamic acid has been demonstrated to inhibit phytopathogen *Fusarium oxysporum* f.sp. *niveum in vitro* [7,22]. Therefore, we speculate that cinnamic acid is the primary player of interest in the tested root exudates, and it is directly responsible for the protection of soybean roots from the attack of *C. parasiticum* in the maize/soybean intercropping system. Since plant roots can exude a broad range of compounds, a more

thorough search for more efficient phytotoxins from soybean and/ or maize needs to be carried out in the future.

The *PR* related genes included pathogenesis-related, antimicrobial protein genes *PR1* (PR1a precursor), *PR2* (β 1–3 endoglucanase), *PR3* (chitinase class I), *PR4* (wound-induced protein), *PR10* (ribonuclease-like protein), *PR12* (defending-like protein), *PPO* (polyphenol oxidase) and *PAL* (phenylalanine ammonia-lyase) that are involved in phytoalexin biosynthesis [33,34]. Here we shown that the defense responses elicited by inoculation of pathogen and intercropping with maize. In legumes, many genes encoding PR

Table 2. Concentrations of phenolic acids (μg·pot^{-1}) in root exudates from different treatments at the flowering stage as detected by HPLC.

Treatment		Ferulic acid	Gallic acid	*P*-coumaric acid	Cinnamic acid	Salicylic acid
	MS	-	6.53±1.59c	-	-	-
LP	MC	12.78±2.34b	29.08±0.80a	13.06±1.91b	20.98±2.51b	7.51±1.41a
	ISC	22.13±3.24a	14.72±2.84b	20.33±2.78a	25.11±2.52a	7.88±2.72a
	MS	-	11.12±2.88b	-	-	-
HP	MC	13.72±1.05a	13.09±1.58b	8.51±0.26a	14.41±2.86b	11.00±2.21a
	ISC	13.18±1.73a	21.65±2.99a	12.86±2.28a	20.59±2.78a	9.32±0.47a

Note: LP, 15 μM P; HP, 500 μM P. MS, soybean monoculture; MC, maize monoculture; ISC, maize/soybean intercropping. -, not detectable. All the data are the mean of four replicates ± SE. Data with the same letter represent no significant differences among the different cultivation modes at the same P level as determined by Duncan's multiple range test or a *t*-test ($P<0.05$).

□ Cinnamic acid ☑ Salicylic acid
▦ Gallic acid ■ Ferulic acid ▦ P-coumaric acid

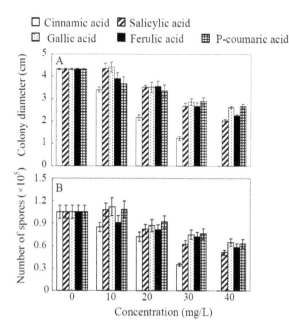

Figure 5. Effect of 5 different phenolic acids on the growth and sporulation of *C. parasiticum*. A, colony diameter, B, number of spores. Each bar represents the mean of four replicates ± SE.

proteins have been shown to be upregulated in plants following inoculation with several pathogens [33]. The upregulation of defense-related genes, such as *PR1* and *PR2* was associated with lesion limitation in soybean seedling roots inoculated with *P. sojae* and to the high level of partial resistance of cultivar Conrad [33]. Defense-related genes are upregulated in both compatible and incompatible interactions, but more rapidly and/or to higher levels in plant varieties with *PR* gene resistance [34]. In addition to producing allelopathic compounds directly, intercropped plants can promote specific plant defense reactions in neighboring plants to protect against phytopathogens [2,31]. Consistent with this, the results herein show that direct interactions with maize roots significantly induce the expression of important soybean *PR* genes and increase the activities of corresponding PR enzymes in soybean roots (Fig. 3, Figure C in File S1). This elicitation, via root exudates or signal compounds for specific plant defense reactions can predispose soybean plants to an early response to attack by root pathogens [24,31]. Infecting soybean roots after pre-treatment with maize enhances PR gene expression and enzyme activity, and therefore, might improve soybean's capability to defend against disease invasion. Furthermore, salicylic acid is a reported enhancer of plant defense pathways [35]. In this study, more salicylic acid was measured in root exudates from the maize/soybean intercropping system than in monoculture, which suggests that salicylic acid or a related component might also contribute to disease control through promotion of defense pathways.

Although adding additional P fertilizer in field and sand culture experiments increased plant growth, it also promoted disease incidence and index of red crown rot (Table 1 and Fig. 2). Likewise, the application of P has been shown to increase the severity of diseases in many plant species [18,36]. Here, we demonstrate that higher P addition is associated with more severe soybean red crown rot (Table 1 and Fig. 2). Furthermore, high P induces colony *C. parasiticum* growth and proliferation *in vitro* (Fig. 4). Plant roots and phytopathogens most likely compete directly for resources (e.g. P) in the rhizosphere, and therefore low

P supply could inhibit phytopathogen growth, especially if the roots more effectively acquire the soil P. Consequently, optimal fertilizer application should be carefully considered, especially in acid soils where P is limiting.

To the best of our knowledge, this study is the first report to demonstrate that intercropping with maize can promote disease resistance in soybean to red crown rot in a root interaction dependent manner. Direct root interactions of soybean with maize may inhibit pathogen growth and reproduction, while enhancing expression of *PR* genes and the activities of corresponding enzymes in host plants. This study suggests that soybean and maize intercropping is an effective tool to sustainably control soil-borne disease and should be incorporated in agricultural management practices.

Supporting Information

File S1 Supporting figures. Figure A. Planting arrangement of diagram and actual pictures in monoculture or intercropping of soybean and maize. **A**, Planting arrangement in the field. **B**, Three root barriers were set-up in sand culture, either with a solid barrier (left) to eliminate root interactions and exudates movement, a nylon mesh barrier (middle) to prevent root intermingling of two species while permitting root exudates exchange, and no barrier (right) to allow roots and exudates to completely interact. **Figure B.** Soybean growth and grain yield as affected by red crown rot. HP: 80 kg P_2O_5 ha^{-1} added as calcium superphosphate, LP: no P fertilizer added. The reduction of growth parameters were calculated as follows: reduction (%) = (yield/biomass of healthy plants- yield/biomass of infected plants) yield/biomass of healthy plants ×100. Each bar represents the mean of four replicates ± SE. **Figure C.** Disease severity of soybean red crown rot in sand culture (Second round). A, disease incidence, B, disease index, C, CFU. LP, 15 μM P; HP, 500 μM P. All of the data are the means of four replicates ±SE. Bars with different letter(s) vary significantly among treatments as determined by Duncan's multiple range test ($P<0.05$). **Figure D.** Activities of the enzyme PPO (A) and PAL (B) in soybean roots. LP, 15 μM P; HP, 500 μM P. Except for control, all the roots were inoculated with *C. parasiticum* (see Materials and Methods for details). The solid barrier eliminated root contact and exudates movement, the nylon mesh (30 μm) barrier prevented root intermingling for the two species while permitting root exudates exchange, and no root barrier permitted roots and exudates to completely interact. Each bar represents the mean of three replicates ± SE. **Figure E.** HPLC scan of soybean and/or maize root exudates. (**a and b**) HPLC scan of soybean root exudates, (**c and d**) HPLC scan of maize root exudates, (**e and f**) HPLC scan of the root exudates from maize/soybean intercropping, (**g**) HPLC analysis of eight standard phenolic acids. LP, 15 μM P (a, c and e); HP, 500 μM P (b, d and f). 1, Gallic acid; 2, *P*-coumaric; 3, Phthalic acid; 4, Vanillic acid; 5, Syringic acid; 6, Ferulic acid; 7, Salicylic acid; 8, Cinnamic acid.

File S2 Supporting tables. Table S1. Real time PCR primers designed for this study. **Table S2.** AMF colonization rate and nodule number of soybean in the field.

Acknowledgments

We thank Mingfang Guan, Haiyan Zhang, Kai Li, Hui Jiang, Zuotong He, Hanxiang Wu, Yiyong Zhu and Renshen Zeng for their helps in field and lab experiments; Zide Jiang and Shuxian Li for guidance on pathogen analysis, and Thomas C Walk for critical comments.

Author Contributions

Conceived and designed the experiments: XG HL. Performed the experiments: XG MW RX. Analyzed the data: XG XW HK HL.

Contributed reagents/materials/analysis tools: RP. Wrote the paper: XG HL.

References

1. Ratnadass A, Fernandes P, Avelino J, Habib R (2012) Plant species diversity for sustainable management of crop pests and diseases in agroecosystems: a review. Agronomy for Sustainable Development 32, 273–303.
2. Zhu Y, Chen H, Fan J, Wang Y, Li Y, et al. (2000) Genetic diversity and disease control in rice. Nature 406, 718–722.
3. Browning JA, Frey KJ (1969) Multiline cultivars as a means of disease control. Annual Review of Phytopathology 7, 355–382.
4. Smedegaard-Petersen V (1985) The limiting effect of disease resistance on yield. Annual Review of Phytopathology 23, 475–490.
5. Yu JQ (1999) Allelopathic suppression of Pseudomonas solanacearum infection of tomato (*Lycopersicon esculentum*) in a tomato-chinese chive (*Allium tuberosum*) intercropping system. Journal of Chemical Ecology 25, 2409–2417.
6. Ren LX, Su SM, Yang XM, Xu YC, Huang QW, et al. (2008) Intercropping with aerobic rice suppressed Fusaium wilt in watermelon. Soil Biology Biochemistry 40, 834–844.
7. Hao WY, Ren LX, Ran W, Shen QR (2010) Allelopathic effects of root exudates from watermelon and rice plants on *Fusarium oxysporum* f.sp. *niveum*. Plant and Soil 336, 485–497.
8. Gómez RO, Zavaleta M, González H, Livera M, Cárdenas S (2003) Allelopathy and microclimatic modification of intercropping with marigold on tomato early blight disease development. Field Crops Research 83, 27–34.
9. Bell DK, Sobers EK (1966) A peg, pod and root necrosis of peanuts caused by a species of *Calonectria*. Phytopathology 56, 1361–1364.
10. Kuruppu P, Schneider R, Russin J (2004) Factors affecting soybean root colonization by *Calonectria ilicicola* and development of red crown rot following delayed planting. Plant Disease 88, 613–619.
11. Guan M, Pan R, Gao X, Xu D, Deng Q, et al. (2010) First report of red crown rot caused by *Cylindrocladium parasiticum* on soybean in Guangdong, Southern China. Plant Disease 94, 485.
12. Li L, Zhang F, Li X, Peter C, Sun J, et al (2003) Interspecific facilitation of nutrient uptake by intercropped maize and faba bean. Nutrient Cyclling in Agroecosystems 65, 61–71.
13. He Y, Ding N, Shi JC, Wu M, Liao H, et al. (2013) Profiling of microbial PLFAs: Implications for interspecific interactions due to intercropping which increase phosphorus uptake in phosphorus limited acidic soils. Soil Biology Biochemistry 57, 625–634.
14. Ae N, Arihara J, Okada K, Yoshihara T, Johansen C (1990) Phosphorus uptake by pigeon pea and its role in cropping systems of Indian subcontinent. Science 248, 477–480.
15. Li C, He X, Zhu S, Zhou H, Wang Y, et al. (2009) Crop diversity for yield increase. PLoS ONE 4, e8049.
16. Fang SQ, Gao X, Deng Y, Chen XP, Liao H (2011) Crop root behavior coordinates phosphorus status and neighbors: from field studies to three-dimensional in situ reconstruction of root system architecture. Plant Physiology 155, 1277–1285.
17. Fang S, Clark RT, Zheng Y, Iyer-Pascuzzia AS, Weitzf JS, et al. (2013) Genotypic recognition and spatial responses by rice roots. Proceedings of the National Academy of Sciences, USA 110, 2670–2675.
18. Gao X, Lu X, Wu M, Zhang HY, Pan RQ, et al. (2012) Co-inoculation with rhizobia and AMF inhibited soybean red crown rot: from field study to plant defense-related gene expression analysis. PLoS ONE 7, e33977.
19. Wang XR, Pan Q, Chen FX, Yan XL, Liao H (2011) Effects of co-inoculation with arbuscular mycorrhizal fungi and rhizobia on soybean growth as related to root architecture and availability of N and P. Mycorrhiza 21, 173–181.
20. Liao H, Wan HY, Shaff J, Wang XR, Yan XL, et al. (2006) Phosphorus and aluminum interactions in soybean in relation to aluminum tolerance. Exudation of specific organic acids form different regions of the intact root system. Plant Physiology 141, 674–684.
21. Song YY, Zeng RS, Xu JF, Li J, Shen X, et al. (2010) Interplant communication of tomato plants through underground common mycorrhizal networks. PLoS ONE 5, e13324.
22. Ling N, Huang QW, Guo SW, Shen QR (2011) Paenibacillus polymyxa SQR-21 systemically affects root exudates of watermelon to decrease the conidial germination of *Fusarium oxysporum* f.sp. *niveum*. Plant and Soil 341, 485–493.
23. Li L, Li SM, Sun JH, Zhou LL, Bao XG, et al. (2007) Diversity enhances agricultural productivity via rhizosphere phosphorus facilitation on phosphorus-deficient soils. Proceedings of the National Academy of Sciences, USA 104, 11192–11196.
24. Trenbath BR (1993) Intercropping for the management of pests and diseases. Field Crops Research 34, 381–405.
25. Meynard JM, Dore T, Lucas P (2003) Agronomic approach: cropping systems and plant disease. Comptes Rendus Biologies 326, 37–46.
26. Hoffland E, Findenegg G, Nelemans J, van den Boogaard R (1992) Biosynthesis and root exudation of citric and malic acids in phosphate-starved rape plants. New Phytologist 122, 675–680.
27. Bardgett RD, Denton CS, Cook R (1999) Below-ground herbivory promotes soil nutrient transfer and root growth in grassland. Ecology Letters 2, 357–360.
28. Bais HP, Prithiviraj B, Jha AK, Ausubel FM, Vivanco JM (2005) Mediation of pathogen resistance by exudation of antimicrobials from roots. Nature 434, 217–221.
29. Bais HP, Vepachedu R, Gilroy S, Callaway RM, Vivanco JM (2003) Allelopathy and exotic plant invasion: from molecules and genes to species interactions. Science 301, 1377–1380.
30. Weir TL, Park SW, Vivanco JM (2004) Biochemical and physiological mechanisms mediated by allelochemicals. Current Opinion in Plant Biology 7, 472–479.
31. Bais HP, Weir TL, Perry LG, Gilroy S, Vivanco JM (2006) The role of root exudates in rhizosphere interactions with plants and other organisms. Annual Review of Plant Biology 57, 233–266.
32. Wu HS, Liu DY, Ling N, Bao W, Ying RR, et al. (2009) Influence of root exudates of wantermelon on *Fusarium oxysporum* f. sp. *niveum*. Soil Biology Biochemistry 73, 1150–1156.
33. Robert GU, Martha ER (2010) Defense-related gene expression in soybean leaves and seeds inoculated with *Cercospora kikuchii* and *Diaporthe phaseolorum* var. *Meridionalis*. Physiological and Molecular Plant Pathology 75, 64–70.
34. Van Loon LC, Rep M, Pirterse CMJ (2006) Significance of inducible defense-related proteins in infected plants. Annual Review of Phytopathology 44, 135–162.
35. Wees S, Swart E, Pelt J, Loon L, Pieterse C (2000) Enhancement of induced disease resistance by simultaneous activation of salicylate- and jasmonate-dependent defense pathways in Arabidopsis thaliana. Proceedings of the National Academy of Sciences, USA 97, 8711–8716.
36. Dordas C (2008) Role of nutrients in controlling plant disease in sustainable agriculture: a review. Agronomy for Sustainable Development 28, 33–46.

10

Sustainable Management in Crop Monocultures: The Impact of Retaining Forest on Oil Palm Yield

Felicity A. Edwards[1]*, David P. Edwards[2,3], Sean Sloan[2], Keith C. Hamer[1]

1 School of Biology, University of Leeds, Leeds, West Yorkshire, United Kingdom, 2 Centre for Tropical Environmental and Sustainability Science (TESS) and School of Marine and Tropical Biology, James Cook University, Cairns, Queensland, Australia, 3 Department of Animal and Plant Sciences, University of Sheffield, Sheffield, South Yorkshire, United Kingdom

Abstract

Tropical agriculture is expanding rapidly at the expense of forest, driving a global extinction crisis. How to create agricultural landscapes that minimise the clearance of forest and maximise sustainability is thus a key issue. One possibility is protecting natural forest within or adjacent to crop monocultures to harness important ecosystem services provided by biodiversity spill-over that may facilitate production. Yet this contrasts with the conflicting potential that the retention of forest exports dis-services, such as agricultural pests. We focus on oil palm and obtained yields from 499 plantation parcels spanning a total of ≈23,000 ha of oil palm plantation in Sabah, Malaysian Borneo. We investigate the relationship between the extent and proximity of both contiguous and fragmented dipterocarp forest cover and oil palm yield, controlling for variation in oil palm age and for environmental heterogeneity by incorporating proximity to non-native forestry plantations, other oil palm plantations, and large rivers, elevation and soil type in our models. The extent of forest cover and proximity to dipterocarp forest were not significant predictors of oil palm yield. Similarly, proximity to large rivers and other oil palm plantations, as well as soil type had no significant effect. Instead, lower elevation and closer proximity to forestry plantations had significant positive impacts on oil palm yield. These findings suggest that if dipterocarp forests are exporting ecosystem service benefits or ecosystem dis-services, that the net effect on yield is neutral. There is thus no evidence to support arguments that forest should be retained within or adjacent to oil palm monocultures for the provision of ecosystem services that benefit yield. We urge for more nuanced assessments of the impacts of forest and biodiversity on yields in crop monocultures to better understand their role in sustainable agriculture.

Editor: João Pinto, Instituto de Higiene e Medicina Tropical, Portugal

Funding: F.A.E was supported by a Biotechnology and Biological Sciences Research Council doctoral training grant studentship, and D.P.E and S.S. were supported by an Australian Research Council Fellowship. The funders had no role in study design, data collection and analysis, decision to publish, or preparation of the manuscript.

Competing Interests: The authors have declared that no competing interests exist.

* E-mail: bs08f2a@leeds.ac.uk

Introduction

More than 50% of the global land area that is purportedly suitable for agriculture has already been converted to farmland [1]. Moreover, by 2050, projections suggest that an increase of one billion hectares in agricultural land is required to feed a growing population and to meet increasing consumption per capita [2], much of which will come at the expense of natural habitat in the tropics [3]. Following agricultural development, the landscape is often left with highly fragmented patches of natural habitat that create sharp habitat boundaries with agriculture, and with remaining patches of natural habitat showing varying degrees of degradation and isolation [4,5]. The simplification of vegetation structure and altered environmental conditions within the agricultural matrix often prove too extreme for much native biodiversity to persist, and valuable ecosystem services may also be threatened by the loss of natural habitats [4,6–8]. Consequently, agricultural expansion is one of the key threats to biodiversity [1,2], and there is an increasing strain between conserving biodiversity and maximising agricultural production [8–10].

Many crops are highly dependent on functional interactions provided by biodiversity, such as soil nutrient supply, pollination, and biological pest control [11–14]. Integration of remnant natural habitat features such as forest fragments, riparian strips, and hedgerows within agricultural landscapes is advocated as a means to enhance ecosystem services and thus yield, in addition to providing conservation benefits to native biodiversity, within sustainable landscapes [15–21]. While there is a large literature on how the retention of natural habitat can encourage biodiversity and ecosystem services, there is a lack of knowledge of the degree to which remnant habitat might negatively affect yield. The spill-over of biodiversity from natural habitats to agricultural land can negatively alter species diversity and food web interactions [22,23], with ecosystem dis-services potentially arising as a consequence of providing reservoir populations of insect or fungal pests, crop raiders, invasive weeds, or predators and parasites of beneficial species [12,23].

Retaining natural habitat remnants within agricultural landscapes also reduces the land available for growing crops, and so may constitute an opportunity cost to local production as well as potentially increasing the demand for converting land elsewhere to agriculture [1]. Landscape-scale planning for agricultural sustainability and conservation therefore hinges on whether or not remnant habitat features provide a net benefit for agricultural

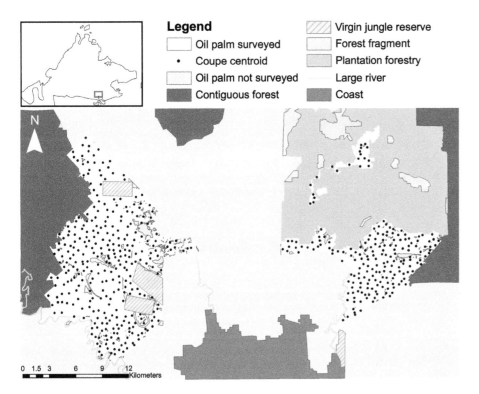

Figure 1. Different land-use types within the study area. The inset shows Sabah, Northeast Borneo, and the red box denotes the study area.

production, for conservation, or for both. This is a particularly important issue in the tropics, where conversion to agriculture consumed 1.4% of the tropical forest biome between 2000 and 2005 [24]. To date, research on the relationship between natural vegetation cover and crop yield in the tropics had focused on two agro-forestry crops: coffee [8,17,18,25,26] and cocao [27,28,29]. Both coffee and cocao plantations consist of a mix of crop plants and (non)-native shade trees, which results in an agro-forestry matrix that is comparatively hospitable to forest species (e.g., [30]), and can enhance spill-over from forest and resulting ecosystem services. Consequently, these studies found that close proximity to forest improved pollinator bee numbers [31] and thus coffee yields by up to 20% [18] compared to locations 1,400–1,600 m from forest, and that distance to forest had a marginal positive effect on yield in cacao plantations [27], which have increasing numbers of predatory ant and spider species with higher densities of native shade trees [28]. Furthermore, exclusion experiments showed that bird and bat predation, and the extent of forest cover were important in controlling pest populations and thus positively impacting yield [8,29].

To our knowledge, the impact of forest on yield has not been assessed in the context of tropical crop monocultures, in which a single crop species is planted in stands that do not contain non-crop trees or other crop species, yet the majority of crop expansion within the tropics now creates monocultures of sugar cane, soya, oil palm, and even cocao. Oil palm *Elaeis guineensis* is one of the world's highest yielding and most financially lucrative monoculture crops [32]. As such, it is expanding very rapidly, with production increasing by >5.5 million ha between 2001 and 2011 [33] and with the majority of this expansion occurring at the expense of hyperdiverse tropical rainforest in Southeast Asia [34]. Unlike coffee and cocao plantations, which can retain high levels of within-plantation biodiversity, whole-sale forest conversion to oil palm results in dramatic local extinctions of most forest-

dwelling species [35–37]. To reduce the environmental footprint of oil palm, The Roundtable for Sustainable Palm Oil (RSPO), via the high conservation value (HCV) forest protocol [38,39], and conservation scientists (e.g., [40,41]) have both highlighted the potential benefits of creating oil palm landscapes that retain forest remnants and riparian strips within plantations, but the net effect of such management on oil palm yield is not known [42].

In this study, we explore the impacts of the local extent of forest cover and the proximity to forest on oil palm yields in Sabah, Malaysian Borneo, where palm oil production covers 19% of the state land area [43] and where there is increasing pressure for further expansion. We thus assess whether the retention of forest within and adjacent to oil palm plantations has a positive, negative or neutral impact on oil palm yield, with the aim of informing sustainable land-use planning.

Materials and Methods

Study Area

Our study landscape spans 49.5 km×29.8 km (total area = 1474 km^2 or 147,400 ha) in Sabah, Malaysian Borneo (Figure 1). The landscape comprises >91,000 ha of contiguous oil palm plantations owned by multiple companies, plus a single > 28,000 ha block of plantation forestry (*Eucalyptus* spp., Teak, *Acacia* spp.; Sabah Softwoods Bhd.) (Figure 1). All of the soils within our study oil palm plantations are Acrisols, as defined by the World Reference Base for Soil Resources [44]. However, these soils also contain other main soil components (e.g., Luvisols, Cambisols, etc.) and they have a mixture of alluvium, mudstone, sandstone and igneous rock as parent material [45]; these are combined into ten soil groups (Table S1, Figure S1). Study oil palm plantations also span an elevational range from 10 to 379 m a.s.l. (Figure S2).

Surrounding these plantations are two areas of contiguous lowland dipterocarp forest >100,000 ha in size, which were not bounded by our study area: to the west and north is the Yayasan Sabah (YS) logging concession and to the east is the Ulu Kalumpang forest reserve (itself contiguous with Tawau Hills National Park). Surrounding contiguous forests have both undergone at least two rotations of selective logging [32,46]. To the south of our study area is a coastline of tidal mangrove creeks, >2 km from the nearest oil palm coupe.

We focus on the oil palm of a single company—Sabah Softwoods Bhd. (we thank Sabah Softwoods Bhd. for providing data, logistical support and site access), a subsidiary of the state-owned Yayasan Sabah Group—with ≈23,000 ha of plantings (Figure 1, in white). Oil palm plantings are separated into three separate zones, which are 2.5 to 9.3 km apart, partitioned by other oil palm plantations between the western and eastern blocks and by plantation forestry between the two eastern blocks (Figure 1). Each zone is sub-divided into discrete parcels known as coupes (n_{total} = 499), which vary in size from 3 to 89 ha (mean±SE: 45±0.7 ha) and which are planted with a density of 100 palms per ha [36].

The Sabah Softwoods oil palm plantations border both contiguous areas of forest, plus numerous isolated forest fragments, increasing in size from tiny patches to large fragments of dipterocarp forest. Forest fragments are divided into Virgin Jungle Reserves (VJRs), which are large (n = 4; mean±SE: 813.95±197.6 ha), were gazetted prior to industrial-scale logging, and thus contain mostly primary forest; whereas privately owned patches (herein 'private fragments') tend to be smaller (n = 307, 11.5±4.2 ha, range = 0.01 to 886 ha), to have been selectively logged at least once (the precise logging history of each fragment is unknown) and open to other disturbances (e.g., hunting). Forest fragments were typically retained within plantations due to their steepness and/or unfavourable underlying substrate.

Oil Palm Yields

Yield data were fresh fruit bunch (FFB) weights (metric tonnes) per hectare for individual coupes from 2008 to 2010. Sabah Softwoods employees visit each oil palm tree within a coupe to harvest ripe fruit bunches and cut decaying fronds twice per month. Bunches are collected into trailors and weighed at the depot. We were provided with the total weight of fruit bunches collected in each coupe on a yearly basis. Oil palm age varied across coupes, from 3 to 15 years old, and because yield varies with age of an oil palm [47] we used the *deviation from the mean expected yield by age* (i.e., observed yield - mean yield for the age of palm) as our indication of yield per coupe. A positive value indicates greater yield than expected, while a negative value indicates a lower yield than expected, given the age of the oil palm. Observed yield data were used from all 499 coupes in 2010. Expected yield was calculated from two yield-by-age curves: firstly, generated from the subset of coupes for which data were provided in 2008 (n = 240 coupes) and 2009 (n = 400; yldSS), and secondly from Butler *et al.* [47] using their average FFB curve (yldB; Figure S3).

Quantifying Extent of Forest Cover and Proximity to Forest

Forest coverage maps were supplied by Sabah Softwoods, and supplemented with additional maps obtained from the literature [43,48] and Google Earth images from 2009. The extent of dipterocarp forest cover surrounding and within each oil palm coupe was calculated within circles of radii 100 m, 250 m, 500 m and 1,000 m from the centroid of each coupe. Radii thus span a range of spatial scales relevant to different taxonomic groups, as determined by observations of species' movements between forest and oil palm [49]. From these four radii, an inverse distance-weighted measure of forest-cover area as a proportion of the 1000-m radius circle area F_{IDW} was calculated, giving greater weight to forest area closer to a coupe centroid than forest further away [50,51], using the formula:

$$F_{IDW} = \sum_{i=1}^{i=4} \frac{fi}{d+1}$$

where fi is the proportion of forest within a buffer ring (0–100 m,

Table 1. The range and mean (±SE) of oil palm yield, elevation, and nearest distance to different forest classes, forestry plantations, large rivers and other (not within Sabah Softwoods Bhd.) oil palm plantations within 499 oil palm coupes in Sabah, Malaysian Borneo.

Measure	Maximum	Minimum	Mean	SE
2010 oil palm yield (mt ha^{-1})	33.46	0.12	16.82	0.39
Elevation (m.a.s.l.)	393.53	7.83	127.51	3.11
Forest cover (%) within radii:				
100 m	36.00	0.00	0.18	0.08
250 m	70.00	0.00	1.43	0.24
500 m	83.00	0.00	3.74	0.38
1000 m	79.00	0.00	6.43	0.51
Distance (km) to nearest:				
Contiguous forest	14.63	0.12	5.03	0.15
Virgin forest reserve (VJR)	20.71	0.05	5.93	0.19
Privately owned fragment	3.89	0.03	0.84	0.03
Plantation forestry	26.95	0.09	13.35	0.41
Large river	16.06	0.20	5.79	0.16
Other oil palm	8.66	0.04	2.96	0.10

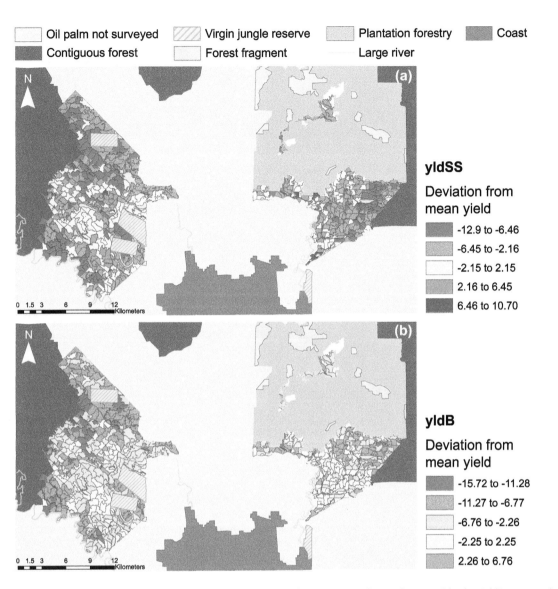

Oil palm not surveyed	Virgin jungle reserve	Plantation forestry	Coast
Contiguous forest	Forest fragment	Large river	

yldSS

Deviation from mean yield

- -12.9 to -6.46
- -6.45 to -2.16
- -2.15 to 2.15
- 2.16 to 6.45
- 6.46 to 10.70

yldB

Deviation from mean yield

- -15.72 to -11.28
- -11.27 to -6.77
- -6.76 to -2.26
- -2.25 to 2.25
- 2.26 to 6.76

Figure 2. The variation in oil palm yield with adjacent land-uses across the study area. Oil palm yield is measured as the mean deviation from yield-by-age curves (a) generated from the study area data (yldSS), and (b) published by Butler et al. [47] (yldB). Yield is quantified as the fresh fruit bunch weight per hectare (mt ha^{-1}).

100–250 m, 250–500 m, and 500–1,000 m) and d (m) is the mean distance of a buffer ring.

Dipterocarp forest included three qualitatively different classes that differed in size and/or logging history, and thus vegetation composition and species communities (e.g., [46,52,53], namely (i) contiguous forest, (ii) Virgin Jungle Reserves, and (iii) private fragments. To account for this variation, we also assessed proximity to these dipterocarp forest classes by calculating, from each coupe centroid, the shortest distance to each class. We also calculated distance to plantation forestry, which directly borders some oil palm coupes and which in this study area has more bird biodiversity than local oil palm [54,55], largely due to the secondary forest understorey that develops under plantation trees. In addition, we included the distance to the nearest surrounding oil palm (i.e. not owned by Sabah Softwoods Bhd.) since a coupe located within a large expanse of oil palm monoculture could benefit if dis-services such as pest infestations originate from within forest or could be disadvantaged if they develop within oil palm.

Finally, we evaluated the proximity of the nearest large river from each coupe centroid, the mean elevation across the coupe, and the dominant soil type by area (mean dominant soil coverage was 96.4%±0.01 SE of coupe area), because these environmental variables have the potential to influence oil palm growth and yield. Elevation (m a.s.l.) was calculated from a digital elevation model at 90 m resolution [56]. Soil types were grouped into ten categories (see above; Table S1) and were assessed using a regional soil survey map at 1:250000 scale [45].

Statistical Analysis

We used Generalised Least Square models (GLS) to firstly test whether the distance-weighted proportional area of forest affected oil palm yield at the coupe level. The distance-weighted measure of forest cover was square-root transformed to reduce the influence of two outliers. Secondly, we used a GLS to test whether proximity of a coupe centroid to the nearest dipterocarp forest class (contiguous forest; VJR; private fragment) affected oil palm yield.

Table 2. The estimates and parameter coefficients from the minimum adequate generalised least square models testing the effects of forest cover and forest proximity on oil palm yield across the study landscape in Sabah, Malaysian Borneo.

Model	Parameter	Estimate	SE	T	P
Forest cover (yldSS*)					
	(Intercept)	−3.1284	0.6915	−4.5240	0.0000
	forest cover	20.2703	13.3163	1.5222	0.1286
Forest cover (yldB$)					
	(Intercept)	−0.6051	1.0429	−0.5802	0.5620
	forest cover	13.1835	12.7962	1.0303	0.3034
	elevation	**−0.0162**	**0.0041**	**−3.9334**	**0.0001**
	tree plantation	**−0.0002**	**0.0001**	**−3.0541**	**0.0024**
Forest proximity (yldSS)					
	(Intercept)	−1.2576	5.7548	−0.2185	0.8271
	contiguous forest	−0.0042	0.0423	−0.0985	0.9216
Forest proximity (yldB)					
	(Intercept)	−0.9506	1.2752	−0.7455	0.4563
	elevation	**−0.0143**	**0.0041**	**−3.4587**	**0.0006**
	tree plantation	**−0.0002**	**0.0001**	**−2.2356**	**0.0258**

* yldSS – yield estimate derived from the yield-by-age curve generated from Sabah Softwoods coupes.
$yldB - yield estimate derived from the Butler et al.'s [43] average FFB yield-by-age curve.
Bold indicates significance at $P < 0.001$.

Distance to the nearest forest class was square-root transformed to account for the likely declining effect of forest and the associated reduction of biodiversity spill-over at increasing distances [27]. Additionally, the area of the nearest private fragment was also included as a covariate in proximity models, because different sized fragments could export different levels of services or dis-services. In both cases, the minimum adequate model was achieved by a model selection process comparing nested models [57]. All models included proximity to tree plantation, proximity to large river, proximity to other oil palm plantation, mean elevation and dominant soil type as fixed effects. All models also included a correlation structure using the latitude and longitude of the coupe centroids to account for spatial autocorrelation [58]. Lastly, using our model residuals with 1000 repetitions, we performed a Monte-Carlo permutation test for Moran's I statistic (moran.mc within spdep package) to test whether our results were influenced by spatial autocorrelation (i.e., that the correlation structure had effectively accounted for impacts of space). All spatial analyses were run in ArcGIS 10.0 [59] and all statistical analyses were run in R 2.15.2 [60].

Results

Oil palm coupes within the landscape spanned a range of distances to forest and degrees of forest cover (Table 1), with the percentage of forest cover at 1000 m ranging from 0 to 79% and distances to forest classes from 30 m and 20.7 km (Table 1), indicating a perfect landscape within which to test the impacts of forest on oil palm yield. Across the study area, there was also a large variation in oil palm yield, spanning over an order of magnitude from 0.12 to 33.46 mt ha^{-1} (Table 1), with a strong correlation between yield and oil palm age ($r^2 = 0.88$). However, having accounted for the increase in yield with palm age (see **Materials and Methods**), the spatial distribution of oil palm yield in relation to forest cover showed no clear visual pattern, with

a mix of high yield oil palm both close and far from major blocks of forest (Figure 2a, b), and with the same visual pattern for lower yields.

Yield Response to Forest Cover

The distance-weighted area of forest cover was retained by the minimum adequate model (MAM), but it was not a significant predictor when yield was derived from either yield-by-age curves: i) the yield-by-age curve generated using Sabah Softwoods coupes (yldSS; GLS: $t_{499} = 1.52$, $P = 0.13$), and ii) Butler et al.'s [47] average FFB yield-by-age curve (yldB; $t_{499} = 1.03$, $P = 0.30$) (Table 2). The environmental variables of elevation and distance to nearest forestry plantation were found to be significant predictors when yield was derived from Butler et al.'s [47] average FFB yield-by-age curve (yldB; elevation: $t_{499} = -3.93$, $P < 0.01$, plantation: $t_{499} = -3.05$, $P < 0.01$) (Table 2). All model residuals had no spatial autocorrelation ($P \geq 0.39$).

Yield Response to Forest Proximity

Proximity to any of the three classes of dipterocarp forest (contiguous, VJR, or private fragment) did not have a significant effect on oil palm yield when considering yield derived from either yield-by-age curves (Table 2). Instead environmental variables were more important predictors when oil palm yield was derived from Butler et al.'s [47] average FFB yield-by-age curve. Increasing elevation (Figure 3a; $t_{499} = -3.46$, $P < 0.01$) and increasing distance from tree plantation (Figure 3b; $t_{499} = -2.24$, $P = 0.03$) both had a significant negative effect on yield (Table 2). Proximity to large river or other oil palm plantation, size of private fragment, and soil type were not significant predictors of yield when using either yield-by-age curve. All model residuals had no spatial autocorrelation ($P \geq 0.06$).

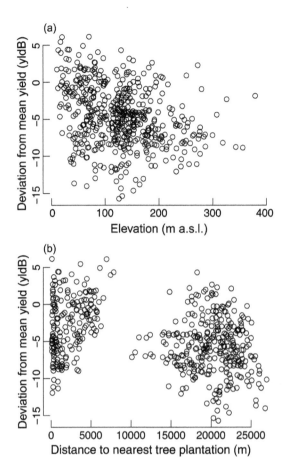

Figure 3. The relationship between oil palm yield and (a) elevation (m a.s.l.), and (b) distance to nearest non-native tree plantation. Oil palm yield was measured as the mean deviation from the yield-by-age curve generated from Butler et al. [47] (yldB), and is quantified as the fresh fruit bunch weight per hectare (mt ha^{-1}).

Discussion

Agricultural expansion in the tropics is a key driver of the global biodiversity crisis. Pressure to mitigate threats from agriculture and improve sustainability has encouraged suggestions that the retention of natural habitat patches within and adjacent to tropical agriculture would result in the export of ecosystem services [14,42,61–63], including to oil palm [40,41]. Yet the potential for spill-over of biodiversity from these features into the agricultural landscape [31,49], and in turn, whether this impacts upon crop yields positively or negatively has only received attention in the context of coffee and cocao agro-forestry plantations [17,18,25–28]. Our study is thus the first to focus on the link between forest and crop yield in a tropical monoculture crop, focusing specifically on oil palm, which is rapidly expanding at the expense of forest and highly lucrative. Spill-over from forest is difficult to quantify [64], especially across large scales and when there are various taxa that may spill-over to different degrees and have contrasting impacts. In this study, we instead assess the impacts of the extent of local forest cover and of forest proximity on oil palm yield directly; we therefore did not focus on biodiversity *per se*, and a precise link between biodiversity and yield is absent.

Using both forest cover and proximity metrics, we found that the retention of dipterocarp forest had no significant effect on yield

in oil palm monocultures, whereas the environmental variables of elevation and proximity to tree plantations did. These results provide a cautionary note for arguments that forest retention within monoculture landscapes can enhance ecosystem service provisioning and thus improve crop yields [14,42,61–63]. They also do not support concerns that ecosystem dis-services, such as increased pest populations or mammal crop raiders, are a major issue resulting from the protection of HCV forests under the RSPO. Because we did not directly measure either ecosystem benefits or dis-services, we do not rule out that these are occurring. Rather, our results suggest that either there is an equal balance between ecosystem service benefits and dis-services, resulting in a net neutral impact on yield, or that there is no spill-over occurring. Across our monoculture landscape, it is likely to be a combination of these possibilities, with the former more likely close to forest where species are known to spill-over into oil palm, and the latter more likely far from forest.

Our results suggest that there is no economic rationale for greater forest protection within and adjacent to oil palm monocultures. However, we acknowledge that riparian forest strips and larger fragments may have other important roles. They could provide hydrological and erosion prevention benefits, which might have longer-term benefits that cannot be quantified by focusing only on a single year of oil palm yield. These features could also provide biological benefits, harbouring some biodiversity [36,52,53] or by acting as stepping-stones and corridors for dispersal of species through the oil palm matrix [19,49], which could be vital for retaining meta-population dynamics.

The optimum growing conditions of oil palm (*Elaeis sp.*) are in lowland wet tropics of <1000 m elevation [65]: the negative effect of increasing elevation on yield is thus not surprising. This result highlights the limitation for future expansion of oil palm, especially in regions such as Southeast Asia where many of the prime locations have already been developed, and less optimum areas are already being considered and converted for oil palm development [66]. Proximity to tree plantations may provide some positive spillover, for example pest predation by birds, which are supported in greater numbers in tree plantations than oil palm [54,55]. In other agricultural systems multi-cropping has been found to be beneficial ([14] and references there in, [67]), and this is an important future direction for optimal agricultural landscape design. However, these results should be interpreted with caution, because elevation and proximity to tree plantation are positively correlated (Pearson's correlation: r = 0.12, p = 0.02), with lower lying areas of higher oil palm yield also closer on average to tree plantations.

In this study, we did not consider the potential impacts of different management activities, such as the use of pesticides or permitting the growth of understory vegetation, or of palm condition (e.g. pest abundance, disease, or structural damage) on yield, which represent important next steps to disentangle drivers of yield change [42]. With the exception of VJRs, which have only been lightly logged in patches, all of the forests in the study area have been selectively logged on an intensive, industrial scale. It is plausible that proximity to primary, unlogged forest could impact differently upon yield. However, this seems unlikely because previous work in the region has shown the retention of high levels of biodiversity, including most primary forest species [46,68,69], and high functional diversity [70,71] within contiguous blocks of logged forests. It is also possible that ecological services or dis-services from forest could affect palm oil quality, and hence price. Finally, we only focused on Southeast Asia and on one monoculture crop, and there could be different relationships between forest and yield in other tropical biomes, where oil palm is now expanding rapidly [72], or with other crops such as soya and sugar cane.

Conclusion

Our results show a neutral effect of forest on oil palm yield. Consequently, dipterocarp forests appear neither to export sufficient ecosystem service benefits to result in a net increase in yield nor to export sufficient ecosystem dis-services to result in a net reduction of yield within oil palm plantations. We thus observe no evidence to support arguments for the retention of forest for the provision of ecosystem services explicitly for yield benefits within oil palm monocultures [42,62]. Many arguments have been made for implementing an integrated framework of agricultural design, which considers biodiversity conservation, ecosystem services and agricultural output [42,73,74]. These are to be warmly welcomed, but in light of our study the proposed benefits of such designer landscapes within monocultures should avoid couching arguments for forest retention in the context of yield benefits. We finish by urging for more empirical assessments of the impacts of forest and biodiversity on crop monoculture yields to better understand their potential role in sustainable agriculture: we fear that by resting arguments for the retention of forest on improved oil palm yield, there could be unintended consequences such as the clearance of retained forest patches and thus the removal of refugia for biodiversity if no such empirical support were to emerge.

Supporting Information

File S1   Table S1. Description of soil types found across the study area. Figure S1. Distribution of soil types across the study area. Figure S2. Distribution of elevation across the study area. Figure S3. Oil palm yield-by-age curves.

Acknowledgments

We thank Sabah Softwoods for providing data, logistical support and site access, along with Glen Reynolds and the Royal Society's Southeast Asian Rainforest Research Program (SEARRP) for additional logistical support, and Yayasan Sabah, Danum Valley Management Committee, the Sabah Biodiversity Council, the State Secretary, Sabah Chief Minister's Department, and the Economic Planning Unit of the Prime Minister's Department for permission to conduct research in Sabah.

Author Contributions

Conceived and designed the experiments: DPE FAE. Performed the experiments: DPE FAE. Analyzed the data: FAE SS. Wrote the paper: FAE DPE SS KCH.

References

1. Green RE, Cornell SJ, Scharlemann JPW, Balmford A (2005) Farming and the fate of wild nature. Science 307: 550–555.
2. Tilman D (2001) Forecasting agriculturally driven global environmental change. Science, 292: 281–284.
3. Gibbs HK, Ruesch AS, Achard F, Clayton MK, Holmgren P, et al. (2010) Tropical forests were the primary sources of new agricultural land in the 1980s and 1990s. Proc Natl Acad Sci USA 107: 16732–16737.
4. Tscharntke T, Klein AM, Kruess A, Steffan-Dewenter I, Thies C (2005) Landscape perspectives on agricultural intensification and biodiversity - ecosystem service management. Ecol Lett 8: 857–874.
5. Ribeiro MC, Metzger JP, Martensen AC, Ponzoni FJ, Hirota MM (2009) The Brazilian Atlantic Forest: How much is left, and how is the remaining forest distributed? Implications for conservation. Biol Conserv 142: 1141–1153.
6. Benton TG, Vickery JA, Wilson JD (2003) Farmland biodiversity: is habitat heterogeneity the key? Trends Ecol Evol 18: 182–188.
7. Hooper DU, Chapin FS, Ewel JJ, Hector A, Inchausti P, et al. (2005) Effects of biodiversity on ecosystem functioning: A consensus of current knowledge. Ecol Monogr 75: 3–35.
8. Karp DS, Mendenhall CD, Sandi RF, Chaumont N, Ehrlich PR, et al. (2013) Forest bolsters bird abundance, pest control and coffee yield. Ecol Lett 16: 1339–1347.
9. Ranganathan J, Krishnaswamy J, Anand MO (2010) Landscape-level effects on avifauna within tropical agriculture in the Western Ghats: Insights for management and conservation. Biol Conserv 143: 2909–2917.
10. Sayer J, Sunderland T, Ghazoul J, Pfund J-L, Sheil D, et al. (2013) Ten principles for a landscape approach to reconciling agriculture, conservation, and other competing land uses. Proc Natl Acad Sci USA 110: 8349–8356.
11. Thies C, Tscharntke T (1999) Landscape structure and biological control in agroecosystems. Science 285: 893–895.
12. Kremen C, Williams NM, Thorp RW (2002) Crop pollination from native bees at risk from agricultural intensification. Proc Natl Acad Sci USA 99: 16812–16816.
13. Sande SO, Crewe RM, Raina SK, Nicolson SW, Gordon I (2009) Proximity to a forest leads to higher honey yield: Another reason to conserve. Biol Conserv 142: 2703–2709.
14. Tscharntke T, Clough Y, Wanger TC, Jackson L, Motzke I, et al. (2012) Global food security, biodiversity conservation and the future of agricultural intensification. Biol Conserv 151: 53–59.
15. Chaplin-Kramer R, O'Rourke ME, Blitzer EJ, Kremen C (2011) A meta-analysis of crop pest and natural enemy response to landscape complexity. Ecol Lett 14: 922–932.
16. Landis DA, Wratten SD, Gurr GM (2000) Habitat management to conserve natural enemies of arthropod pests in agriculture. Ann Rev Entomol 45: 175–201.
17. Klein AM, Steffan-Dewenter I, Tscharntke T (2003) Fruit set of highland coffee increases with the diversity of pollinating bees. Proc R Soc Lond B: Biol Sci 270: 955–961.
18. Ricketts TH, Daily GC, Ehrlich PR, Michener CD (2004) Economic value of tropical forest to coffee production. Proc Natl Acad Sci USA 101: 12579–12582.
19. Koh LP (2008) Can oil palm plantations be made more hospitable for forest butterflies and birds? J Appl Ecol 45: 1002–1009.
20. Tscharntke T, Sekercioglu CH, Dietsch TV, Sodhi NS, Hoehn P, et al. (2008) Landscape constraints on functional diversity of birds and insects in tropical agroecosystems. Ecology 89: 944–951.
21. Woltz JM, Isaacs R, Landis DA (2012) Landscape structure and habitat management differentially influence insect natural enemies in an agricultural landscape. Agriculture Ecosyst Environ 152: 40–49.
22. Tscharntke T, Brandl R (2004) Plant-insect interactions in fragmented landscapes. Ann Rev Entomol 49: 405–430.
23. Zhang W, Ricketts TH, Kremen C, Carney K, Swinton SM (2007) Ecosystem services and dis-services to agriculture. Ecol Econom 64: 253–260.
24. Asner GP, Rudel TK, Aide TM, Defries R, Emerson R (2009) A contemporary assessment of change in humid tropical forests. Conserv Biol 23: 1386–1395.
25. Olschewski R, Klein A-M, Tscharntke T (2010) Economic trade-offs between carbon sequestration, timber production, and crop pollination in tropical forested landscapes. Ecol Complexity 7: 314–319.
26. Olschewski R, Tscharntke T, Benitez PC, Schwarze S, Klein A-M (2006) Economic evaluation of pollination services comparing coffee landscapes in Ecuador and Indonesia. Ecol Soc 11.
27. Clough Y, Barkmann J, Juhrbandt J, Kessler M, Wanger TC, et al. (2011) Combining high biodiversity with high yields in tropical agroforests. Proc Natl Acad Sci USA 108: 8311–8316.
28. Bisseleua HBD, Fotio D, Yede, Missoup AD, Vidal S (2013) Shade tree diversity, cocoa pest damage, yield compensating inputs and farmers' net returns in West Africa. PLoS ONE 8.
29. Maas B, Clough Y, Tscharntke T (2013) Bats and birds increase crop yield in tropical agroforestry landscapes. Ecol Lett 16: 1480–1487.
30. Steffan-Dewenter I, Kessler M, Barkmann J, Bos MM, Buchori D, et al. (2007) Tradeoffs between income, biodiversity, and ecosystem functioning during tropical rainforest conversion and agroforestry intensification. Proc Natl Acad Sci USA 104: 4973–4978.
31. Ricketts TH (2004) Tropical forest fragments enhance pollinator activity in nearby coffee crops. Conserv Biol 18: 1262–1271.
32. Fisher B, Edwards DP, Giam X, Wilcove DS (2011) The high costs of conserving Southeast Asia's lowland rainforests. Frontiers Ecol Environ 9: 329–334.
33. FAOSTAT Statistical databases (2013) Food and Agriculture Organization of the United Nations. Available: http://faostat.fao.org/. Accessed 2013 Jul 01.
34. Wilcove DS, Giam X, Edwards DP, Fisher B, Koh LP (2013) Navjot's nightmare revisited: logging, agriculture, and biodiversity in Southeast Asia. Trends Ecol Evol 28: 531–540.
35. Fitzherbert E, Struebig M, Morel A, Danielsen F, Bruhl C, et al. (2008) How will oil palm expansion affect biodiversity? Trends Ecol Evol 23: 538–545.
36. Edwards DP, Hodgson JA, Hamer KC, Mitchell SL, Ahmad AH, et al. (2010) Wildlife-friendly oil palm plantations fail to protect biodiversity effectively. Conserv Lett 3: 236–242.

37. Fayle TM, Turner EC, Snaddon JL, Chey VK, Chung AYC, et al. (2010) Oil palm expansion into rain forest greatly reduces ant biodiversity in canopy, epiphytes and leaf-litter. Basic Appl Ecol 11: 337–345.

38. Edwards DP, Fisher B, Wilcove DS (2012) High Conservation Value or high confusion value? Sustainable agriculture and biodiversity conservation in the tropics. Conserv Lett 5: 20–27.

39. Edwards DP, Laurance SGW (2012) Green labelling, sustainability and the expansion of tropical agriculture: critical issues for certification schemes. Biol Conserv 151: 60–64.

40. Bhagwat SA, Willis KJ (2008) Agroforestry as a solution to the oil-palm debate. Conserv Biol 22: 1368–1369.

41. Koh LP, Levang P, Ghazoul J (2009) Designer landscapes for sustainable biofuels. Trends Ecol Evol 24: 431–438.

42. Foster WA, Snaddon JL, Turner EC, Fayle TM, Cockerill TD, et al. (2011) Establishing the evidence base for maintaining biodiversity and ecosystem function in the oil palm landscapes of South East Asia. Philos Trans R Soc Lond B: Biol Sci 366: 3277–3291.

43. Reynolds G, Payne J, Sinun W, Mosigil G, Walsh RPD (2011) Changes in forest land use and management in Sabah, Malaysian Borneo, 1990–2010, with a focus on the Danum Valley region. Philos Trans R Soc Lond B: Biol Sci 366: 3168–3176.

44. FAO website. WRB Map of World Soil Resources. Avaliable: http://www.fao.org/nr/land/soils/soil/wrb-soil-maps/wrb-map-of-world-soil-resources/en/ Accessed 2014 Jan 6.

45. Director of National Mapping (1974) Tawau. The Soils of Sabah. Sheet NB 50-15. D.O.S. 3180J. The British Government's Overseas Development Administration U.K.

46. Edwards DP, Larsen TH, Docherty TDS, Ansell FA, Hsu WW, et al. (2011) Degraded lands worth protecting: the biological importance of Southeast Asia's repeatedly logged forests. Proc R Soc Lond B: Biol Sci 278: 82–90.

47. Butler RA, Koh LP, Ghazoul J (2009) REDD in the red: palm oil could undermine carbon payment schemes. Conserv Lett 2: 67–73.

48. Miettinen J, Shi C, Tan WJ, Liew SC (2012) 2010 land cover map of insular Southeast Asia in 250-m spatial resolution. Remote Sensing Lett 3: 11–20.

49. Lucey JM, Hill JK (2012) Spillover of insects from rain forest into adjacent oil palm plantations. Biotropica 44: 368–377.

50. Peterson EE, Sheldon F, Darnell R, Bunn SE, Harch BD (2011) A comparison of spatially explicit landscape representation methods and their relationship to stream condition. Freshwater Biol 56: 590–610.

51. Rheinhardt R, Brinson M, Meyer G, Miller K (2012) Integrating forest biomass and distance from channel to develop an indicator of riparian condition. Ecol Indicators 23: 46–55.

52. Benedick S, Hill JK, Mustaffa N, Chey VK, Maryati M, et al. (2006) Impacts of rain forest fragmentation on butterflies in northern Borneo: species richness, turnover and the value of small fragments. J Appl Ecol 43: 967–977.

53. Hill JK, Gray MA, Khen CV, Benedick S, Tawatao N, et al. (2011) Ecological impacts of tropical forest fragmentation: how consistent are patterns in species richness and nestedness? Philos Trans R Soc Lond B: Biol Sci 366: 3265–3276.

54. Sheldon FH, Styring A, Hosner PA (2010) Bird species richness in a Bornean exotic tree plantation: A long-term perspective. Biol Conserv 143: 399–407.

55. Styring AR, Ragai R, Unggang J, Stuebing R, Hosner PA, et al. (2011) Bird community assembly in Bornean industrial tree plantations: Effects of forest age and structure. Forest Ecol Manag 261: 531–544.

56. Jarvis A, Reuter HI, Nelson A, Guevara E (2008) Hole-filled SRTM for the globe Version 4. CGIAR-CSI SRTM 90 m Database, http://srtm.csi.cgiar.org.

57. Zuur AF, Ieno EN, Walker NJ, Saveliev AA, Smith GM (2009) Mixed Effects Models and extensions in ecology with R. Springer, New York, U.S.A.

58. Dormann CF, McPherson JM, Araujo MB, Bivand R, Bolliger J, et al. (2007) Methods to account for spatial autocorrelation in the analysis of species distributional data: a review. Ecography 30: 609–628.

59. ESRI (2011) ArcGIS Desktop: Release 10. Redlands, CA: Environmental Systems Research Institute.

60. R Development Core Team (2011). R: a language and environment for statistical computing. In R Foundation for Statistical Computing. Vienna, Austria.

61. Fischer J, Lindenmayer DB, Manning AD (2006) Biodiversity, ecosystem function, and resilience: ten guiding principles for commodity production landscapes. Frontiers Ecol Environ 4: 80–86.

62. Koh LP (2008a) Birds defend oil palms from herbivorous insects. Ecol Appl 18: 821–825.

63. Perfecto I, Vandermeer J (2010) The agroecological matrix as alternative to the land-sparing/agriculture intensification model. Proc Natl Acad Sci USA 107: 5786–5791.

64. Kremen C (2005) Managing ecosystem services: what do we need to know about their ecology? Ecol Lett 8: 468–479.

65. Corley RHV, Tinker PBH (2003) The oil palm, 4 edn. Blackwell Science Ltd, Oxford.

66. Wicke B, Sikkema R, Dornburg V, Faaij A (2011) Exploring land use changes and the role of palm oil production in Indonesia and Malaysia. Land Use Policy 28: 193–206.

67. Perfecto I, Vandermeer JH, Bautista GL, Nunez GI, Greenberg R, et al. (2004) Greater predation in shaded coffee farms: The role of resident neotropical birds. Ecology 85: 2677–2681.

68. Berry NJ, Phillips OL, Lewis SL, Hill JK, Edwards DP, et al. (2010) The high value of logged tropical forests: lessons from northern Borneo. Biodiv Conserv 19: 985–997.

69. Woodcock P, Edwards DP, Fayle TM, Newton RJ, Khen CV, et al. (2011) The conservation value of South East Asia's highly degraded forests: evidence from leaf-litter ants. Philos Trans R Soc Lond B: Biol Sci 366: 3256–3264.

70. Edwards FA, Edwards DP, Hamer KC, Davies RG (2013) Impacts of logging and conversion of rainforest to oil palm on the functional diversity of birds in Sundaland. Ibis 155: 313–326.

71. Senior MM, Hamer K, Bottrell S, Edwards DP, Fayle T, et al. (2013) Trait-dependent declines of species following conversion of rain forest to oil palm plantations. Biodiv Conserv 22: 253–268.

72. Garcia-Ulloa J, Sloan S, Pacheco P, Ghazoul J, Koh LP (2012) Conserv Lett 5: 366–375.

73. Schroth G, McNeely JA (2011) Biodiversity conservation, ecosystem services and livelihoods in tropical landscapes: towards a common agenda. Environ Manag 48: 229–236.

74. Phalan B, Bertzky M, Butchart SHM, Donald PF, Scharlemann JPW, et al. (2013) Crop expansion and conservation priorities in tropical countries. PLoS ONE, 8: e51759.

11

Early Root Overproduction Not Triggered by Nutrients Decisive for Competitive Success Belowground

Francisco M. Padilla[1]*[¤], **Liesje Mommer**[1,2], **Hannie de Caluwe**[1], **Annemiek E. Smit-Tiekstra**[1],
Cornelis A. M. Wagemaker[3], **N. Joop Ouborg**[3], **Hans de Kroon**[1]

1 Experimental Plant Ecology, Institute for Water and Wetland Research, Radboud University Nijmegen, Nijmegen, The Netherlands, 2 Nature Conservation and Plant Ecology, Wageningen University, Wageningen, The Netherlands, 3 Molecular Ecology, Institute for Water and Wetland Research, Radboud University Nijmegen, Nijmegen, The Netherlands

Abstract

Background: Theory predicts that plant species win competition for a shared resource by more quickly preempting the resource in hotspots and by depleting resource levels to lower concentrations than its competitors. Competition in natural grasslands largely occurs belowground, but information regarding root interactions is limited, as molecular methods quantifying species abundance belowground have only recently become available.

Principal Findings: In monoculture, the grass *Festuca rubra* had higher root densities and a faster rate of soil nitrate depletion than *Plantago lanceolata*, projecting the first as a better competitor for nutrients. However, *Festuca* lost in competition with *Plantago*. *Plantago* not only replaced the lower root mass of its competitor, but strongly overproduced roots: with only half of the plants in mixture than in monoculture, *Plantago* root densities in mixture were similar or higher than those in its monocultures. These responses occurred equally in a nutrient-rich and nutrient-poor soil layer, and commenced immediately at the start of the experiment when root densities were still low and soil nutrient concentrations high.

Conclusions/Significance: Our results suggest that species may achieve competitive superiority for nutrients by root growth stimulation prior to nutrient depletion, induced by the presence of a competitor species, rather than by a better ability to compete for nutrients per se. The root overproduction by which interspecific neighbors are suppressed independent of nutrient acquisition is consistent with predictions from game theory. Our results emphasize that root competition may be driven by other mechanisms than is currently assumed. The long-term consequences of these mechanisms for community dynamics are discussed.

Editor: Frederick R. Adler, University of Utah, United States of America

Funding: F.M.P. was supported by a postdoc grant (Spanish Ministry of Education) and Stipendium Bottelier (KNBV), and L.M. by a Veni grant (016091116, NWO). The funders had no role in study design, data collection and analysis, decision to publish, or preparation of the manuscript.

Competing Interests: The authors have declared that no competing interests exist.

* E-mail: f.padilla@ual.es

¤ Current address: Agronomy Department, Universidad de Almería, La Cañada, Almería, Spain

Introduction

Co-occurring plant species frequently share space and compete belowground for essential soil nutrients [1,2]. Competition theory predicts that plant species win competition for a shared resource by more quickly preempting the resource supply in hotspots, as a result of greater root plasticity [1,3–5], and by depleting resource levels to lower concentrations than their competitors [6–8]. In competition studies with two species, the winner takes the share of the inferior species if resource availability is finite. This results in a competitive replacement where the superior species grows at the expense of the inferior [1,9–11], with the total aboveground yield of the mixture being intermediate to that of the monocultures [9–11]. Mixtures can draw more resources and will produce more biomass than the average of the monocultures ("overyield") if species occupy different niches, such as different rooting depths,

take up different nutrient sources, or if they segregate in phenology [10,12–19].

This classical model of resource competition and plasticity to nutrients does not take into account responses to neighbors independent of responses to nutrients [20,21]. Game theory predicts that plants should allocate a much greater share of their resources to roots than in the absence of competition, in order to prevent competitors from capturing the nutrients [22]. Evidence from pot experiments with individual plants is accumulating that such responses exist, independent of nutrient acquisition [23–27], but to what extent they affect the competition between plant populations of different species has not been examined so far.

Testing these predictions requires that root investments of different species are quantified in mixtures but such information is rarely available [28], as molecular methods quantifying species abundance belowground have only recently become available [29,30]. Results of plant competition have traditionally been analyzed by aboveground responses, despite that up to 80% of the plant community biomass may be belowground [31–33]. Today it is still unknown how aboveground responses are mirrored belowground [20], and therefore we are missing the contribution of a critical component involved in plant competition [28].

Here, two common West-European grassland perennials, *Plantago lanceolata* L. and *Festuca rubra* L., were led to compete in large containers in a facility specifically designed to study root growth under near-natural conditions for two growing seasons. The 55 cm deep containers in which the communities were grown contained a deep nutrient-rich soil layer (28–42 cm depth; Fig. 1A), with a high concentration of humus-rich soil, to test how these species competed for nutrients placed at depth. We combined monocultures and 50/50 mixtures to test the expectation from resource competition theory that the superior competitor takes resources at the expense of the inferior competitor, leading to a replacement of one species by another [10,11,34]. In particular, we expected the species developing the densest roots per unit of soil volume (i.e., *F. rubra*), quickly taking up the available nutrients in the nutrient-rich soil layer, acquiring a greater fraction of the nutrient supply rate and depleting the soil to the lowest nutrient concentrations (i.e. having the lowest R*), to win the competition [6,35]. The species with lower root densities (i.e., *P. lanceolata*) would only be expected to win if it would forage more effectively than its competitor for the nutrients in the nutrient-rich layer, or, following game theoretical predictions, if this species would pre-empt belowground space at the expense of its competitor and independent of nutrients. To address belowground responses, minirhizotron images were taken on a monthly basis and root mass in mixtures was determined at final harvest by applying a recent molecular method to quantify root mass of different species in mixed samples [29,30].

Methods

Species and experimental setup

The two species investigated commonly co-occur in hay meadows in central-northern Europe. The research was conducted in the Phytotron of the Radboud University Nijmegen (http://www.ru.nl/phytotron) in containers with separate units of $50(w) \times 50(l) \times 70(h)$ cm each. It is situated under a transparent rain shelter (high-quality commercial greenhouse film, 90% transparency) and open at all sides in order to allow natural weather conditions, except for some wire-netting. Plants were thus grown under near-ambient growth conditions except for watering. Monocultures of *F. rubra* and *P. lanceolata*, or mixtures of both species (1:1 proportion), were planted in June 2008 in a replacement design. Assignment of the planting treatments occurred randomly to a total of 11 units, resulting in 3–4 replicates. Planted seedlings were raised in the greenhouse for four weeks before transplanting. Seeds from local provenance (forelands of the river Rhine, near Nijmegen, the Netherlands) were first germinated in Petri dishes and then transferred to small pots containing the same background soil until transplant. Thirty-six seedlings (6×6) were then planted in each unit giving a plant density of 144 m^{-2}, but only the area of the inner 4×4 plants $(32 \times 32$ cm$)$ was used for further measurements. Interplant distant was 8 cm but distance from the edge plants to the rim was 5 cm. During the growing season, plants were irrigated 2 L unit^{-1} three times a week with tap water through an automatic irrigation system (PRIVA, de Lier, The Netherlands). In winter, watering was supplied manually once a week.

The bottom of each unit was filled with a five-cm layer of coarse gravel covered with weed cloth. Soil depth from surface to gravel

stones was 55 cm, divided in a 14-cm nutrient-rich layer consisting of black soil placed at 28 cm depth, and the remaining of the profile being filled up with a mixture of the same nutrient-rich black soil and nutrient-poor riverine sand (1:3; v:v) resulting in a poor sandy background soil. Soil nutrients were measured with an autoanalyzer (Bran+Luebbe, Norderstedt, Germany) after nutrients were extracted by diluting 20 g of freshly mixed soil samples in 50 mL of demineralized water and shaking for 1 h. At the start of the experiment, the nutrient-rich soil contained 26.4 ± 1.5 g kg^{-1} organic matter, and available nutrients were as follow: 256.4 ± 65.1 mg kg^{-1} nitrate (NO$_3^-$), 11.2 ± 2.2 mg kg^{-1} ammonium (NH$_4^+$) and 4.1 ± 0.3 mg kg^{-1} phosphate (PO$_4^{-3}$), whereas these values were 9.6 ± 0.1, 60.8 ± 0.1, 2.5 ± 0.0 and 1.5 ± 0.3, respectively, for the nutrient-poor background soil.

Each unit had separate drainage at the bottom and holes to insert a minirhizotron tube (6.4 cm inner diameter×50 cm length) horizontally with the top of the tube at 10 cm depth, and two soil suctions cups (Rhizosphere Research Products, Wageningen, The Netherlands) at 7 and 35 cm depth for collecting soil solution for analysis.

Measurements

In late August 2009, after two growing seasons, standing shoot biomass in the inner area was harvested by clipping 2 cm above soil surface. Root mass density was estimated by soil cores (20 mm diameter, four sub-replicates in each plot) in the inner area down to four soil layers (0–14, 14–28, 28–42, 42–55 cm depth). Distances to surrounding individual plants were equal. Roots per soil increment were collected after carefully rinsing them with tap water. Fresh weight was determined immediately with a micro-balance (Sartorius AG, Goettingen, Germany). Up to 100 mg of fresh roots was then stored at $-80°$C for later molecular analyses. Shoot and root dry weights were determined after drying samples at 70°C for 48 hours. Species abundance belowground in mixtures was quantified only in the top soil layer (0–14 cm) and in the intermediate nutrient-rich layer (28–42 cm), thus processing $\approx 70\%$ of the total root biomass. On these samples, genomic DNA extracts were subjected separately to quantitative real time polymerase chain reactions (RT-PCR) with primers for non-coding species-specific markers [29]. Analyses were performed on the basis of 100 mg fresh root mass, and recalculated in terms of dry weight as this was highly correlated with the fresh weight $(R^2 = 0.89, P < 0.001)$.

Root images from minirhizotron tubes at 10 cm depth $(21.6 \times 7.0$ cm, 300 dpi; CI-600 Root Scanner, CID Inc., Camas, WA, USA) captured the rooting area of four individuals in a row (either of the same species in monocultures, or half of each species in mixtures). Images were taken every 37 days on average, except in winter (Nov–Feb). Roots were digitized and analyzed using the WinRhizoTron V. 2005a software (Regents Inc., Quebec, Canada) for root length production. In mixtures, analyses were separated by species as the different color of newly-formed roots enabled species distinction: dark red to brown roots for *F. rubra*, pale grey to white roots for *P. lanceolata* (Fig. 2).

Soluble nutrients in the soil solution at two depths (7 and 35 cm; poor and nutrient-rich soil, respectively) were monitored every 65 days on average except in winter, by sampling soil water through porous soil suction cups and analyzing the extracted water solution for available nitrate with an autoanalyzer (Bran+Luebbe, Norderstedt, Germany).

Calculations and data analysis

In our replacement design, where total plant density in the mixture was equal to the plant density used in the monoculture of

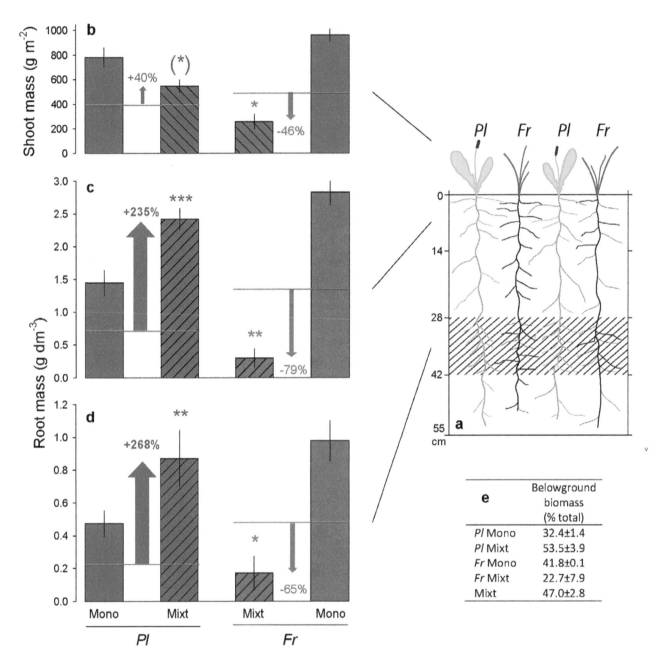

Figure 1. Experimental setup and biomass data. Planting scheme (**a**); shoot (**b**) and root mass in the poor top (**c**) and rich bottom layer (**d**); percentage belowground biomass at harvest (**e**), in *Festuca rubra* (*Fr*) and *Plantago lanceolata* (*Pl*) monocultures and mixtures. Horizontal lines in **b–d** show expected values for mixtures in case of competitive-equivalence (i.e., 50% of monocultures, or a relative yield of 0.5), and arrows depict the percentage deviation. Asterisks show significant differences between observed and expected values after t-tests. Data are means ± SE, N = 3–4. (*) P<0.06; * P<0.05; ** P<0.01; *** P<0.001.

each component, the competitive ability of the species was given by the Relative Yield [9,10,36], calculated as the ratio between the observed yield of a species in mixture and the yield of this same species in monoculture. These calculations find their origin in Lotka-Volterra competition theory [10], which is the central concept in competition and coexistence theory [37,38]. Relative Yields of 0.5 for both species reflect the situation of competitive equivalence with intraspecific competition equal to interspecific competition. If plants compete for a finite resource, a superior competitor is expected to take a larger proportion of the shared resource resulting in a higher Relative Yield, at the expense of an inferior competitor which will develop a proportionally lower relative yield. Belowground, the root Relative Yields are expected to deviate particularly in the nutrient-rich layer with the superior competitor developing a much larger root mass than the inferior competitor. As root investments will pay-off in nutrient uptake and growth, the root Relative Yields in the nutrient-rich layer are expected to be similar to the Relative Yields aboveground.

As a finite resource is partitioned among competing species, the sum of the relative yields (Relative Yield Total, RYT) is not expected to deviate from unity. RYT>1 or overyielding is only expected in the case of niche differentiation, i.e. when both

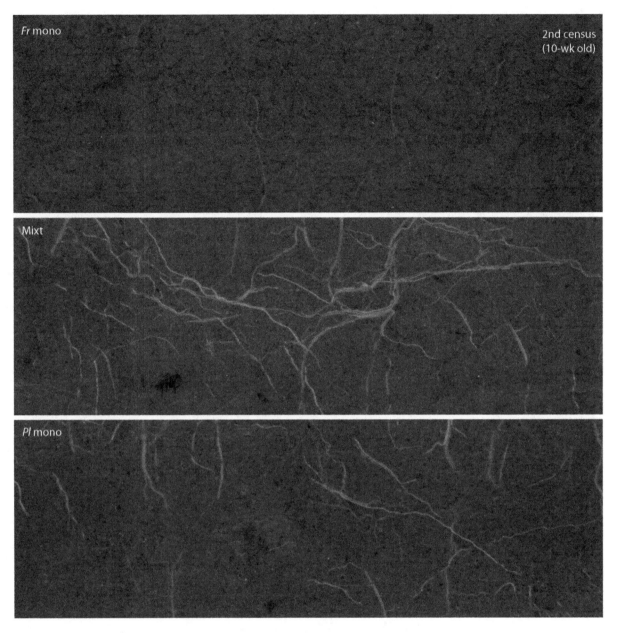

Figure 2. Minirhizotron images at 10 cm depth taken ten weeks after the start of the experiment. Note the abundance of *P. lanceolata* (*Pl*, white roots) and absence of *F. rubra* (*Fr*, brown roots) in mixtures images compared to the respective monocultures.

competitors have access to partly unique resources, as in the case of species with different rooting depths [13]. In such cases intraspecific competition is greater than interspecific competition for both species which is the criterion for species coexistence [38]. Significance of Relative Yields is tested by comparing the observed values with those expected from monocultures representing the null-expectation of competitive equivalence (RY = 0.5), calculated as ½ of the monoculture values.

If the species are involved in a competitive game, one of the species is expected to overinvest in roots (Relative Yield >>0.5) at the expense of the other species (Relative Yield <<0.5). Such investment will take place similarly in the nutrient-poor topsoil as in the nutrient-rich deep soil layer. Depending on the extent of dominance and suppression, RYT may appear larger than 1. As

root investments are expected to be altered, Relative Yields belowground will not reflect Relative Yields aboveground.

Root and shoot masses were estimated on the basis of above and below-ground biomass values per area and core soil volume, respectively. To compare whether values of root mass observed in mixtures for each species deviated from the expected values, we ran t-test.

Significance of differences over time in soil solution nutrients were tested by ANCOVA, using diversity of species and soil depth as fixed factors, and days after plantation as covariate. Conventional tests aimed at testing temporal trends (RM-ANOVA and MANOVA) could not be applied because the sphericity assumption was not met. In ANCOVA, differences in the temporal pattern between factors were considered significant when the interaction(s) between factor(s) and 'days after plantation' result was significant. When a factor or interaction resulted significant, pair-wise comparisons were performed using the Sidak correction for multiple comparisons.

Differences between expected and observed root length density from minirhizotron images were explored by linear regression, by plotting expected against observed values deviating from the null 1:1 expectation (i.e., expected = observed). Differences between observed and expected values of root lengths in mixtures on specific dates for each species were tested by t-tests, using Sidak correction for multiple comparisons. All analyses were run with PASW Statistics 18 (SPSS Inc., Chicago, IL, USA).

Results

Festuca was clearly projected as the superior species in nutrient competition: its root length densities in monoculture were 1.4–2.1 times higher than of *Plantago* (Fig. 1, C and D), and soil nutrient solution measurements throughout the study period showed that *Festuca* monocultures more quickly took-up nutrients in the rich soil layer and depleted them to a lower concentration than *Plantago* monocultures (Fig. 3). Nitrate, the most limiting nutrient in these soils and the most differentiating nutrient between the rich and poor layer (Table 1), was much more available in the deeper nutrient-rich layer than in the poor top at the beginning of the experiment, but differences between both layers levelled off as the experiment progressed. In the deep-rich layer, nitrate in *Festuca* monocultures became significantly lower than in *Plantago* monocultures, while in the poor-top layer, availability of nitrate did not differ among communities. Soil nitrate measurements (Fig. 3) further showed that *Festuca* monocultures depleted soil nitrate more rapidly than *Plantago* monocultures from the very beginning of the experiment, although the root densities were still low. If the species behave similarly in mixture than in monoculture, it is to be expected that *Festuca* will more quickly take up available nitrate and develop more root mass at the expense of *Plantago*.

However, results from the mixtures immediately contrasted with this expectation: *Plantago*, rather than *Festuca*, won the competition belowground. Moreover, *Plantago* did not competitively replace the inferior *Festuca* but strongly overproduced roots and did so both in deep-rich and the poor-top soil layer. This dominance and suppression belowground became established early in the experiment, prior to and not as a result of soil nutrient depletion.

Where *Festuca* was severely reduced in mixtures (72% less root mass than expected from monoculture; t-test obs. *vs.* exp., P = 0.001 top layer, P = 0.038 bottom layer; Fig. 1, C and D), roots of *Plantago* did not simply take the space from which *Festuca* was ousted. Rather, *Plantago* overcompensated and produced on average a massive 252% more root mass in mixtures than expected from monocultures (t-test obs. *vs.* exp., P<0.001 top layer, P = 0.009 bottom layer; Fig. 1, C and D). Root overproduction of *Plantago* in mixtures was so overwhelming that this species had as much (t-test, P = 0.069 bottom layer) or even higher (t-test, P = 0.015 top layer) root biomass in mixtures than in its monocultures, with only half the number of plants.

One basic tenet of resource competition is that a superior competitor takes resources at the expense of an inferior species, resulting in a differentiation of their relative yields from the null-expectation of competitive equivalence (Relative Yield, RY = 0.5) [6]. As both species have access to the same pool of limiting resources, the sum of relative yields (Relative Yield Total, RYT) is not expected to differ from unity [10,34]. However, with belowground RY of 1.66 and 0.13 for *Plantago* and *Festuca*, respectively, our results significantly deviate from these expectations. A RYT significantly higher than unity can be expected if species have access to different resources, as in species with different rooting depths [13,18,19,39], but this was not the case in

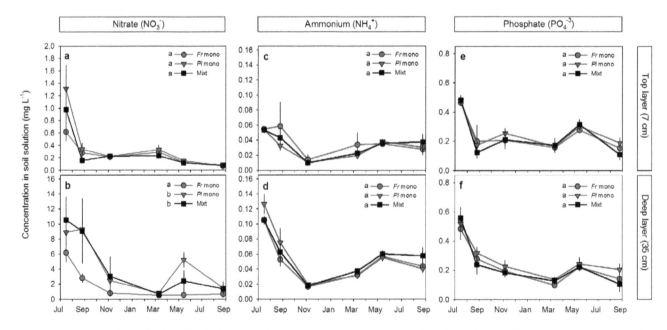

Figure 3. Nutrients dynamics in soil solution over time. Nitrate (**a, b**), ammonium (**c, d**) and phosphate (**e, f**) concentration in *Festuca rubra* and *Plantago lanceolata* monocultures and in mixtures of the two species, at 7 and 35 cm depth. In nitrate, different letters in legends show significant differences between species over time, after ANCOVA$_{layer \times species}$. Data are means ± SE, N = 3–4. No significant second and third order interactions involving species and time were detected (Table 1), meaning that similarities/differences between species were consistent all over the experimental period. Soil nutrient concentrations were derived from regular sampling of soil water over the course of the experiment through porous suction cups that had been placed in the soil layers.

Table 1. ANCOVA results for available nutrients in soil solution.

Variable	Source	d.f.	F-value
Nitrate	Species x Depth x Time	2	1.277^{ns}
	Species x Depth	2	**4.140***
	Species x Time	2	1.606^{ns}
	Depth x Time	1	**30.285*****
	Species	2	**5.027****
	Depth	1	**80.600*****
	Time	1	**42.499*****
	Error	120	
Ammonium	Species x Depth x Time	2	0.878^{ns}
	Species x Depth	2	1.521^{ns}
	Species x Time	2	0.569^{ns}
	Depth x Time	1	2.957^{ns}
	Species	2	0.322^{ns}
	Depth	1	**15.480*****
	Time	1	**14.675*****
	Error	118	
Phosphate	Species x Depth x Time	2	0.075^{ns}
	Species x Depth	2	0.087^{ns}
	Species x Time	2	0.085^{ns}
	Depth x Time	1	2.802^{ns}
	Species	2	0.055^{ns}
	Depth	1	2.266^{ns}
	Time	1	**40.007*****
	Error	117	

Diversity of species (*F. rubra* monoculture, *P. lanceolata* monoculture and mixture of the two species) and soil depth were fixed factors, and time after plantation was a covariate.
*P<0.05;
**P<0.01;
***P<0.001;
^{ns}P>0.09. Bold shows significant effects.

our experiment where both species in mixture had similar mean rooting depths (i.e., 7.4±1.0 cm; $F_{1,10}$ = 0.001, P = 0.973).

Root overproduction of *Plantago* occurred similarly in top (poor soil) and bottom (rich soil) layers, despite their very different nutrient availability (Fig. 3). Overproduction of *Plantago* roots solely in the rich layer could have been interpreted as a nutrient-induced (foraging) response, but the fact that the same degree of overproduction was found in the poor-top layer suggest that the overproduction of *Plantago* was not related to differences in soil nutrients but induced by the presence of the competitor species.

Non-destructive observations from minirhizotron tubes located in the top layer revealed when root densities of the species started to diverge. Disentangling the roots on the images by color (dark red to brown roots for *Festuca*, pale grey to white roots for *Plantago*, Fig. 2) showed that the root overproduction of *Plantago* was initiated immediately after the start of the experiment (Fig. 4a). Already at first census (four weeks after plantation onwards), *Plantago* produced 3× more root length in mixtures than expected from its monocultures. At this time, root length densities were only 20% of the densities developed after the two growing seasons. Soil nitrate concentrations of the rich layer were still >6-fold higher

than later in the experiment and not significantly different between mixtures and *Plantago* monocultures (Fig. 3). These differences in root length were maintained until the end of the experiment. Likewise, immediately after the start of the experiment, *Festuca* produced less root length in mixtures than expected from its monocultures, despite the relatively high soil nutrient concentrations in the mixed soil, and differences remained over the two growing seasons of the experiment (Fig. 4b). Importantly, these results suggest that root overproduction in *Plantago* and suppression in *Festuca* preceded soil nitrate depletion and that they were not the result of higher soil nitrate uptake by *Plantago*.

Discussion

By analysing the root responses of two common perennials in a straightforward competition experiment, two surprising results were apparent. Firstly, the species projected as the better competitor for nutrients based on the monocultures (*Festuca rubra*) did not win the competition. Soil nutrient and root growth analyses through time revealed a sequence of events that deviated from what may be expected in resource competition. Dominance and suppression were established very early in the experiment, irrespective of soil nutrient availability. Root growth of the superior species (*Plantago lanceolata*) was stimulated in mixtures, and root growth of the inferior species severely reduced, *prior to* nutrient depletion. *Plantago* did not win because it took up a larger proportion of the shared resources after which it was able to develop more roots [1,22,40], but because its root growth was immediately stimulated in the presence of *Festuca*. Our results suggest that a species may win competition for nutrients for different reasons than is currently assumed.

Secondly, the massive root overproduction of *Plantago* in mixture and the overyielding belowground (RYT>>1) is inconsistent with niche differentiation as in such case it is to be expected that both species develop Relative Yields larger than 0.5 [13,34,38]. In our experiment an RYT>>1 was reached by severe suppression of the inferior species combined with disproportional root growth of the superior species. As disproportional root growth did not only occur in the nutrient-rich layer, it is not a reflection of a better ability to forage for nutrient-rich hotspots of this species. Rather, root overproduction in the presence of another species and independent of nutrients follow game theoretical predictions. Game theory also predicted that investments belowground increase in competition for nutrients, as observed in mixtures relative to monocultures. We discuss the mechanisms and consequences of these two results below.

Mechanisms of belowground competition

Except that we have been able to rule out a differential response to soil nutrients, we do not know what mechanisms have been driving this strong dominance and suppression early in this competition experiment. There has been a lot of attention in recent years to the effects of species-specific communities of soil pathogens affecting coexistence and production of plant communities [41–43]. *Plantago* growth is known to be sensitive to its own conditioned soils due to accumulation of self-harming fungi [44] and root pathogens [45], suggesting the presence of negative plant-soil feedback [46] in monoculture of this species. Mixtures would have been a better environment for *Plantago* roots to grow as self-harming biota would have been diluted. A recent plant–soil feedback experiment showed that *Plantago* monocultures developed 3.2-fold more biomass in the presence of *Festuca* soil biota compared to soil biota of its own, but the reverse was also true: *Festuca* monocultures grew 2.5-fold more biomass on *Plantago* soil

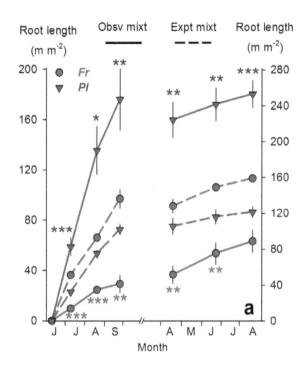

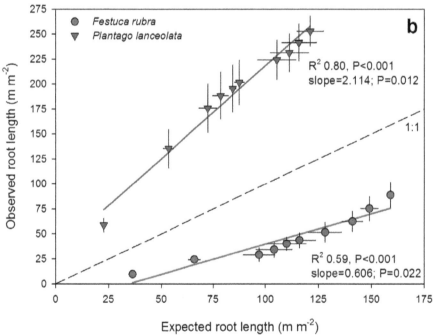

Figure 4. Root growth observed through minirhizotron tubes. (a) Root length production over time (m m^{-2} image) of *Festuca rubra* (*Fr*) and *Plantago lanceolata* (*Pl*) in mixtures, obtained from minirhizotron observations at 10 cm depth. Solid lines are for observed values, dashed lines for expected values from monocultures (½ of monocultures). On each date, t-test were run separately to detect significant differences between observed and expected values in each species. *P*-values were then adjusted using the Sidak correction for multiple comparisons. After correction, * $P<0.009$; ** $P<0.002$; *** $P<0.001$. (b) Linear regression of expected *versus* observed root length of *Plantago* and *Festuca* in mixtures over the whole experiment, and null expectation expected = observed (1:1). Significance of deviation of slopes from unity is shown by p-values. Data are means ± SE, N = 3–4.

than soil of its own [47]. However, it should be noted that soils in the current experiment were not conditioned purposely and, therefore, it is unlikely that species-specific soil biota solely explain the observed root responses. Moreover, differential root growth

developed very early in the experiment well before species–specific soil communities were likely built up [48].

Root growth suppression, apparent already at low densities and independent of local soil nutrient concentrations, is reminiscent of allelopathy or chemical interference [49–51]. Release of chemical substances may be expected to quickly reduce root growth, as in the case of *Festuca*, but to our knowledge immediate overproduction by the superior species, as observed in *Plantago*, is a phenomenon that has hitherto been unassociated with allelopathy. Some studies have suggested that root exudates can stimulate root growth in response to interspecific neighbors [26,52–55], which would imply that *Plantago* root growth was stimulated by *Festuca*. These phenomena have not been previously described for these species; phytotoxic effects have been attributed only to root exudates of *Festuca rubra* [56].

Facilitation by roots of certain species through the release of organic acids by roots of leguminous species may account for some cases of root growth stimulation in mixtures [57]. However, it seems unlikely that such facilitative mechanism can explain root growth stimulation in our non-leguminous system [20]. As discussed below, there is no sign of facilitation aboveground in our system as overyielding aboveground was not detected.

Interpreting root overproduction

Consistent with game theoretical predictions [22], *Plantago* won by rapid investment in roots in the presence of *Festuca*, at the same time suppressing *Festuca* and preventing it from acquiring soil resources, resulting in a much larger root investment of *Plantago* than predicted on the basis of classical resource competition. However, from a game theoretical perspective, pertinent questions remain. Firstly, why did *Plantago* win and not *Festuca*? With its higher root densities, fine roots and high nutrient uptake rates, the grass species *Festuca* had a much better starting position to compete for nutrients. For our species pair, well-known traits conferring competitive ability belowground [1,4,6] could not predict the competitive outcome. Further research has to unravel the root traits that have predictive power and even then, the outcome may well depend on specific combinations of species and soils.

Secondly, as competition is a process taking place among individuals, why did *Plantago* not overproduce to a similar extent in monoculture? Craine [22] suggested that the best solution for a plant is to alter root allocation in proportion to the root length density of competitors. This exactly seems to have occurred in our experiment: plants with lower root densities (*Plantago*) overproduced roots strongly and won competition from plants with already high root densities (*Festuca*). As root densities in *Plantago* monocultures are lower than in *Festuca* monocultures, as is generally true for forbs versus grasses, a similar competitive game between *Plantago* individuals may have resulted in less overproduction. In other words, making substantially more roots may have paid-off only in competition with *Festuca* individuals, not with other *Plantago* individuals.

Consequences for plant competition and coexistence

We do not know how common this belowground competitive mechanism is in plant communities. However, if it is widespread it may have easily gone unnoticed in many competition experiments. The reason is that, aboveground, competitive relationships among our species appeared to conform to the resource competition model. Similar to numerous other experiments with only aboveground information (e.g., [9]; see refs there), Relative Yields

of 0.70 and 0.27 for *Plantago* and *Festuca*, respectively, would project *Plantago* as the winner replacing *Festuca* in resource competition (Fig. 1B), further confirmed by an aboveground RYT similar to unity (0.97). Due to inherent difficulties in quantifying the roots of different species, our experiment is one of the first to compare competitive interactions aboveground with those belowground. Doing so revealed that apparently classical competitive relationships aboveground were combined by unexpected responses belowground.

If our results for these two common plant species are representative for a wider group of plants, the implications for long-term competitive superiority and coexistence may be profound. Competitive games are predicted to generate a "Tragedy of the Commons" where plants invest more to the acquisition of a limiting recourse than is optimal in the absence of competition [22]. Likewise, *Plantago* individuals invested 65% more biomass in their roots in mixtures than in monoculture (percentage total biomass increased from 32.4 to 53.5; Fig. 1E). If *Plantago* roots had not overproduced but only replaced the roots of *Festuca* in mixture (belowground RY 0.87 rather than 1.66), this increase in root investment would only have been 16% (percentage total biomass increase from 32.4 to 37.7). Although the investment pays-off in terms of immediate competitive gain, such major root investment may compromise biomass production in the long run, reminiscent of a Tragedy of the Commons. Interestingly, in a two-species *Plantago-Festuca* mixture within a long-term biodiversity experiment [58], *Plantago* initially dominated the mixture aboveground as in our experiment. But over the 11 years of study *Plantago* never outcompeted *Festuca* and after eight years *Festuca* even gained in abundance (J. van Ruijven, pers. comm.). This trajectory suggests that the overinvestment of *Plantago* in roots may have compromised its competitive ability in the long run.

Consistent with our results, there are indications from biodiversity studies that root mass is increased in species mixtures [53,57,59] and that this higher root biomass may already develop prior to positive effects of biodiversity on aboveground production [59]. Moreover, evidence is increasing that interactions in multi-species communities are driven by species-specific soil biota giving opportunities for local coexistence [41–43,47,60,61], whereas opportunities for niche partitioning for nutrients seem to be limited [28,62]. Future work should demonstrate to what extent the root responses seen in our experiment also play a role in more diverse plant communities.

Acknowledgments

We thank N. Anten, F. Berendse, J. van Ruijven, R. Pierik, E. Visser and H. Huber for discussion, two anonymous reviewers for thought-provoking comments on earlier drafts, and G. Bögemann, C. Ruiz, G. van der Weerden and the Botanical Garden staff for help.

Author Contributions

Conceived and designed the experiments: FMP HdK. Performed the experiments: FMP HdC AES-T. Analyzed the data: FMP. Contributed reagents/materials/analysis tools: LM AES-T CAMW NJO. Wrote the paper: FMP LM HdK.

References

1. Casper BB, Jackson RB (1997) Plant competition underground. Annual Review of Ecology and Systematic 28: 545–570.
2. Goldberg DE, Barton AM (1992) Patterns and consequences of interspecific competition in natural communities - a review of field experiments with plants. The American Naturalist 139: 771–801.
3. Grime JP (1994) The role of plasticity in exploiting environmental heterogeneity. In: Caldwell MM, Pearcy RW, editors. Exploitation of environmental

heterogeneity by plants: ecophysiological processes above- and below-ground. San Diego, CA, , USA: Academic Press, Inc. pp. 1–19.
4. Robinson D, Hodge A, Griffiths BS, Fitter AH (1999) Plant root proliferation in nitrogen-rich patches confers competitive advantage. Proceedings of the Royal Society of London Series B-Biological Sciences 266: 431–435.

5. Craine JM, Fargione J, Sugita S (2005) Supply pre-emption, not concentration reduction, is the mechanism of competition for nutrients. New Phytologist 166: 933–940.

6. Fargione J, Tilman DD (2006) Plant species traits and capacity for resource reduction predict yield and abundance under competition in nitrogen-limited grassland. Functional Ecology 20: 533–540.

7. Tilman D (1982) Resource competition and community structure. Princeton, New Jersey: Princeton University Press. 296 p.

8. Craine JM (2005) Reconciling plant strategy theories of Grime and Tilman. Journal of Ecology 93: 1041–1052.

9. Fransen B, de Kroon H, Berendse F (2001) Soil nutrient heterogeneity alters competition between two perennial grass species. Ecology 82: 2534–2546.

10. de Wit CT (1960) On competition. Verslagen van landbouwkundige onderzoekingen 66: 1–82.

11. Olde Venterink H, Gusewell S (2010) Competitive interactions between two meadow grasses under nitrogen and phosphorus limitation. Functional Ecology 24: 877–886.

12. Berendse F (1981) Competition between plant-populations with different rooting depths. 2. Pot experiments. Oecologia 48: 334–341.

13. Berendse F (1983) Interspecific competition and niche differentiation between Plantago lanceolata and Anthoxanthum odoratum in a natural hayfield. Journal of Ecology 71: 379–390.

14. Piper JK (1998) Growth and seed yield of three perennial grains within monocultures and mixed stands. Agriculture Ecosystems & Environment 68: 1–11.

15. Roscher C, Schumacher J, Weisser WW, Schmid B, Schulze ED (2007) Detecting the role of individual species for overyielding in experimental grassland communities composed of potentially dominant species. Oecologia 154: 535–549.

16. Li L, Li SM, Sun JH, Zhou LL, Bao XG, et al. (2007) Diversity enhances agricultural productivity via rhizosphere phosphorus facilitation on phosphorus-deficient soils. Proceedings of the National Academy of Sciences of the United States of America 104: 11192–11196.

17. Polley HW, Wilsey BJ, Tischler CR (2007) Species abundances influence the net biodiversity effect in mixtures of two plant species. Basic and Applied Ecology 8: 209–218.

18. Fargione J, Tilman D (2005) Niche differences in phenology and rooting depth promote coexistence with a dominant C 4 bunchgrass Oecologia 143: 598–606.

19. Levine JM, HilleRisLambers J (2009) The importance of niches for the maintenance of species diversity. Nature 461: 254–257.

20. de Kroon H (2007) How do roots interact? Science 318: 1562–1563.

21. Cahill JF, McNickle GG, Haag JJ, Lamb EG, Nyanumba SM, et al. (2010) Plants integrate information about nutrients and neighbors. Science 328: 1657–.

22. Craine JM (2006) Competition for nutrients and optimal root allocation. Plant and Soil 285: 171–185.

23. Chen BJW, During HJ, Anten NPR (2012) Detect thy neighbor: Identity recognition at the root level in plants. Plant Science 195: 157–167.

24. O'Brien EE, Brown JS, Moll JD (2007) Roots in space: a spatially explicit model for below-ground competition in plants. Proceedings of the Royal Society B-Biological Sciences 274: 929–934.

25. O'Brien EE, Gersani M, Brown JS (2005) Root proliferation and seed yield in response to spatial heterogeneity of below-ground competition. New Phytologist 168: 401–412.

26. Semchenko M, John EA, Hutchings MJ (2007) Effects of physical connection and genetic identity of neighbouring ramets on root-placement patterns in two clonal species. New Phytologist 176: 644–654.

27. Semchenko M, Zobel K, Hutchings MJ (2010) To compete or not to compete: an experimental study of interactions between plant species with contrasting root behaviour. Evolutionary Ecology 24: 1433–1445.

28. de Kroon H, Hendriks M, van Ruijven J, Ravenek J, Padilla FM, et al. (2012) Root responses to nutrients and soil biota: drivers of species coexistence and ecosystem productivity. Journal of Ecology 100: 6–15.

29. Mommer L, Wagemaker CAM, De Kroon H, Ouborg N (2008) Unravelling below-ground plant distributions: a real-time polymerase chain reaction method for quantifying species proportions in mixed root samples. Molecular Ecology Resources 8: 947–953.

30. Mommer L, Dumbrell AJ, Wagemaker CAM, Ouborg NJ (2011) Belowground DNA-based techniques: untangling the network of plant root interactions. Plant and Soil 348: 115–121.

31. Poorter H, Niklas KJ, Reich PB, Oleksyn J, Poot P, et al. (2012) Biomass allocation to leaves, stems and roots: meta-analyses of interspecific variation and environmental control. New Phytologist 193: 30–50.

32. Mokany K, Raison RJ, Rokushkin AS (2006) Critical analysis of root : shoot ratios in terrestrial biomes. Global Change Biology 12: 84–96.

33. Canadell J, Jackson RB, Ehleringer JR, Mooney HA, Sala OE, et al. (1996) Maximum rooting depth of vegetation types at the global scale. Oecologia 108: 583–595.

34. Berendse F (1979) Competition between plant-populations with different rooting depths.1. Theoretical considerations. Oecologia 43: 19–26.

35. Wedin D, Tilman D (1993) Competition among grasses along a nitrogen gradient - Initial conditions and mechanisms of competition. Ecological Monographs 63: 199–229.

36. van den Bergh JP, Elberse WT (1970) Yields of monocultures and mixtures of 2 grass species differing in growth habit. Journal of Applied Ecology 7: 311–320.

37. Begon M, Townsend CA, Harper JL (2005) Ecology: From Individuals to Ecosystems: Wiley.

38. Chesson P (2000) Mechanisms of maintenance of species diversity. Annual Review of Ecology and Systematics 31: 343–366.

39. Cardinale BJ, Wright JP, Cadotte MW, Carroll IT, Hector A, et al. (2007) Impacts of plant diversity on biomass production increase through time because of species complementarity. Proceedings of the National Academy of Sciences of the United States of America 104: 18123–18128.

40. Goldberg DE (1990) Components of resource competition in plant communities. In: Grace J, Tilman D, editors. Perspectives in Plant Competition: Academic Press. pp. 27–49.

41. Petermann JS, Fergus AJF, Turnbull LA, Schmid B (2008) Janzen-Connell effects are widespread and strong enough to maintain diversity in grasslands. Ecology 89: 2399–2406.

42. Maron JL, Marler M, Klironomos JN, Cleveland CC (2011) Soil fungal pathogens and the relationship between plant diversity and productivity. Ecology Letters 14: 36–41.

43. Schnitzer SA, Klironomos JN, HilleRisLambers J, Kinkel LL, Reich PB, et al. (2011) Soil microbes drive the classic plant diversity–productivity pattern. Ecology 92: 296–303.

44. Bever JD (2002) Negative feedback within a mutualism: host-specific growth of mycorrhizal fungi reduces plant benefit. Proceedings of the Royal Society of London Series B-Biological Sciences 269: 2595–2601.

45. Mills KE, Bever JD (1998) Maintenance of diversity within plant communities: Soil pathogens as agents of negative feedback. Ecology 79: 1595–1601.

46. Bever JD (1994) Feedback between plants and their soil communities in an old field community. Ecology 75: 1965–1977.

47. Hendriks M, Mommer L, de Caluwe H, Smit-Tiekstra AE, van der Putten WH, et al. (2013) Independent variations of plant and soil mixtures reveal soil feedback effects on plant community overyielding. Journal of Ecology. In press.

48. Mitchell CE, Blumenthal D, Jarosik V, Puckett EE, Pysek P (2010) Controls on pathogen species richness in plants' introduced and native ranges: roles of residence time, range size and host traits. Ecology Letters 13: 1525–1535.

49. Schenk HJ (2006) Root competition: beyond resource depletion. Journal of Ecology 94: 725–739.

50. Mahall BE, Callaway RM (1992) Root communication mechanisms and intracommunity distributions of 2 Mojave desert shrubs. Ecology 73: 2145–2151.

51. Inderjit, Wardle DA, Karban R, Callaway RM (2011) The ecosystem and evolutionary contexts of allelopathy. Trends in Ecology and Evolution 26: 655–662.

52. Bartelheimer M, Steinlein T, Beyschlag W (2006) Aggregative root placement: a feature during interspecific competition in inland sand-dune habitats. Plant and Soil 280: 101–114.

53. Brassard BW, Chen HYH, Bergeron Y, Paré D (2011) Differences in fine root productivity between mixed- and single-species stands. Functional Ecology 25: 238–246.

54. Bais HP, Weir TL, Perry LG, Gilroy S, Vivanco JM (2006) The role of root exudates in rhizosphere interations with plants and other organisms. Annual Review of Plant Biology 57: 233–266.

55. Badri DV, Vivanco JM (2009) Regulation and function of root exudates. Plant, Cell & Environment 32: 666–681.

56. Bertin C, Weston LA, Huang T, Jander G, Owens T, et al. (2007) Grass roots chemistry: meta-Tyrosine, an herbicidal nonprotein amino acid. Proceedings of the National Academy of Sciences of the United States of America 104: 16964–16969.

57. Li L, Sun J, Zhang F, Guo T, Bao X, et al. (2006) Root distribution and interactions between intercropped species. Oecologia 147: 280–290.

58. van Ruijven J, Berendse F (2009) Long-term persistence of a positive plant diversity-productivity relationship in the absence of legumes. Oikos 118: 101–106.

59. Mommer L, van Ruijven J, de Caluwe H, Smit-Tiekstra AE, Wagemaker CAM, et al. (2010) Unveiling below-ground species abundance in a biodiversity experiment: a test of vertical niche differentiation among grassland species. Journal of Ecology 98: 1117–1127.

60. Mangan SA, Schnitzer SA, Herre EA, Mack KML, Valencia MC, et al. (2010) Negative plant-soil feedback predicts tree-species relative abundance in a tropical forest. Nature 466: 752–755.

61. Kulmatiski A, Beard KH, Heavilin J (2012) Plant-soil feedbacks provide an additional explanation for diversity-productivity relationships. Proceedings of the Royal Society B-Biological Sciences 279: 3020–3026.

62. von Felten S, Niklaus PA, Scherer-Lorenzen M, Hector A, Buchmann N (2012) Do grassland plant communities profit from N partitioning by soil depth? Ecology 93: 2386–2396.

12

A Meta-Analysis of the Impacts of Genetically Modified Crops

Wilhelm Klümper, Matin Qaim*

Department of Agricultural Economics and Rural Development, Georg-August-University of Goettingen, Goettingen, Germany

Abstract

Background: Despite the rapid adoption of genetically modified (GM) crops by farmers in many countries, controversies about this technology continue. Uncertainty about GM crop impacts is one reason for widespread public suspicion.

Objective: We carry out a meta-analysis of the agronomic and economic impacts of GM crops to consolidate the evidence.

Data Sources: Original studies for inclusion were identified through keyword searches in ISI Web of Knowledge, Google Scholar, EconLit, and AgEcon Search.

Study Eligibility Criteria: Studies were included when they build on primary data from farm surveys or field trials anywhere in the world, and when they report impacts of GM soybean, maize, or cotton on crop yields, pesticide use, and/or farmer profits. In total, 147 original studies were included.

Synthesis Methods: Analysis of mean impacts and meta-regressions to examine factors that influence outcomes.

Results: On average, GM technology adoption has reduced chemical pesticide use by 37%, increased crop yields by 22%, and increased farmer profits by 68%. Yield gains and pesticide reductions are larger for insect-resistant crops than for herbicide-tolerant crops. Yield and profit gains are higher in developing countries than in developed countries.

Limitations: Several of the original studies did not report sample sizes and measures of variance.

Conclusion: The meta-analysis reveals robust evidence of GM crop benefits for farmers in developed and developing countries. Such evidence may help to gradually increase public trust in this technology.

Editor: emidio albertini, University of Perugia, Italy

Funding: This research was financially supported by the German Federal Ministry of Economic Cooperation and Development (BMZ) and the European Union's Seventh Framework Programme (FP7/2007-2011) under Grant Agreement 290693 FOODSECURE. The funders had no role in study design, data collection and analysis, decision to publish, or preparation of the manuscript. Neither BMZ nor FOODSECURE and any of its partner organizations, any organization of the European Union or the European Commission are accountable for the content of this article.

Competing Interests: The authors have declared that no competing interests exist.

* Email: mqaim@uni-goettingen.de

Introduction

Despite the rapid adoption of genetically modified (GM) crops by farmers in many countries, public controversies about the risks and benefits continue [1–4]. Numerous independent science academies and regulatory bodies have reviewed the evidence about risks, concluding that commercialized GM crops are safe for human consumption and the environment [5–7]. There are also plenty of studies showing that GM crops cause benefits in terms of higher yields and cost savings in agricultural production [8–12], and welfare gains among adopting farm households [13–15]. However, some argue that the evidence about impacts is mixed and that studies showing large benefits may have problems with the data and methods used [16–18]. Uncertainty about GM crop impacts is one reason for the widespread public suspicion towards this technology. We have carried out a meta-analysis that may help to consolidate the evidence.

While earlier reviews of GM crop impacts exist [19–22], our approach adds to the knowledge in two important ways. First, we include more recent studies into the meta-analysis. In the emerging literature on GM crop impacts, new studies are published continuously, broadening the geographical area covered, the methods used, and the type of outcome variables considered. For instance, in addition to other impacts we analyze effects of GM crop adoption on pesticide quantity, which previous meta-analyses could not because of the limited number of observations for this particular outcome variable. Second, we go beyond average impacts and use meta-regressions to explain impact heterogeneity and test for possible biases.

Our meta-analysis concentrates on the most important GM crops, including herbicide-tolerant (HT) soybean, maize, and cotton, as well as insect-resistant (IR) maize and cotton. For these crops, a sufficiently large number of original impact studies have been published to estimate meaningful average effect sizes. We estimate mean impacts of GM crop adoption on crop yield, pesticide quantity, pesticide cost, total production cost, and farmer profit. Furthermore, we analyze several factors that may influence outcomes, such as geographic location, modified crop trait, and type of data and methods used in the original studies.

Materials and Methods

Literature search

Original studies for inclusion in this meta-analysis were identified through keyword searches in relevant literature databanks. Studies were searched in the ISI Web of Knowledge, Google Scholar, EconLit, and AgEcon Search. We searched for studies in the English language that were published after 1995. We did not extend the review to earlier years, because the commercial adoption of GM crops started only in the mid-1990s [23]. The search was performed for combinations of keywords related to GM technology and related to the outcome of interest. Concrete keywords used related to GM technology were (an asterisk is a replacement for any ending of the respective term; quotation marks indicate that the term was used as a whole, not each word alone): GM*, "genetically engineered", "genetically modified", transgenic, "agricultural biotechnology", HT, "herbicide tolerant", Roundup, Bt, "insect resistant". Concrete keywords used related to outcome variables were: impact*, effect*, benefit*, yield*, economic*, income*, cost*, soci*, pesticide*, herbicide*, insecticide*, productivity*, margin*, profit*. The search was completed in March 2014.

Most of the publications in the ISI Web of Knowledge are articles in academic journals, while Google Scholar, EconLit, and AgEcon Search also comprise book chapters and grey literature such as conference papers, working papers, and reports in institutional series. Articles published in academic journals have usually passed a rigorous peer-review process. Most papers presented at academic conferences have also passed a peer-review process, which is often less strict than that of good journals though. Some of the other publications are peer reviewed, while many are not. Some of the working papers and reports are published by research institutes or government organizations, while others are NGO publications. Unlike previous reviews of GM crop impacts, we did not limit the sample to peer-reviewed studies but included all publications for two reasons. First, a clear-cut distinction between studies with and without peer review is not always possible, especially when dealing with papers that were not published in a journal or presented at an academic conference [24]. Second, studies without peer review also influence the public and policy debate on GM crops; ignoring them completely would be short-sighted.

Of the studies identified through the keyword searches, not all reported original impact results. We classified studies by screening titles, abstracts, and full texts. Studies had to fulfill the following criteria to be included:

- The study is an empirical investigation of the agronomic and/or economic impacts of GM soybean, GM maize, or GM cotton using micro-level data from individual plots and/or farms. Other GM crops such as GM rapeseed, GM sugarbeet, and GM papaya were commercialized in selected countries [23], but the number of impact studies available for these other crops is very small.
- The study reports GM crop impacts in terms of one or more of the following outcome variables: yield, pesticide quantity (especially insecticides and herbicides), pesticide costs, total

variable costs, gross margins, farmer profits. If only the number of pesticide sprays was reported, this was used as a proxy for pesticide quantity.
- The study analyzes the performance of GM crops by either reporting mean outcomes for GM and non-GM, absolute or percentage differences, or estimated coefficients of regression models that can be used to calculate percentage differences between GM and non-GM crops.
- The study contains original results and is not only a review of previous studies.

In some cases, the same results were reported in different publications; in these cases, only one of the publications was included to avoid double counting. On the other hand, several publications involve more than one impact observation, even for a single outcome variable, for instance when reporting results for different geographical regions or derived with different methods (e.g., comparison of mean outcomes of GM and non-GM crops plus regression model estimates). In those cases, all observations were included. Moreover, the same primary dataset was sometimes used for different publications without reporting identical results (e.g., analysis of different outcome variables, different waves of panel data, use of different methods). Hence, the number of impact observations in our sample is larger than the number of publications and primary datasets (Data S1). The number of studies selected at various stages is shown in the flow diagram in Figure 1. The number of publications finally included in the meta-analysis is 147 (Table S1).

Effect sizes and influencing factors

Effect sizes are measures of outcome variables. We chose the percentage difference between GM and non-GM crops for five different outcome variables, namely yield, pesticide quantity, pesticide cost, total production cost, and farmer profits per unit area. Most studies that analyze production costs focus on variable costs, which are the costs primarily affected through GM technology adoption. Accordingly, profits are calculated as revenues minus variable production costs (profits calculated in this way are also referred to as gross margins). These production costs also take into account the higher prices charged by private companies for GM seeds. Hence, the percentage differences in profits considered here are net economic benefits for farmers using GM technology. Percentage differences, when not reported in the original studies, were calculated from mean value comparisons between GM and non-GM or from estimated regression coefficients.

Since we look at different types of GM technologies (different modified traits) that are used in different countries and regions, we do not expect that effect sizes are homogenous across studies. Hence, our approach of combining effect sizes corresponds to a random-effects model in meta-analysis [25]. To explain impact heterogeneity and test for possible biases, we also compiled data on a number of study descriptors that may influence the reported effect sizes. These influencing factors include information on the type of GM technology (modified trait), the region studied, the type of data and method used, the source of funding, and the type of publication. All influencing factors are defined as dummy variables. The exact definition of these dummy variables is given in Table 1. Variable distributions of the study descriptors are shown in Table S2.

Statistical analysis

In a first step, we estimate average effect sizes for each outcome variable. To test whether these mean impacts are significantly

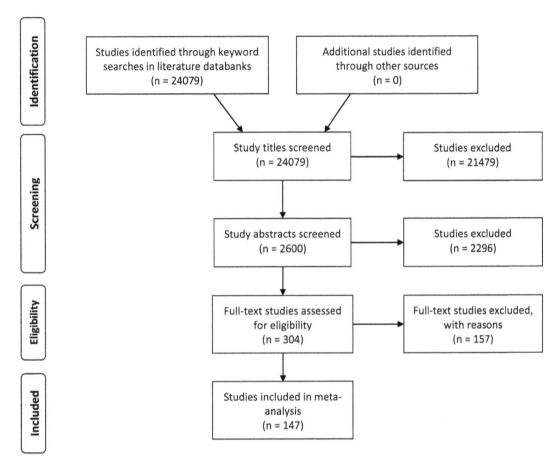

Figure 1. Selection of studies for inclusion in the meta-analysis.

different from zero, we regress each outcome variable on a constant with cluster correction of standard errors by primary dataset. Thus, the test for significance is valid also when observations from the same dataset are correlated. We estimate average effect sizes for all GM crops combined. However, we expect that the results may differ by modified trait, so that we also analyze mean effects for HT crops and IR crops separately.

Meta-analyses often weight impact estimates by their variances; estimates with low variance are considered more reliable and receive a higher weight [26]. In our case, several of the original studies do not report measures of variance, so that weighting by variance is not possible. Alternatively, weighting by sample size is common, but sample sizes are also not reported in all studies considered, especially not in some of the grey literature publications. To test the robustness of the results, we employ a

Table 1. Variables used to analyze influencing factors of GM crop impacts.

Variable name	Variable definition
Insect resistance (IR)	Dummy that takes a value of one for all observations referring to insect-resistant GM crops with genes from *Bacillus thuringiensis* (Bt), and zero for all herbicide-tolerant (HT) GM crops.
Developing country	Dummy that takes a value of one for all GM crop applications in a developing country according to the World Bank classification of countries, and zero for all applications in a developed country.
Field-trial data	Dummy that takes a value of one for all observations building on field-trial data (on-station and on-farm experiments), and zero for all observations building on farm survey data.
Industry-funded study	Dummy that takes a value of one for all studies that mention industry (private sector companies) as source of funding, and zero otherwise.
Regression model result	Dummy that takes a value of one for all impact observations that are derived from regression model estimates, and zero for observations derived from mean value comparisons between GM and non-GM.
Journal publication	Dummy that takes a value of one for all studies published in a peer-reviewed journal, and zero otherwise.
Journal/academic conference	Dummy that takes a value of one for all studies published in a peer-reviewed journal or presented at an academic conference, and zero otherwise.

different weighting procedure, using the inverse of the number of impact observations per dataset as weights. This procedure avoids that individual datasets that were used in several publications dominate the calculation of average effect sizes.

In a second step, we use meta-regressions to explain impact heterogeneity and test for possible biases. Linear regression models are estimated separately for all of the five outcome variables:

$$\%\Delta Y_{hij} = \alpha_h + \mathbf{X}_{hij}\boldsymbol{\beta}_h + \varepsilon_{hij}$$

$\%\Delta Y_{hij}$ is the effect size (percentage difference between GM and non-GM) of each outcome variable h for observation i in publication j, and $\mathbf{X}_{hij}$ is a vector of influencing factors. α_h is a coefficient and $\boldsymbol{\beta}_h$ a vector of coefficients to be estimated; ε_{hij} is a random error term. Influencing factors used in the regressions are defined in Table 1.

Results and Discussion

Average effect sizes

Distributions of all five outcome variables are shown in Figure S1. Table 2 presents unweighted mean impacts. As a robustness check, we weighted by the inverse of the number of impact observations per dataset. Comparing unweighted results (Table 2) with weighted results (Table S3) we find only very small differences. This comparison suggests that the unweighted results are robust.

On average, GM technology has increased crop yields by 21% (Figure 2). These yield increases are not due to higher genetic yield potential, but to more effective pest control and thus lower crop damage [27]. At the same time, GM crops have reduced pesticide quantity by 37% and pesticide cost by 39%. The effect on the cost of production is not significant. GM seeds are more expensive than non-GM seeds, but the additional seed costs are compensated through savings in chemical and mechanical pest control. Average profit gains for GM-adopting farmers are 69%.

Results of Cochran's test [25], which are reported in Figure S1, confirm that there is significant heterogeneity across study observations for all five outcome variables. Hence it is useful to

further disaggregate the results. Table 2 shows a breakdown by modified crop trait. While significant reductions in pesticide costs are observed for both HT and IR crops, only IR crops cause a consistent reduction in pesticide quantity. Such disparities are expected, because the two technologies are quite different. IR crops protect themselves against certain insect pests, so that spraying can be reduced. HT crops, on the other hand, are not protected against pests but against a broad-spectrum chemical herbicide (mostly glyphosate), use of which facilitates weed control. While HT crops have reduced herbicide quantity in some situations, they have contributed to increases in the use of broad-spectrum herbicides elsewhere [2,11,19]. The savings in pesticide costs for HT crops in spite of higher quantities can be explained by the fact that broad-spectrum herbicides are often much cheaper than the selective herbicides that were used before. The average farmer profit effect for HT crops is large and positive, but not statistically significant because of considerable variation and a relatively small number of observations for this outcome variable.

Impact heterogeneity and possible biases

Table 3 shows the estimation results from the meta-regressions that explain how different factors influence impact heterogeneity. Controlling for other factors, yield gains of IR crops are almost 7 percentage points higher than those of HT crops (column 1). Furthermore, yield gains of GM crops are 14 percentage points higher in developing countries than in developed countries. Especially smallholder farmers in the tropics and subtropics suffer from considerable pest damage that can be reduced through GM crop adoption [27].

Most original studies in this meta-analysis build on farm surveys, although some are based on field-trial data. Field-trial results are often criticized to overestimate impacts, because farmers may not be able to replicate experimental conditions. However, results in Table 3 (column 1) show that field-trial data do not overestimate the yield effects of GM crops. Reported yield gains from field trials are even lower than those from farm surveys. This is plausible, because pest damage in non-GM crops is often more severe in farmers' fields than on well-managed experimental plots.

Table 2. Impacts of GM crop adoption by modified trait.

Outcome variable	All GM crops	Insect resistance	Herbicide tolerance
Yield	21.57***	24.85***	9.29**
	(15.65; 27.48)	(18.49; 31.22)	(1.78; 16.80)
n/m	451/100	353/83	94/25
Pesticide quantity	−36.93***	−41.67***	2.43
	(−48.01; −25.86)	(−51.99; −31.36)	(−20.26; 25.12)
n/m	121/37	108/31	13/7
Pesticide cost	−39.15***	−43.43***	−25.29***
	(−46.96; −31.33)	(−51.64; −35.22)	(−33.84; −16.74)
n/m	193/57	145/45	48/15
Total production cost	3.25	5.24**	−6.83
	(−1.76; 8.25)	(0.25; 10.73)	(−16.43; 2.77)
n/m	115/46	96/38	19/10
Farmer profit	68.21***	68.78***	64.29
	(46.31; 90.12)	(46.45; 91.11)	(−24.73; 153.31)
n/m	136/42	119/36	17/9

Average percentage differences between GM and non-GM crops are shown with 95% confidence intervals in parentheses. *, **, *** indicate statistical significance at the 10%, 5%, and 1% level, respectively. *n* is the number of observations, *m* the number of different primary datasets from which these observations are derived.

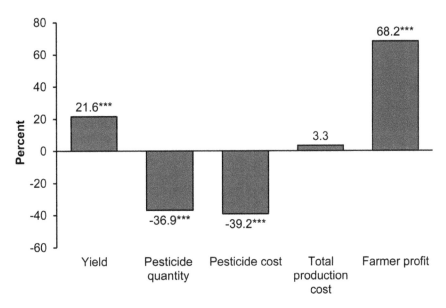

Figure 2. Impacts of GM crop adoption. Average percentage differences between GM and non-GM crops are shown. Results refer to all GM crops, including herbicide-tolerant and insect-resistant traits. The number of observations varies by outcome variable; yield: 451; pesticide quantity: 121; pesticide cost: 193; total production cost: 115; farmer profit: 136. *** indicates statistical significance at the 1% level.

Another concern often voiced in the public debate is that studies funded by industry money might report inflated benefits. Our results show that the source of funding does not significantly influence the impact estimates. We also analyzed whether the statistical method plays a role. Many of the earlier studies just compared yields of GM and non-GM crops without considering possible differences in other inputs and conditions that may also affect the outcome. Net impacts of GM technology can be estimated with regression-based production function models that control for other factors. Interestingly, results derived from regression analysis report higher average yield effects.

Finally, we examined whether the type of publication matters. Controlling for other factors, the regression coefficient for journal publications in column (1) of Table 3 implies that studies published in peer-reviewed journals show 12 percentage points higher yield gains than studies published elsewhere. Indeed, when only including observations from studies that were published in journals, the mean effect size is larger than if all observations are included (Figure S2). On first sight, one might suspect publication bias, meaning that only studies that report substantial effects are accepted for publication in a journal. A common way to assess possible publication bias in meta-analysis is through funnel plots [25], which we show in Figure S3. However, in our case these funnel plots should not be over-interpreted. First, only studies that report variance measures can be included in the funnel plots, which holds true only for a subset of the original studies used here. Second, even if there were publication bias, our mean results would be estimated correctly, because we do include studies that were not published in peer-reviewed journals.

Further analysis suggests that the journal review process does not systematically filter out studies with small effect sizes. The journal articles in the sample report a wide range of yield effects, even including negative estimates in some cases. Moreover, when combining journal articles with papers presented at academic conferences, average yield gains are even higher (Table 3, column 2). Studies that were neither published in a journal nor presented at an academic conference encompass a diverse set of papers, including reports by NGOs and outspoken biotechnology critics.

These reports show lower GM yield effects on average, but not all meet common scientific standards. Hence, rather than indicating publication bias, the positive and significant journal coefficient may be the result of a negative NGO bias in some of the grey literature.

Concerning other outcome variables, IR crops have much stronger reducing effects on pesticide quantity than HT crops (Table 3, column 3), as already discussed above. In terms of pesticide costs, the difference between IR and HT is less pronounced and not statistically significant (column 4). The profit gains of GM crops are 60 percentage points higher in developing countries than in developed countries (column 6). This large difference is due to higher GM yield gains and stronger pesticide cost savings in developing countries. Moreover, most GM crops are not patented in developing countries, so that GM seed prices are lower [19]. Like for yields, studies published in peer-reviewed journals report higher profit gains than studies published elsewhere, but again we do not find evidence of publication bias (column 7).

Conclusion

This meta-analysis confirms that – in spite of impact heterogeneity – the average agronomic and economic benefits of GM crops are large and significant. Impacts vary especially by modified crop trait and geographic region. Yield gains and pesticide reductions are larger for IR crops than for HT crops. Yield and farmer profit gains are higher in developing countries than in developed countries. Recent impact studies used better data and methods than earlier studies, but these improvements in study design did not reduce the estimates of GM crop advantages. Rather, NGO reports and other publications without scientific peer review seem to bias the impact estimates downward. But even with such biased estimates included, mean effects remain sizeable.

One limitation is that not all of the original studies included in this meta-analysis reported sample sizes and measures of variance. This is not untypical for analyses in the social sciences, especially when studies from the grey literature are also included. Future

Table 3. Factors influencing results on GM crop impacts (%).

Variables	(1) Yield	(2) Yield	(3) Pesticide quantity	(4) Pesticide cost	(5) Total cost	(6) Farmer profit	(7) Farmer profit
Insect resistance (IR)	6.58** (2.85)	5.25* (2.82)	-37.38*** (11.81)	-7.28 (5.44)	5.63 (5.60)	-22.33 (21.62)	-33.41 (21.94)
Developing country	14.17*** (2.72)	13.32*** (2.65)	-10.23 (8.99)	-19.16*** (5.35)	3.43 (4.78)	59.52*** (18.02)	60.58*** (17.67)
Field-trial data	-7.14** (3.19)	-7.81** (3.08)	–#	-17.56 (11.45)	-10.69* (5.79)	–#	–#
Industry-funded study	1.68 (5.30)	1.05 (5.21)	37.04 (23.08)	-7.77 (10.22)	–#	–#	
Regression model result	7.38* (3.90)	7.29* (3.83)	9.67 (10.40)	–#	–#	-11.44 (24.03)	-9.85 (24.03)
Journal publication	12.00*** (2.52)		9.95 (6.79)	-3.71 (4.09)	-3.08 (3.30)	48.27*** (15.48)	
Journal/academic conference		16.48*** (2.64)					65.29*** (17.75)
Constant	-0.22 (2.84)	-2.64 (2.86)	-4.44 (10.33)	-16.13 (4.88)	-1.02 (4.86)	8.57 (24.33)	-1.19 (24.53)
Observations	451	451	121	193	115	136	136
R²	0.23	0.25	0.20	0.14	0.12	0.12	0.14

Coefficient estimates from linear regression models are shown with standard errors in parentheses. Dependent variables are GM crop impacts measured as percentage differences between GM and non-GM. All explanatory variables are 0/1 dummies (for variable definitions see Table 1). The yield models in columns (1) and (2) and the farmer profit models in columns (6) and (7) have the same dependent variables, but they differ in terms of the explanatory variables, as shown. *, **, *** indicate statistical significance at the 10%, 5%, and 1% level, respectively. # indicates that the variable was dropped because the number of observations with a value of one was smaller than 5.

impact studies with primary data should follow more standardized reporting procedures. Nevertheless, our findings reveal that there is robust evidence of GM crop benefits. Such evidence may help to gradually increase public trust in this promising technology.

Supporting Information

Figure S1 Histograms of effect sizes for the five outcome variables.

Figure S2 Impacts of GM crop adoption including only studies published in journals.

Figure S3 Funnel plots for the five outcome variables.

Table S1 List of publications included in the meta-analysis.

Table S2 Distribution of study descriptor dummy variables for different outcomes.

Table S3 Weighted mean impacts of GM crop adoption.

Data S1 Data used for the meta-analysis.

Acknowledgments

We thank Sinja Buri and Tingting Xu for assistance in compiling the dataset. We also thank Joachim von Braun and three reviewers of this journal for useful comments.

Author Contributions

Conceived and designed the research: WK MQ. Analyzed the data: WK MQ. Contributed to the writing of the manuscript: WK MQ. Compiled the data: WK.

References

1. Gilbert N (2013) A hard look at GM crops. Nature 497: 24–26.
2. Fernandez-Cornejo J, Wechsler JJ, Livingston M, Mitchell L (2014) Genetically Engineered Crops in the United States. Economic Research Report ERR-162 (United Sates Department of Agriculture, Washington, DC).
3. Anonymous (2013) Contrary to popular belief. Nature Biotechnology 31: 767.
4. Andreasen M (2014) GM food in the public mind–facts are not what they used to be. Nature Biotechnology 32: 25.
5. DeFrancesco L (2013) How safe does transgenic food need to be? Nature Biotechnology 31: 794–802.
6. European Academies Science Advisory Council (2013) Planting the Future: Opportunities and Challenges for Using Crop Genetic Improvement Technologies for Sustainable Agriculture (EASAC, Halle, Germany).
7. European Commission (2010) A Decade of EU-Funded GMO Research 2001–2010 (European Commission, Brussels).
8. Pray CE, Huang J, Hu R, Rozelle S (2002) Five years of Bt cotton in China - the benefits continue. The Plant Journal 31: 423–430.
9. Huang J, Hu R, Rozelle S, Pray C (2008) Genetically modified rice, yields and pesticides: assessing farm-level productivity effects in China. Economic Development and Cultural Change 56: 241–263.
10. Morse S, Bennett R, Ismael Y (2004) Why Bt cotton pays for small-scale producers in South Africa. Nature Biotechnology 22: 379–380.
11. Qaim M, Traxler G (2005) Roundup Ready soybeans in Argentina: farm level and aggregate welfare effects. Agricultural Economics 32: 73–86.
12. Sexton S, Zilberman D (2012) Land for food and fuel production: the role of agricultural biotechnology. In: The Intended and Unintended Effects of US Agricultural and Biotechnology Policies (eds. Zivin, G. & Perloff, J.M.), 269–288 (University of Chicago Press, Chicago).
13. Ali A, Abdulai A (2010) The adoption of genetically modified cotton and poverty reduction in Pakistan. Journal of Agricultural Economics 61, 175–192.
14. Kathage J, Qaim M (2012) Economic impacts and impact dynamics of Bt (*Bacillus thuringiensis*) cotton in India. Proceedings of the National Academy of Sciences USA 109: 11652–11656.
15. Qaim M, Kouser S (2013) Genetically modified crops and food security. PLOS ONE 8: e64879.
16. Stone GD (2012) Constructing facts: Bt cotton narratives in India. Economic & Political Weekly 47(38): 62–70.
17. Smale M, Zambrano P, Gruere G, Falck-Zepeda J, Matuschke I, et al. (2009) Measuring the Economic Impacts of Transgenic Crops in Developing Agriculture During the First Decade: Approaches, Findings, and Future Directions (International Food Policy Research Institute, Washington, DC).
18. Glover D (2010) Is Bt cotton a pro-poor technology? A review and critique of the empirical record. Journal of Agrarian Change 10: 482–509.
19. Qaim M (2009) The economics of genetically modified crops. Annual Review of Resource Economics 1: 665–693.
20. Carpenter JE (2010) Peer-reviewed surveys indicate positive impact of commercialized GM crops. Nature Biotechnology 28: 319–321.
21. Finger R, El Benni N, Kaphengst T, Evans C, Herbert S, et al. (2011) A meta analysis on farm-level costs and benefits of GM crops. Sustainability 3: 743–762.
22. Areal FJ, Riesgo L, Rodríguez-Cerezo E (2013) Economic and agronomic impact of commercialized GM crops: a meta-analysis. Journal of Agricultural Science 151: 7–33.
23. James C (2013) Global Status of Commercialized Biotech/GM Crops: 2013. ISAAA Briefs No.46 (International Service for the Acquisition of Agri-biotech Applications, Ithaca, NY).
24. Rothstein HR, Hopewell S (2009) Grey literature. In: Handbook of Research Synthesis and Meta-Analysis, Second Edition (eds. Cooper, H., Hedges, L.V. & Valentine, J.C.), 103–125 (Russell Sage Foundation, New York).
25. Borenstein M, Hedges LV, Higgins JPT, Rothstein HR (2009) Introduction to Meta-Analysis (John Wiley and Sons, Chichester, UK).
26. Shadish WR, Haddock CK (2009) Combining estimates of effect size. In: Handbook of Research Synthesis and Meta-Analysis, Second Edition (eds. Cooper, H., Hedges, L.V. & Valentine, J.C.), 257–277 (Russell Sage Foundation, New York).
27. Qaim M, Zilberman D (2003) Yield effects of genetically modified crops in developing countries. Science 299: 900–902.

13

Alfalfa (*Medicago sativa* L.)/Maize (*Zea mays* L.) Intercropping Provides a Feasible Way to Improve Yield and Economic Incomes in Farming and Pastoral Areas of Northeast China

Baoru Sun, Yi Peng, Hongyu Yang, Zhijian Li*, Yingzhi Gao*, Chao Wang, Yuli Yan, Yanmei Liu

Key Laboratory of Vegetation Ecology, Northeast Normal University, Changchun, China

Abstract

Given the growing challenges to food and eco-environmental security as well as sustainable development of animal husbandry in the farming and pastoral areas of northeast China, it is crucial to identify advantageous intercropping modes and some constraints limiting its popularization. In order to assess the performance of various intercropping modes of maize and alfalfa, a field experiment was conducted in a completely randomized block design with five treatments: maize monoculture in even rows, maize monoculture in alternating wide and narrow rows, alfalfa monoculture, maize intercropped with one row of alfalfa in wide rows and maize intercropped with two rows of alfalfa in wide rows. Results demonstrate that maize monoculture in alternating wide and narrow rows performed best for light transmission, grain yield and output value, compared to in even rows. When intercropped, maize intercropped with one row of alfalfa in wide rows was identified as the optimal strategy and the largely complementary ecological niches of alfalfa and maize were shown to account for the intercropping advantages, optimizing resource utilization and improving yield and economic incomes. These findings suggest that alfalfa/maize intercropping has obvious advantages over monoculture and is applicable to the farming and pastoral areas of northeast China.

Editor: Wen-Xiong Lin, Agroecological Institute, China

Funding: This work was supported by the National Natural Science Foundation of China (NSFC, 31072080 and 31270444), the program for New Century Excellent Talents in University (NCET-13-0717), Key Science and Technology Program of Jilin Province (20100212, 2012ZDGG008), National Key Technology R&D Program (2011BAD17B04-3-2), and the National Program on Key Basic Research Project (2012CB722202). The funders had no role in study design, data collection and analysis, decision to publish, or preparation of the manuscript.

Competing Interests: The authors have declared that no competing interests exist.

* Email: lizj004@nenu.edu.cn (ZL); gaoyz108@nenu.edu.cn (YG)

Introduction

The farming and pastoral area (FPA) of northeast China (NEC) is an agriculture-based ecozone combining forestry and animal husbandry, and it is an important grain commodity and animal husbandry base. However, it is also a vulnerable eco-environmental zone owing to low vegetation cover, fine sandy soil and strong wind [1–2]. Wind erosion, water erosion and unsustainable production activities (e.g., single cropping and multiple-year continuous cropping) have made the land subject to dust storms in winter and spring. Moreover, cropland soil is being increasingly eroded, causing low soil fertility and reduced crop productivity and quality [3–5]. Together with the growing use of agricultural chemicals, such as fertilizers and herbicides, sustainable agricultural development is now facing many serious challenges [6].

In addition to the cropland, grassland has been degraded due to overgrazing and excessive agricultural reclamation, and grassland degradation is reflected by the reduced grassland production and forb quality and low carrying capacity [7]. "Grain for green" initiatives including reseeding of forage grasses have long been considered as effective ways to restore grassland vegetation and help to balance the ecological system [8]. Moreover, with increasing demand for meat products and high quality forage grass, China continues to give substantial support and increased financial investment to the development of the forage industry. Additionally, the Chinese government has strongly endorsed a proposal for boosting the development of the alfalfa industry so as to ensure the production, processing and sustainable supply of high quality forage grass [9]. Therefore, in the context of food and eco-environmental security and animal husbandry sustainable development, traditional farming patterns in the northeast FPA are being altered; with a tendency to adjust agricultural structure, introduce forage grass into the main crop farming system and establish an intercropping pattern between crops and forage grass, resulting in efficient resource utilization, a friendly ecological environment and good economic benefits [10–11].

Intercropping, the practice of growing two or more crops in proximity, is advantageous due to the differences in ecological characteristics and growth of the intercropped varieties. This can establish a composite population, producing complementary effects and increasing yield and economic incomes per unit area [12]. Furthermore, intercropping can improve soil fertility, alleviate disease and insect harm, and inhibit the growth of weeds [13–16]. Maize, as a principal crop of the northeast FPA, is an important food and forage crop. Its grain is an important fodder with high energy, known as "queen feed" [17]. Alfalfa is a

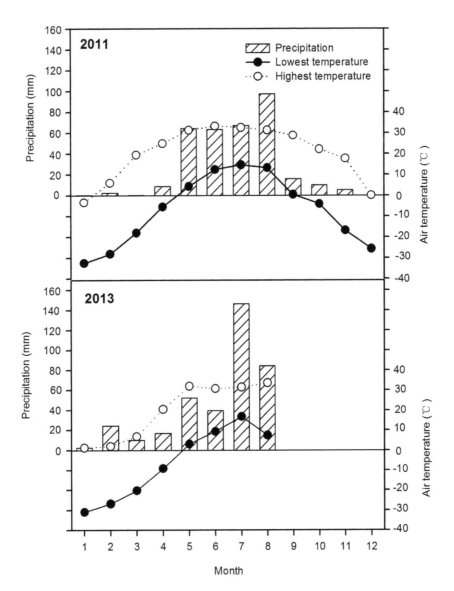

Figure 1. Monthly precipitation (bar) and air temperature (curve) of the experimental site in 2011 and 2013.

leguminous forb that has been prioritized in the development of the PFA of NEC; suitable because of its high yield, rich content and high quality of protein, abundant vitamins and minerals, good palatability and high digestibility [18]. Moreover, as a perennial, alfalfa supplies soil cover throughout the year; providing wind resistance, fixing soil and improving the environment of the planting area. Its large root system can significantly improve soil fertility and physico-chemical properties, leading to the win-win relationship between utilization and conservation [19]. Therefore, it is predictable that intercropping alfalfa with maize can not only guarantee regional food security and meet the nutritional requirements of forage industry, but also provide eco-environmental protections, and it is a promising cropping pattern in the future development of this region.

In China, maize has been traditionally cultivated in even rows. More recently, in attempts to improve maize yield agricultural scientists have experimented with alternative cultivation strategies. Indeed, a study has shown that in the main agricultural region of Jilin Province, planting maize in alternating wide and narrow rows can achieve up to 10% improved yield over the conventional

pattern [20]. Previous intercropping studies focused mainly on food crop combinations [21–22], but few studies have investigated the possibility of intercropping alfalfa with maize, a strategy combining annual food crop with perennial forage crop together. Those studies that have addressed alfalfa/maize intercropping focused on improving group yield, forage quality, soil fertility and the environment [23–24]. To our knowledge, no study has investigated whether intercropping alfalfa with maize in alternating wide and narrow rows could provide sustained high yield and economic incomes while taking investment and environmental factors into account, and whether there are some constraints limiting its popularization such as crop management and the acceptability of local farmers. Hence, a field experiment was conducted to explore the following questions: (1) Compared to an even row approach, does an alternating wide and narrow row planting approach in maize improve yield in the FPA of NEC? (2) Can intercropping alfalfa with maize provide sustained high yield and economic incomes? Which intercropping mode is best, considering farmer incomes and land management?

Materials and Methods

Experimental site

The study was conducted between 2007 and 2013 at the Grassland Ecosystem Field Station of the Northeast Normal University at Songnen Grassland (123° 44′ E and 44° 40′ N, 137.8–144.8 masl), a typical FPA of NEC. This area is characterized by a semi-arid and temperate continental monsoon climate with a mean annual temperature of 4.6–6.4°C, an annual accumulated temperature ($\geq 10°C$) of 2546–3375°C, mean annual precipitation of 300–400 mm (86% of precipitation occurring from May to September) and mean annual evaporation of 1500–2000 mm. The frost-free period lasts approximately 140 days, from the end of April to early October. The two experimental years contrasted each other in terms of precipitation. In 2011, the annual total precipitation was 335 mm and mostly occurred in the growing season (308 mm), whereas the year 2013 was a year with a higher precipitation amount (376 mm from January to August), better seasonal distribution and pronounced peak in July. Air temperature in 2011 and 2013 showed a similar dynamic with the maximum and minimum air temperatures of 33°C and −32°C, respectively (Figure 1). The soil type at the site is light chernozem with deep soil layers. The plough layer consists of organic C (17.24 ± 1.76 g kg^{-1}), total N (0.98 ± 0.15 g kg^{-1}), rapidly available P (5.88 ± 0.65 mg kg^{-1}) and rapidly available K (140.70 ± 11.75 mg kg^{-1}), with an initial soil pH of 7.46 ± 0.04.

Experiment materials and design

Alfalfa variety *Medicago sativa* L. cv. Dongmu No. 1 was used throughout this study. This variety was bred by Northeast Normal University to adapt readily to drought and cold and is now the principally cultivated variety in the study area. Alfalfa generally turns green in mid-April, continues to grow until end of October and can be harvested three times per year. The hay yield can reach up to 7500–10,000 kg ha^{-1} in the second year under the rainfed condition. *Zea mays* L. cv. Zhengdan 958 was chosen as the maize test variety. This variety matures at about 128 days and is widely cultivated by local farmers due to the high and stable yield and a great resistance to lodging and disease. The differences of alfalfa and maize in growth dynamics (Figure S4) make them easy to form temporal and spatial complementarity and promote the efficient utilization of light, water and nutrients.

At the beginning of the experiment, the whole field was fully ploughed to ensure uniform soil conditions. The experiment was conducted in a completely randomized block design with four blocks each containing five cropping patterns: maize monoculture in even (65 cm) rows (MME); maize monoculture in alternating wide (90 cm) and narrow rows (40 cm) (MMW); alfalfa monoculture in even (30 cm) rows (MA); MMW intercropped with alfalfa, with one row of alfalfa in the wide rows (23.1% alfalfa in intercropping area) (IMA1); MMW intercropped with alfalfa, with two rows of alfalfa in the wide rows (46.2% alfalfa in intercropping area) (IMA2). In the maize planting patterns, every three rows of maize were defined as one belt (1.3 m wide) with three belts in each treatment. Each plot had an area of 46.8 m^2 (3.9 m × 12 m), with 50 cm spacing between each plot and 1 m separating each block.

To establish the intercropping system, alfalfa was sown in early July 2007 at a seeding rate of 15 kg ha^{-1} and had been allowed to grow for 4 years before data collection in 2011. Every year maize was sown in early May with 26 cm separating each plant, and was irrigated with 75 mm before its sowing to ensure good germination. In all planting patterns, 135 kg P ha^{-1} and 90 kg K ha^{-1} were applied for alfalfa, and 225 kg N ha^{-1}, 120 kg P ha^{-1} and 60 kg K ha^{-1} were applied for maize. The commercial fertilizer used were: nitrogen, urea (46% nitrogen content); phosphate, diammonium phosphate (46% phosphorus content, 18% nitrogen content); potash, potassium chloride (60% potassium content). All fertilizers required by alfalfa were spread in the soil at the time of sowing. For maize, all of the phosphate and potash, and half of the nitrogen fertilizers were spread in the soil at the time of sowing and the remaining nitrogen fertilizers applied during the big flare opening period. Weeds were regularly controlled using a hand hoe, and pest and disease of alfalfa or maize were separately controlled timely with the idea of minimizing the pesticide application effects on the non-target crop.

Data collection

The data was collected from 2011. Unfortunately, there was considerably small snowfall and the air temperature was relatively high in the winter of 2011 (Figure 1), which weakened alfalfa resistance to cold and freezing [25]. In the early spring of 2012, the sprout of alfalfa was promoted due to the continuously high air temperature from 21th to 30th in March with the maximum 16.8°C, whereas the air temperature dramatically decreased from 31th March to 6th in April with the minimum −9.0°C (Figure S1). This unexpected cold snap made a serious freezing injury to the sprouting alfalfa due to its lowest cold resistence at that time [26]. Consequently, alfalfa achieved a low turning green rate. In order to ensure the sustainable production of alfalfa, we stopped data collection in 2012 and restarted sampling in 2013 when alfalfa turned a good recovery in its growth and development.

Light intensity was determined using a ST-80C illuminometer (Photoelectric Instrument Factory of Beijing Normal University, China) in 2011. Four layers for maize or alfalfa were selected in each treatment: (1) the reference layer, above the canopy; (2) the bottom layer, close to the soil surface; (3) the intermediate layer, at the point of 1/2 plant height and (4) the top layer, 10 cm below the top of the plant. Based on the light intensity, light transmission was calculated for each of the layers. Maize and alfalfa leaf area index (LAI) was measured using a LAI-2000 plant canopy analyzer (LI-COR, Inc., USA) in 2011 and 2013. From the first flowering stage of alfalfa, both light intensity and LAI were tested from 10:00–11:00 on a sunny day. The measurement was repeated every 15 days at three different positions within each testing belt.

Using time domain reflectometry (TDR 100, Campbell Scientific Inc., Logan, Utah, USA), soil water content (SWC) at depth of 0–20 cm was measured four times for all the treatments in each year (2011: 10th June, 24th July, 10th September and 4th October; 2013: 6th June, 12th July, 26th August and 30th September). The measuring time was corresponding to different developing stages of crop, that was the first, second and third flowering stage of alfalfa and the maturity stage of maize, respectively. Specifically, SWC was tested at ten different representative positions between the alfalfa or maize rows in the monoculture treatments for each plot. As to intercropping treatments, SWC was measured between the alfalfa rows (for IMA2), between the alfalfa and maize rows and between the maize rows with ten different representative positions in each belt of plots and then all measured data were averaged as the SWC condition of the intercropping system.

The final harvest of maize was taken in early October (2011: 5th October; 2013: 1th October). In each planting pattern, the second belt was selected for grain yield determination. Fresh weight was recorded before maize grains were oven-dried at 65°C to a constant weight and dry weight recorded. Water content and grain yield were calculated based on dry weight. Alfalfa was cut three times (2011: 11th June, 25th July and 11th September; 2013: 7th

June, 13th July and 27th August). For each alfalfa cut, fresh weight and dry weight were determined, and water content and hay yield calculated.

Data calculations

Light transmission.   Light transmission (LT) was calculated using equation (1);

$$LT_m = \frac{LI_m}{LI_a} \qquad (1)$$

where m is the canopy layer (including the top, intermediate and bottom layers) of either alfalfa or maize, LT_m is the light transmission at layer m, LI_m is the light intensity at layer m, and LI_a is the light intensity above the canopy [27].

Output value per unit area.   The output value per unit area (OVPUA) was calculated according to equation (2);

$$OVPUA = P_m \times Y_m + P_a \times Y_a \qquad (2)$$

where for each planting pattern, P_m and P_a denote the price of maize grain and alfalfa hay respectively. The price expressed in USD is based on the exchange rate of 630 ¥ 100 USD^{-1} in 2011 and 613 ¥ 100 USD^{-1} in 2013, thus P_m and P_a are respectively 361.90 USD t^{-1} and 380.95 USD t^{-1} in 2011 and 342.58 USD t^{-1} and 358.89 USD t^{-1} in 2013 based on local market values; Y_m and Y_a denote the yield of maize grain and alfalfa hay respectively [28].

Land equivalent ratio.   Land equivalent ratio (LER) is an index that adopts yield as a comparison parameter to evaluate land use efficiency of different cultivated patterns relative to monoculture. However, the value of LER is not necessarily related to yield. The equation is defined as follows;

$$LER = (Y_{ia}/Y_{ma}) + (Y_{im}/Y_{mm}) \qquad (3)$$

where Y_{ia} and Y_{im} are the respective yields of alfalfa and maize in the total intercropped area, and Y_{ma} and Y_{mm} are the yields of monocultured alfalfa and maize. An LER greater than 1.0 reveals an intercropping advantage and the favors of intercropping on crops growth and yield, while an LER less than 1.0 indicates an intercropping disadvantage and the negative affections of inter-cropping on crops growth and yield [29–31].

Aggressivity.   Aggressivity (A_{ac}) measures the relative re-source competitiveness of two intercropped species;

$$A_{ac} = Y_{ia}/(Y_{ma} \times P_a) - Y_{im}/(Y_{mm} \times P_c) \qquad (4)$$

where A_{ac} is the aggressivity of alfalfa relative to maize in the intercropping system, P_a and P_c are the intercropping area proportions occupied by alfalfa and maize respectively, while the meanings of other symbols are the same as those used for the LER equation. If A_{ac} is greater than 0, the competitive ability of alfalfa exceeds that of maize in intercropping; otherwise, maize has greater competitiveness [12,23].

Statistical analysis

Normal distribution and homogeneous variances were tested for all the data with Shapiro-Wilk test [32] using SPSS 17.0 software (SPSS Inc., Chicago, IL, USA), and light transmission was biquadrate after reciprocal transformed to achieve normal distribution. One-way ANOVA was performed to examine the effects of cropping patterns on light transmission and soil water content (SWC). Repeated measures ANOVA in a general linear model (GLM) were conducted to assess the effects of planting modes on LAI, yield and output value per unit area (OVPUA), with year and alfalfa flowering stage as the repeated measures. The results were reported using the Greenhouse-Geisser correction when Mauchly's test of sphericity was violated. If the interaction between factors was significant, one-way ANOVA was conducted to evaluate the effects of cropping pattern or alfalfa flowering stage and significant differences of means were compared with Duncan's multiple-compare range test; while the effects of year were tested by independent-samples t test. Significant level was set at $P < 0.05$.

Results

Maize light transmission

With the exception of the first flowering stage of alfalfa (early June), no significant difference was found in light transmission at the top of maize among treatments (Figure 2A). At the stage of the first flowering of alfalfa, light transmission of monoculture maize (MME and MMW) was significantly higher than that of intercropped maize (IMA1 and IMA2) ($P < 0.0001$), whereas there was no significant difference between monoculture modes or between intercropping modes. This can be accounted for by the fact that during flowering, alfalfa grew taller than maize, which had an overshadowing effect. On the whole, however, no significant difference was observed for average light transmission at the top of maize, regardless of cropping pattern (Figure 2A inset).

During continuous growth of maize, there was a tendency for a reduction in both intermediate and bottom light transmission for all treatments. This was particularly evident before maize entered the big flare opening stage (24th July), with little change in light transmission occurring after this point (Figure 2B and C). In the first flowering stage of alfalfa, both intermediate and bottom light transmission of monoculture maize (MME and MMW) were higher than for intercropped maize (IMA1 and IMA2); however, the opposite pattern was observed in the resting period, when light transmission of IMA1 and IMA2 maize was higher than monoculture maize (Figure 2B and C). In addition, there were significant differences for intermediate and bottom light transmis-sion between intercropping and monoculture maize at the second alfalfa flowering stage (24th July) (intermediate: $P < 0.0001$; bottom: $P = 0.003$) and after the second cutting of alfalfa (8th August) (intermediate: $P = 0.002$; bottom: $P = 0.005$). Significant differences were also found for intermediate and bottom light transmission between intercropped maize (IMA1 and IMA2) at the second flowering stage of alfalfa, but differences between monocultures of maize were never significant. However, for maize intermediate or bottom canopy layers, no significant difference in average light transmission was observed, regardless of planting strategy (Figure 2A inset).

Alfalfa light transmission

Throughout the alfalfa growing season, there was no evident variance in top light transmission in any treatment, whereas the intermediate and bottom light transmission dynamics of all treatments showed a bimodal curve (Figure 2D, E and F), which can be attributed to alfalfa being cut twice (11th June and 25th July). Before cutting, alfalfa was flowering and therefore the canopy had high closure and lower light transmission. Upon

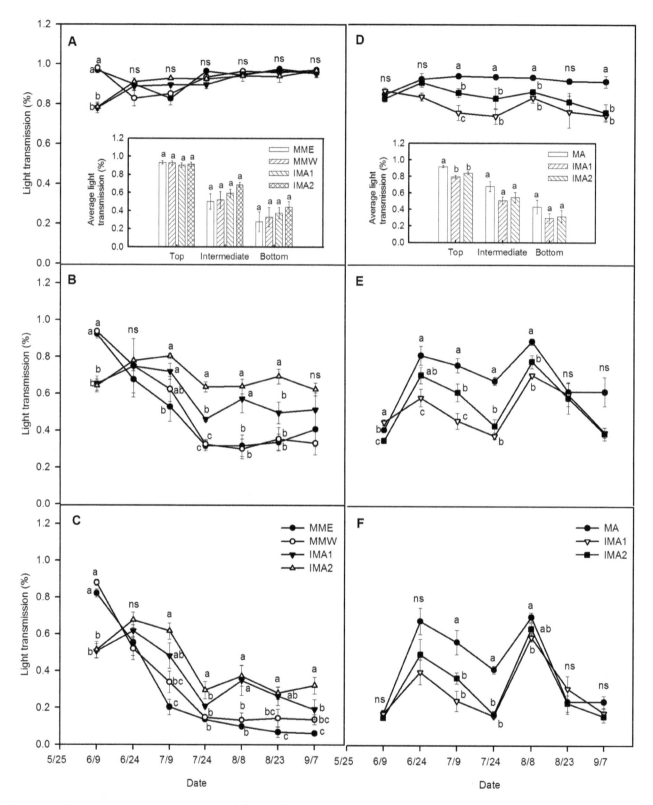

Figure 2. Light transmission dynamics of maize and alfalfa at different layers under monoculture and intercropping. The inset figures show the average light transmission of maize and alfalfa in different planting patterns during the vegetation period. A = top light transmission of maize, B = intermediate light transmission of maize, C = bottom light transmission of maize, D = top light transmission of alfalfa, E = intermediate light transmission of alfalfa, F = bottom light transmission of alfalfa. MME = monoculture maize in even rows, MMW = monoculture maize in alternating wide and narrow rows, MA = alfalfa monoculture, IMA1 = maize intercropped with one row of alfalfa in the wide rows, IMA2 = maize intercropped with two rows of alfalfa in the wide rows. Different letters for the same date indicate significant difference at $P < 0.05$ probability level, and ns represents no difference between treatments. Values = means ± SE.

cutting, light transmission increased. Subsequent alfalfa regrowth produced new canopy closure and decreased light transmission.

With the exception of the first, second and sixth measuring times, the top layer light transmission of monoculture alfalfa (MA) was significantly higher than that of intercropped alfalfa at all recorded time points (third: $P = 0.003$; forth: $P < 0.0001$; fifth: $P = 0.004$; seventh: $P = 0.010$). The season average light transmission of MA was significantly higher than that of intercropped alfalfa. There was no significant difference between the two intercropping modes (Figure 2D and inset). In the intermediate and bottom layers, light transmission showed complex patterns with time. In the intermediate layer, light transmission of MA was significantly higher than in the two intercropping models at the third, fourth and fifth measuring times (third: $P = 0.003$; forth: $P < 0.0001$; fifth: $P = 0.003$). With the exception for the first and third measuring times, there was no significant difference between the light transmissions of the two intercropping modes (Figure 2E). In the bottom layer, for the third and fourth measuring times, light transmission of MA was significantly higher than the two intercropping modes (third: $P = 0.006$; forth: $P < 0.0001$) and light transmission of the two intercropping modes showed no significant difference (Figure 2F). For intermediate and bottom layers, there were no significant differences in the season average light transmission among treatments (Figure 2D inset).

Soil water content (SWC)

SWC of all cropping patterns displayed a strong seasonal dynamic, with a peak in July and August (Figure 3). Irrespective of the growing stage, the difference of SWC between MME and MMW was not significant in both 2011 and 2013. Compared to monoculture, intercropping significantly reduced the SWC, and it was more evident in 2013 than 2011 (Figure 3). In 2011, with the exception of 24th July and 4th October, the SWC of IMA1 and IMA2 was significantly decreased compared to MMW, but with no significant difference compared with MA (both: $P < 0.0001$) (Figure 3A); while in 2013, the SWC of IMA1 and IMA2 was significantly lower than that of MMW as well as MA except for 12th July (all: $P < 0.0001$) (Figure 3B). The differences between treatments were also reflected by seasonal average SWC: the values of IMA1 and IMA2 were significantly lower than that of MMW and MME in both 2011 and 2013, while there was no significant difference between IMA1 and IMA2 in both years. For the MA treatment, the seasonal average SWC was significantly lower than that of MMW and MME in 2011, while there was no significant difference among treatments MMW, MME and MA in 2013 (Figure 3 3A inset and 3B inset).

Leaf area index (LAI)

There was no significant difference in maize LAI between MME and MMW (Figure 4). Compared to MMW, the LAI of intercropped maize (IMA1 and IMA2) was significantly reduced ($P < 0.0001$), and no significant difference was observed between the two intercropping patterns (Table 1; Figure 4). Regarding alfalfa LAI, the values of IMA1 and IMA2 were significantly higher than that of MA ($P = 0.009$), but with no significant difference between IMA1 and IMA2 (Table 1; Figure 5A). Meanwhile, alfalfa LAI was significantly affected by its flowering stage ($F = 32.648, P < 0.0001$), LAI in the first flowering stage was significantly higher than that in the second and third flowering stages ($P < 0.0001$) (Figure 5B).

Yield and output value per unit area (OVPUA)

Although both grain yield and OVPUA of MMW were higher (6.8% in 2011 and 6.5% in 2013) than the corresponding values of

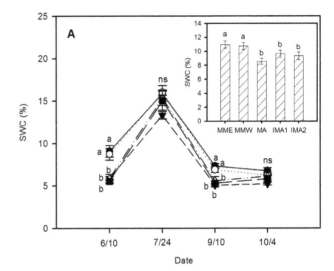

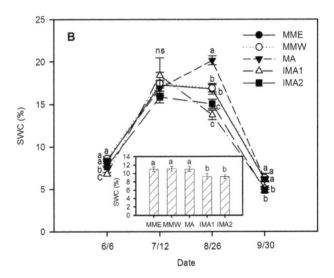

Figure 3. Soil water content comparisons of different cropping patterns in 2011 (A) and 2013 (B). The inset figures show the average soil water content in different planting patterns during the vegetation period. SWC = soil water content. The other symbols are the same as for Figure 2.

MME, the difference between the two cultivated patterns was not statistically significant (Table 2). Compared to monoculture, alfalfa hay yield in the intercropping treatments increased significantly, while maize grain yield in the same treatments was reduced dramatically in both years (all: $P < 0.0001$). The corresponding parameters in the IMA1 cropping pattern were altered in a greater extent than that in IMA2. Additionally, maize grain yield of IMA1 in 2013 was significantly higher than that in 2011 ($P = 0.028$) (Table 1 and 2).

The comprehensive benefits for total yield and OVPUA were significantly affected by the interaction between cropping pattern and year (Table 1). In 2011, both total yield and OVPUA of IMA1 and IMA2 were significantly enhanced compared to MA (both: $P < 0.0001$), while no significant increase was found compared to MMW. In 2013, both total yield and OVPUA of the two intercropping patterns were significantly higher than that of MA as well as MMW (both: $P < 0.0001$). In terms of year effects, both total yield and OVPUA of MA in 2013 was significantly

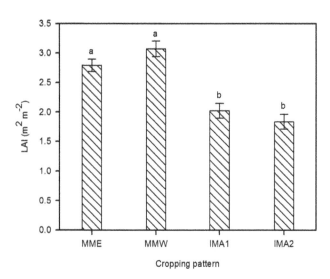

Figure 4. Leaf area index comparisons of maize at the harvest stage under monoculture and intercropping. Significant differences between different cropping patterns are indicated by lower case letters ($P < 0.05$). The other symbols are the same as for Figure 2.

higher than that in 2011 (total yield: $P = 0.038$, OVPUA: $P < 0.0001$), and there was also a significant increase in the total yield of IMA1 in 2013, as compared to 2011 ($P < 0.0001$) (Table 2).

Furthermore, land equivalent ratios (LERs) of IMA1 and IMA2 were 1.27 and 1.23 in 2011 and 1.12 and 1.08 in 2013, respectively, demonstrating that both intercropping strategies were advantageous, and that the IMA1 pattern was superior to IMA2. The calculated aggressivity (A_{ac}) values for IMA1 and IMA2 were 2.71 and 1.40 in 2011 and 1.31 and 0.65 in 2013, respectively, demonstrating that the resource competitiveness of alfalfa was greater than that of maize in the two intercropping systems (Table 2).

Discussion

Maize monoculture in wide and narrow rows vs. in even rows

Liu et al. [20] reported that in the main agricultural region of Jilin Province, compared to even rows, maize planted in alternating wide and narrow rows increased group light transmission, photosynthetic potential, leaf area and grain yield by more than 10%. The results presented here are in agreement with these findings; both grain yield and output value of MMW were enhanced by 6.8% in 2011 and 6.5% in 2013 relative to MME (Table 2). This was mainly attributable to the improved spatial structure of the group, which increased light transmission (Figure 2), improved maize growth conditions and promoted the formation of edge effect [33]. At harvest time, maize achieved a greater LAI and dry matter accumulation (Figure 4 and S3). Therefore, in the PFA of NEC, maize cropped in alternating wide and narrow rows also had a superior economic benefit compared to maize planted in even rows. We recommend that this approach should be popularized and put into widespread use.

The advantages of intercropping alfalfa with maize

Intercropping plays an important part in traditionally intensive agriculture and has captured attention for its efficient utilization of limited resources [21,34]. Among numerous agricultural intercropping modes, legume/cereal intercropping has been most

Table 1. Results of repeated measures ANOVA on maize leaf area index (LAI) and yield, alfalfa yield and comprehensive benefit analysis of total yield and output value per unit area (OVPUA), with cropping pattern (CP) as the independent variable and year (Y) as the repeated measure.

Factors	Maize			Alfalfa		Comprehensive benefit analysis		
	Df	LAI	Yield	Df	Yield	Df	Total yield	OVPUA
CP	3	26.691*	49.157**	2	430.879**	4	16.343**	15.063**
Y	1	0.349 ns	21.174**	1	0.248 ns	1	32.515**	8.052*
CP × Y	3	0.846 ns	9.024**	2	3.985 ns	4	10.850**	10.699**

Df = degrees of freedom, ns = no significant difference, * $p < 0.05$, ** $p < 0.01$

successful, with a long history and several apparent advantages [12,35–36]. Common patterns include intercropping peanuts [37], soybeans [38] or cowpeas [39] with maize. However, few studies have investigated the potential advantages of intercropping alfalfa with maize [24,40–41], and there has been no systematic study attempting to identify whether it can provide sustained high yield and economic incomes while taking investments and environmental factors into consideration or analyze the constraints limiting its popularization. Furthermore, previous intercropping studies have planted their maize in even rows [23,28]. To our knowledge, no field data are available for intercropping alfalfa with maize in alternating wide and narrow rows.

Owing to differences in the traits, growth and development characteristics of alfalfa and maize (Figure S4), the alfalfa/maize intercropping system resulted in temporal and spatial complementarity, which optimized resource utilization and promoted intercropping advantages [27]. The details are presented as follows:

First, as a perennial forb, alfalfa turns green in early spring, grows fast and mainly covers the soil by late April, whereas maize is sown in early or mid-May. We found that when alfalfa was entering the first flowering stage, maize was still a seedling with canopy lower than that of alfalfa. Thus alfalfa and maize occupy complementary spatial and temporal niches, resulting in complementarity in light interception (Figure 2) and facilitating the circulation and diffusion of air (especially CO_2) in the composite population. This result is consistent with many other studies [42–43]. The increase of light transmission of intercropped alfalfa produces an edge effect and enhances the LAI of alfalfa (Figure 5), thereby significantly improving the hay yield of alfalfa in the first cut, especially when using the IMA1 strategy (Figure S2). Furthermore, the hay yield of alfalfa in the first cut accounted for more than 50% of the total hay yield [19]; therefore, the increase in hay yield in the first cut made a great contribution to total hay yield (Figure S2).

Second, between alfalfa first cut and third flowering stage, maize achieves a period of peak growth, became taller than alfalfa; occupying a more advantageous position in the intercropping system so as to make full use of light, heat and other resources. This effect is particularly evident immediately after the second cutting of alfalfa, when intermediate and bottom light transmission of intercropped maize is dramatically enhanced (Figure 2B and C); producing better growing conditions for intercropped maize and accelerating dry matter accumulation (Figure S3). As a result, intercropped maize has an opportunity for recovery [44], partly compensating for the reduction in the maximum dry matter and grain yield caused by competition with alfalfa. In addition, many studies have demonstrated that there is nitrogen transfer from legumes to cereals in intercropping systems [45–46]. Especially, alfalfa, a perennial leguminous forb, has a strong ability to fix nitrogen, and its fixed and transferred nitrogen contributed as much as 30% to the total N accumulated in the associated grass [47–48]. Therefore, we speculate that there would be nitrogen transfer from alfalfa to maize in the alfalfa/maize intercropping system, which can enhance soil nitrogen availability, improve soil physical and chemical properties and be responsible for the improved growth and development of intercropped maize. It should be noted that the relatively favorable precipitation and allocation in the growing season in 2011 and 2013 was also beneficial in alleviating the growth inhibition of intercropped maize caused by water deficit (Figure 1). Consequently, the combined effects of these factors narrowed the grain yield gap between intercropped maize and monoculture maize and promoted the formation of intercropping advantages.

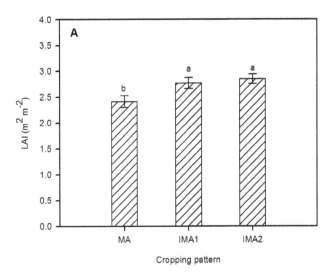

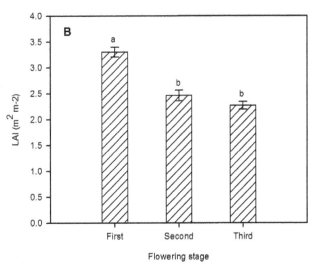

Figure 5. Leaf area index comparisons of alfalfa under monoculture and intercropping (A) and at different flowering stages (B). Significant differences between different alfalfa flowering stages are indicated by lower case letters (P <0.05). The other symbols are the same as for Figure 2 and 4.

Third, alfalfa can continue to grow for around one month after harvesting of maize (Figure S4), providing enough time for alfalfa roots to store adequate carbohydrates to overwinter [23]. Thus, the temporal and spatial differentiation in alfalfa/maize intercropping system avoided wasting light, heat, water, air and other natural resources, prolonged photosynthetic effective time, increased photosynthetic effective area, and ultimately enhanced group yield (Table 2).

Although intercropping alfalfa with maize occupies largely complementary aboveground ecological niches, there is belowground competition. Alfalfa has deep strong roots, that can penetrate>10 m into the soil but mainly proliferate at a soil depth of 0–60 cm. The roots of maize are shallow and mostly distributed at a soil depth of 0–40 cm [49]. Despite some differentiation in root distribution, alfalfa and maize will compete in the shallow soil layers where most water and nutrients are distributed. Moreover, alfalfa has a higher evaporation coefficient and requires more water than maize [18,50]. Therefore, when water is limited, intercropped alfalfa competes with maize for water in the shallow

Table 2. Comparisons of total yield and OVPUA of different planting patterns.

Treatment	Yield (t ha^{-1})			OVPUA (USD ha^{-1})			LER	A$_{ac}$
	Maize	Alfalfa	Total	Maize	Alfalfa	Total		
2011								
MME	11.00±0.33a	-	11.00±0.33a	3982.31±120.85	-	3982.31±120.85a	-	-
MMW	11.75±0.33a	-	11.75±0.33a	4253.31±120.38	-	4253.31±120.38a	-	-
MA	-	7.85±0.31a	7.85±0.31b	-	2991.92±118.56	2991.92±118.56b	-	-
IMA1	5.87±0.34 (7.63±0.44b)	6.08±0.53 (26.35±2.29b)	11.95 ± 0.30a	2124.17±122.92	2316.91±201.23	4441.08±116.95a	1.27	2.70
IMA2	3.71±0.64 (6.89 ± 1.19b)	7.21±0.34 (15.63±0.74c)	10.92±0.67a	1342.17±231.66	2747.93±129.75	4090.10±246.77a	1.23	1.40
2013								
MME	10.68±0.29a	-	10.68±0.29a	3656.77±100.28	-	3656.77±100.28a	-	-
MMW	11.37±0.23a	-	11.37±0.23ab	3894.69±79.08	-	3894.69±79.08a	-	-
MA	-	11.76±0.05a	11.76 ± 0.05b*	-	4221.72±18.90	4221.72±18.90b*	-	-
IMA1	7.18±0.18 (9.34±0.23b*)	5.79±0.27 (25.06±1.17b)	12.97±0.19c*	2460.38 ± 61.59	2075.89±97.10	4536.26±67.64c	1.12	1.31
IMA2	4.77±0.15 (8.87 ± 0.27b)	7.79±0.11 (16.88±0.23c)	12.56 ± 0.05c	1633.92±50.89	2795.73±38.22	4429.65±14.12c	1.08	0.65

MME = monoculture maize in even rows, MMW = monoculture maize in alternating wide and narrow rows, MA = alfalfa monoculture, IMA1 = maize intercropped with one row of alfalfa in the wide rows, IMA2 = maize intercropped with two rows of alfalfa in the wide rows. Values in the parentheses are yields based on the whole of the intercropping area, including the areas occupied by both maize and alfalfa, and are equal to the yields of maize or alfalfa divided by their respective area proportion. The intercropping area proportions of maize and alfalfa were respectively 76.9% and 23.1% in the IMA1 treatment, while the intercropping area ratios occupied by alfalfa and maize were 53.8% and 46.2% in the IMA2 treatment. Different letters in the same column following different cropping patterns, and * denotes significant difference between years (P <0.05). Value = mean ± S.

layers to meet its growth and development besides utilizing deep groundwater [28], which contributed to the promotion of LAI and hay yield of intercropped alfalfa at the flowering stage (Figure 5 and S1; Table 2).

However, the strong competition of alfalfa with maize significantly reduced SWC of the intercropping system, especially at the first flowering stage of alfalfa and between its second and third flowering stage when maize was in the period of seedling and peak growth with high water demand (Figure 3). The water stress restrained the growth and development of intercropped maize, particularly in the early growth stage (first 80 days post-emergence) (Figure S3), and significantly declined the LAI and grain yield at harvest time (Figure 4; Table 2).

The competitiveness for resources of intercropped species differs due to competition and complementary in the intercropping system. Generally, cereals are considered to have a competitive advantage over legumes and have a decisive influence on the total yield in annual legume/cereal intercropping systems [31,51–52]. However, Zhang et al. [23] studied an alfalfa/maize intercropping system and found that alfalfa had much greater competitiveness than maize, and that alfalfa yield dramatically influenced the total yield of the intercropping system. Our results are consistent with this study. In our two intercropping modes (IMA1 and IMA2), the competitiveness for resources of alfalfa was much stronger than that of maize, and alfalfa hay yield was significantly improved whereas maize grain yield was significantly reduced. In addition, any reduction in maize yield was more than offset by increased alfalfa yield (Table 2). Thus both intercropping systems have consistently accomplished a successful tradeoff between complementarity and competition, enhanced land use capability and achieved a significant yield and output value advantage over monoculture, except that in 2011 the two intercropping patterns achieved a similar yield and OVPUA compared to MMW (Table 2).

When investment is considered, both IMA1 and IMA2 can improve economic benefits in a great extent relative to MMW in both years. This is because alfalfa is a perennial legume that does not require ploughing, sowing and fertilization after the establishment. Compared to maize, planting alfalfa can reduce costs by at least 584 USD ha^{-1} (based on the fertilizer level and seeding rate in our study and in which fertilizer, seed and farming labor savings were 376, 113 and 95 USD ha^{-1}). In this way, relative to MMW, the IMA1 mode not only increased output value (+187.77 USD ha^{-1} in 2011 and +641.57 USD ha^{-1} in 2013), but also reduced investment (−135 USD ha^{-1}), thus enhanced economic benefits by 322.77 USD ha^{-1} and 776.57 USD ha^{-1} in 2011 and 2013, respectively. Similarly, the final economic benefit of IMA2 was also enhanced by 106.79 USD ha^{-1} in 2011 and 804.96 USD ha^{-1} in 2013, as compared to MMW.

The higher total yield and output value of alfalfa/maize intercropping in 2013 than 2011 (Table 2) could be attributed to the following two aspects. First, the favorable higher amount and better distribution of precipitation in 2013, especially in the growing season, improved the SWC (Figure 3). Thus water competition between alfalfa and maize was minimized, nutrient availability to crops was enhanced [53] and crops obtained a superior condition for growth and development. Meanwhile, the improved growth conditions made alfalfa reach the flowering stages earlier, especially for the second and third times. Thus the co-growth time of alfalfa and maize with strong competition was shortened, and maize obtained a more unconstrained condition to utilize resource and grow. Second, with the intercropping year's increase, alfalfa roots descend into deeper soil layers [49] and could extract water at greater depths when the upper soil horizons

get drier, which would improve the growth conditions of intercropped maize; at the same time, it has been proved that alfalfa can fix and transfer more nitrogen to the associated grass as time goes on [54], thus, it is likely that alfalfa could transfer more nitrogen to intercropped maize and improve its growth and development.

Therefore, the comprehensive benefit of alfalfa intercropped with maize was not completely stable and had a certain variation with the changes of rainfall and planting years. It should be noted that the economic benefit of intercropping is also strongly correlated with local market prices of alfalfa and maize [28]. In addition, environmental stress (e.g. pest, disease and the freezing injury to alfalfa in early spring) should also be taken into account to evaluate the valorization of the comprehensive benefits of the intercropping system, and it could severely reduce crops productivity and the economic returns. However, compared to monoculture, intercropping enhanced crops resistance to stress and reduced the management costs and economic losses per unit area [14–16]. Overall, both of the two intercropping strategies have potential to improve economic incomes and are superior to monoculture. This result is in accordance with many other studies and manifests the advantages of intercropping [16,28,55]. Moreover, the intercropping mode of IMA1 was superior to that of IMA2, regardless of land use efficiency, total yield or output value.

Nevertheless, the maize/alfalfa intercropping system still remains relatively unpopular for a number of reasons. First, because of the distinct cultural requirements of each crop, it is difficult to manage the two crops together with existing farm machinery. In this study, all crop management was performed using manual labor. Sowing, fertilizer application, weed control and harvest were conducted separately due to growth and management differences between the two crops. It was also required to manage pest and disease control independently for each crop - when either crop suffered from pests or disease, appropriate pesticides and safeguards were selected to minimize influence on the other crop. It is clear that managing the intercropping system is more complicated and inconvenient than management of a monoculture system. In order to simplify management of this intercropping system, it will be necessary to integrate a multidisciplinary body of knowledge to develop efficient machinery and cautiously advocate the popularization of this intercropping system on a moderate scale to realize yield and economic advantages [56]. Secondly, although the 383,000 km^2 FPA and NEC areas [1] are well suited for the popularization of this intercropping system, local farmers have traditionally utilized a single cropping system and are likely unaware of intercropping systems, especially an intercropping system mixing a food crop and forage legume. Under current conditions, these farmers are more likely to select traditional planting strategies with simple and convenient management schedules and are little concerned about sustainable food production, animal husbandry or eco-environmental security (local investigation and personal communications). Therefore, the dissemination of efficient intercropping technologies and expert technical guidance, as well as financial support of government will necessarily play an important role in putting alfalfa/maize intercropping into practice in the FPA of NEC [56]. Finally, further research is required to assess the long-term benefits of the composite crop population and its responses to rainfall, planting years and environmental stresses (pest, disease and freezing injury), in order to avoid agronomic risks and economic loss.

Conclusions

Based on the above analyses, we conclude that alfalfa/maize intercropping has obvious advantages in grain yield and economic incomes; it guarantees regional food security and provides superior forage. Therefore, this intercropping strategy can help maximize use of limited land resources and promote sustainable development of agriculture and animal husbandry. Moreover, nitrogen fixation and transfer from alfalfa to maize can improve soil fertility and reduce fertilizer investment [19]. Furthermore, alfalfa hay yield increases continuously throughout the first five years [18], providing sustainable economic benefits.

With rising demand for meat, egg, milk and nutrient balance, China is giving increasing importance to the development of animal husbandry and investing the forage industry [9]. Therefore, alfalfa market prices are likely to increase. In this way, planting alfalfa will not only meet the demands of the animal husbandry industry, but also promote the rapid development of local economies. In addition, the multiple-year coverage of alfalfa on the soil can alleviate wind erosion and water erosion, improving the environment of planting areas [19]. In conclusion, there are clear and significant economic, social and ecological benefits in alfalfa/maize intercropping, and maize intercropped with one row of alfalfa was identified as the optimal strategy.

Supporting Information

Figure S1 Daily variation dynamics of air temperature from February to April in 2012. T-highest = highest temperature in a day, T-lowest = lowest temperature in a day.

Figure S2 Alfalfa hay yield at three different flowering stages under monoculture and intercropping. MA = alfalfa monoculture, IMA1 = maize intercropped with one row of alfalfa in the wide rows, IMA2 = maize intercropped with two rows of alfalfa in the wide rows. Different lower case letters for the same flowering stage in one year indicate significant difference, and significant differences of alfalfa total hay yield for one year between different cropping patterns are indicated by different capital letters (P <0.05). Values = means ± SE.

Figure S3 Accumulation dynamics of maize aboveground dry matter under monoculture and intercropping. MME = monoculture maize in even rows, MMW = monoculture maize in alternating wide and narrow rows. The other symbols are the same as for Figure S2.

Figure S4 Growth dynamics of alfalfa and maize in the intercropping system.

Acknowledgments

Special thanks to Baotian Zhang for field help and Scott Diloreto for English check.

Author Contributions

Conceived and designed the experiments: ZL YG. Performed the experiments: BS CW YY YL. Analyzed the data: BS YP HY YG. Contributed reagents/materials/analysis tools: ZL. Wrote the paper: BS YG.

References

1. Zhang HX, Shao MA, Zhang XC (2004) The resuming of weak ecological environment and sustainable development in farming-pasture zone of northeastern China. J Arid Land Resour Env 18: 129–134.
2. Luo CP, Xue JY (1995) Ecologically vulnerable characteristics of the farming-pastoral zigzag zone in northern China. J Arid Land Resour Env 9: 1–7.
3. Zhou DW, Lu WX, Xia LH, Wu FZ, Li JD, et al. (1999) Grassland degradation and soil erosion in the eastern ecotone between agriculture and animal husbandry in northern China. Resour Sci 21: 57–61.
4. Zhao LP, Wang HB, Liu HQ, Wang YL, Liu SX, et al. (2006) Mechanism of fertility degradation of black soil in corn belt of Song Liao Plain. Acta Pedol Sinica 43: 79–84.
5. He LN, Liang YL, Gao J, Xiong YM, Zhou MJ, et al. (2008) The effect of continuous cropping on yield, quality and soil enzymes activities in solar green house. J Northwest A F Univ (Nat. Sci. Ed.) 36: 155–159.
6. Zhen Z, Bo WJ, Wu GL, Luo XC, Zheng YH (2012) Important effect of the organic fertilizer on soil fertility and yield of crop: a case study in Zhende organic farm, Henan, China. J Eng Stud 4: 19–25.
7. Lin JX, Wang JF, Li XY, Zhang YT, Xu JT, et al. (2011) Effects of saline and alkaline stresses in varying temperature regimes on seed germination of Leymus chinensis from the Songnen Grassland of China. Grass Forage Sci 66: 578–584.
8. Tian X, Yang YF (2009) Current situation of grassland degradation and its management options in farming-pasturing ecotone in western Jilin Province and eastern Inner Mongolia. Chinese J Ecol 28: 152–157.
9. Lu XS (2012) The opportunity, connotation and future of Chinese grassland agriculture. The Second China Grassland Agriculture Conference. pp. 7–10.
10. Ren JZ (2002) Establishment of on agro-grassland systems for grain storage—A thought on restructure of agricultural framework in Western China. Acta Prataculturae Sinica 11: 1–3.
11. Zhu TC, Li ZJ, Zhang WZ, Liang CZ, Yang HJ, et al. (2002) A preliminary report on the cereal-forage rotation system in the plain of Northeast China. Acta Prataculturae Sinica 12: 34–43.
12. Li L, Sun JH, Zhang FS, Li XL, Yang SC, et al. (2001) Wheat/maize or wheat/soybean strip intercropping I. Yield advantage and interspecific interactions on nutrients. Field Crop Res 71: 123–137.
13. Hauggaard-Nielsen H, Ambus P, Jensen ES (2001) Interspecific competition, N use and interference with weeds in pea-barley intercropping. Field Crop Res 70: 101–109.
14. Miriti JM, Kironchi G, Esilaba AO, Heng LK, Gachene CKK, et al. (2012) Yield and water use efficiencies of maize and cowpea as affected by tillage and cropping systems in semi-arid Eastern Kenya. Agr Water Manage 115: 148–155.
15. Trenbath BR (1993) Intercropping for the management of pests and diseases. Field Crops Res 34: 381–405.
16. Rusinamhodzi L, Corbeels M, Nyamangara J, Giller KE (2012) Maize-grain legume intercropping is an attractive option for ecological intensification that reduces climatic risk for smallholder farmers in central Mozambique. Field Crop Res 136: 12–22.
17. Chen YX, Zhang X, Chen J, Zhou DW (2009) The maize proper harvesting methods in ecotone between agriculture and animal husbandry in Northeast China. J Agr Mech Res 31: 113–117.
18. Wang X, Ma YX, Li J (2003) The nutrient content and main biological characteristics of alfalfa. Pratacultural Sci 10: 39–41.
19. Li ZJ, Guo JX, Zhang YS, Wu ZY (2003) The role and status of alfalfa industry in the structure adjustment of planting industry in Jilin province. J Jilin Agr Sci 28: 40–46.
20. Liu WR, Feng YC, Zheng JY, Liu FC, Zhu XL, et al. (2003) Yield and benefit analysis of maize planted in wide and narrow rows. J Maize Sci 11: 63–65.
21. Lesoing GW, Francis CA (1999) Strip intercropping effects on yield and yield components of corn, grain sorghum, and soybean. Agron J 91: 807–813.
22. Gilbert RA, Heilman JL, Juo ASR (2003) Diurnal and seasonal light transmission to cowpea in sorghum-cowpea intercrops in Mali. Agron Crop Sci 189: 21–29.
23. Zhang GG, Yang ZB, Dong ST (2011) Interspecific competitiveness affects the total biomass yield in an alfalfa and corn intercropping system. Field Crop Res 124: 66–73.
24. Wang T, Zhu B, Xia LZ (2012) Effects of contour hedgerow intercropping on nutrient losses from the sloping farmland in the Three Gorges Area, China. J Mt Sci 9: 105–114.
25. Yin XJ, Cui GW (2006) Causes and preventing techniques of alfalfa freezing injury in northern cold regions. Feed Review 4:31–33.
26. Sun QZ, Wang YQ, Hou XY (2004) Alfalfa winter survival research summary. Pratacultural Sci 21:21–25.
27. Bedoussac L, Justes E (2010) Dynamic analysis of competition and complementarity for light and N use to understand the yield and the protein content of a durum wheat-winter pea intercrop. Plant Soil 330: 37–54.
28. Smith MA, Carter PR (1989) Strip intercropping corn and alfalfa. J Prod Agric 11: 345–353.
29. Anil L, Park J, Phipps RH, Miller FA (1998) Temperate intercropping of cereals for forage: a review of the potential for growth and utilization with particular reference to the UK. Grass Forage Sci 53: 301–317.

30. Chu GX, Shen QR, Gao JL (2004) Nitrogen fixation and N transfer from peanut to rice cultivated in aerobic soil in an intercropping system and its effect on soil N fertility. Plant Soil 263: 17–27.

31. Lithourgidis AS, Viachostergios DN, Dordas CA, Damalas CA (2011) Dry matter yield, nitrogen content, and competition in pea-cereal intercropping systems. Eur J Agron 34: 287–294.

32. Shapiro SS, Wilk MB (1965) An analysis of variance test for normality (complete samples). Biometrika 52:591–611.

33. Liu AN, Liu ZG, Zhou XG, Meng ZJ, Chen JP (2005) Study on edge effect and ecological effect in system of winter wheat intercropping with cotton. J Mt Agr Biol 24: 471–476.

34. Trenbath BR (1976) Plant interactions in mixed crop communities. In: Papendick RI, Sanchez PA, Triplett GB, editors. Multiple cropping. ASSA, CSSA, and SSSA. Madison Wis. pp. 129–170.

35. Willey RW (1990) Resource use in intercropping systems. Agr Water Manage 17: 215–231.

36. Mandal BK, Das D, Saha A, Mohasin M (1996) Yield advantage of wheat (Triticum aestivum) and chickpea (Cicer arietinum) under different spatial arrangements in intercropping. Ind J Agron 41: 17–21.

37. Inal A, Gunes A, Zhang F, Cakmak I (2007) Peanut/maize intercropping induced changes in rhizosphere and nutrient concentrations in shoots. Plant Physiol Biochem 45: 350–356.

38. Prasad RB, Brook RM (2005) Effect of varying maize densities on intercropped maize and soybean in Nepal. Exp Agr 41: 365–382.

39. Ghanbari A, Dahmardeh M, Siahsar BA, Ramroudi M (2010) Effect of maize (Zea mays L.)-cowpea (Vigna unguiculata L.) intercropping on light distribution, soil temperature and soil moisture in arid environment. J Food Agr Environ 8: 102–108.

40. Liebman M, Graef RL, Nettleton D, Cambardella CA (2011) Use of legume green manures as nitrogen sources for corn production. Renew Agr Food Syst 27: 180–191.

41. Guldan SJ, May T, Martin CA, Steiner RL (1998) Yield and forage quality of interseeded legumes in a high-desert environment. J Sustain Agr 12: 85–97.

42. Tsubo M, Walker S (2002) A model of radiation interception and use by a maize-bean intercrop canopy. Agr Forest Meteorol 110: 203–215.

43. Awal MA, Koshi H, Ikeda T (2006) Radiation interception and use by maize/peanut intercrop canopy. Agr For Meteorol 139: 74–83.

44. Li L, Sun JH, Zhang FS, Li XL, Rengel Z, et al. (2001) Wheat/maize or wheat/soybean strip intercropping II. Recovery or compensation of maize and soybean after wheat harvesting. Field Crop Res 71: 173–181.

45. Shen QR, Chu GX (2004) Bi-directional nitrogen transfer in an intercropping system of peanut with rice cultivated in aerobic soil. Biol Fertil Soils 40: 81–87.

46. Li YF, Ran W, Zhang RP, Sun SB, Xu GH (2009) Facilitated legume nodulation, phosphate uptake and nitrogen transfer by arbuscular inoculation in an upland rice and mung bean intercropping system. Plant Soil 315: 285–296.

47. Ta TC, Faris MA (1987) Species variation in the fixation and transfer of nitrogen from legumes to associated grasses. Plant Soil 98: 265–274.

48. Yang SX, Yang ZZ (1992) A study on superiorities in mixed cropping of alfalfa and Siberian wildrye. Sci Agr Sin 25: 63–68.

49. Zhang GG, Zhang CY, Yang ZB, Dong ST (2013) Root distribution and N acquisition in an alfalfa and corn intercropping system. J Agr Sci 5: 128–142.

50. Sun HR (2003) Alfalfa's pre-blossoming transpiration coefficients and comparison of the water consumption coefficients between alfalfa and maize in terms of economic yield. Acta Agrestia Sinica 4: 346–349

51. Misra AK, Acharya CL, Rao AS (2006) Interspecific interaction and nutrient use in soybean/sorghum intercropping system. Agron J 98: 1097–1108.

52. Wahla IH, Ahmad R, Ehsanullah AA, Jabbar A (2009) Competitive functions of components crops in some barley based intercropping systems. Intl J Agr Biol (Pakistan) 11: 69–71.

53. Jensen JR, Bernhard RH, Hansen S, McDonagh J, MØberg JP, et al. (2003) Productivity in maize based cropping systems under various soil-water-nutrient management strategies in semi-arid Alfisol environment in East Africa. Agr Water Manage 59: 217–237.

54. Goodman PJ (1988) Nitrogen fixation transfer and turnover in upland and lowland grass-clover awards, using ^{15}N isotope dislution. Plant Soil 112: 247–254.

55. Mucheru-Muna M, Pypers P, Mugendi D, Kung'u J, Mugwe J, et al. (2009) A staggered maize-legume intercrop arrangement robustly increases crop yields and economic returns in the highlands of Central Kenya. Field Crop Res 115: 132–139.

56. Yu Y, He Y (2009) The investigation and popularization of the efficient mode of grain-economic intercropping. Shanxi J Agr Sci 5: 90–93.

14

Expression of Cry1Ab and Cry2Ab by a Polycistronic Transgene with a Self-Cleavage Peptide in Rice

Qichao Zhao[1**❾**]**, Minghong Liu**[1**❾**]**, Miaomiao Tan**[1]**, Jianhua Gao**[2]**, Zhicheng Shen**[1]*

1 State Key Laboratory of Rice Biology, Institute of Insect Sciences, Zhejiang University, Hangzhou, China, 2 College of Life Science, Shanxi Agricultural University, Taigu, China

Abstract

Insect resistance to *Bacillus thuringiensis* (Bt) crystal protein is a major threat to the long-term use of transgenic Bt crops. Gene stacking is a readily deployable strategy to delay the development of insect resistance while it may also broaden insecticidal spectrum. Here, we report the creation of transgenic rice expressing discrete Cry1Ab and Cry2Ab simultaneously from a single expression cassette using 2A self-cleaving peptides, which are autonomous elements from virus guiding the polycistronic viral gene expression in eukaryotes. The synthetic coding sequences of Cry1Ab and Cry2Ab, linked by the coding sequence of a 2A peptide from either foot and mouth disease virus or porcine teschovirus-1, regardless of order, were all expressed as discrete Cry1Ab and Cry2Ab at high levels in the transgenic rice. Insect bioassays demonstrated that the transgenic plants were highly resistant to lepidopteran pests. This study suggested that 2A peptide can be utilized to express multiple Bt genes at high levels in transgenic crops.

Editor: Mario Soberón, Instituto de Biotecnología, Universidad Nacional Autónoma de México, Mexico

Funding: This research was supported by The National Natural Science Foundation of China (31021003, http://www.nsfc.gov.cn/) received by Zc Shen. The funders had no role in study design, data collection and analysis, decision to publish, or preparation of the manuscript.

Competing Interests: The authors have declared that no competing interests exist.

* Email: zcshen@zju.edu.cn

❾ These authors contributed equally to this work.

Introduction

Since first demonstrated in transgenic tobacco in 1987, Bt crystal toxin genes have been widely utilized in transgenic crops for pest management [1,2]. However, due to the widespread application of Bt toxins, several major insect pests, including fall armyworm (*Spodoptera frugiperda*), dimondback moth (*Plutella xylostella*), maize stalk borer (*Busseola fusca*), cotton bollworm (*Helicoverpa armigera*), western corn rootworm (*Diabrotica virgifera*) and cabbage looper (*Trichoplusia ni*) have developed resistance to Bt toxins in field or greenhouse, which threatens the long-term utilization of Bt crops in the future [3–5].

Several strategies have been proposed and/or deployed to cope with the development of insect resistance to Bt toxins, including high-dose/refuge strategy, discovery of novel insecticidal genes with novel modes of actions, modification of used Bt genes and Bt gene stacking [6–8]. Bt gene stacking strategy introduces more than one Bt genes into plant and has been demonstrated to be an effective way to delay the development of insect resistance to Bt toxins [9–11]. Expression of multiple genes using conventional approaches has several potential limitations, most notably imbalanced expression among different genes and a large T-DNA size required to include multiple genes [12]. Efficient and stoichiometric expression of discrete proteins may be achieved by a polycistronic system involving self-cleaving peptides such as the 2A peptide from foot and mouth disease virus (F2A) or porcine teschovirus-1 (P2A) [12,13]. When expressed in eukaryotic system,

nascent 2A peptide can interact with the ribosome exit tunnel to dictate a stop-codon-independent termination at the final proline codon of 2A peptide [14]. Subsequently translation is reinitiated on the same proline codon. When linked by a 2A peptide coding sequence, different genes are co-expressed from a single open reading frame.

2A self-cleaving peptides have been extensively studied previously [15]. In this study, the DNA encoding F2A or P2A was used to link two potent insecticidal Bt genes, the truncated *Cry1Ab* encoding the N-terminal 648 amino acids of active Cry1Ab endotoxin (Genbank:AAG16877.1) and the full-length *Cry2Ab* encoding Cry2Ab endoxin (Genbank:AAA22342.1) with 634 amino acid residues, to generate a polycistronic gene for co-expression in transgenic rice. Analysis of the obtained transgenic rice lines revealed that discrete Cry1Ab and Cry2Ab were indeed co-expressed at a level approximately comparable to transgenic plants expressing traditional monocistronic Bt genes. Insect bioassays demonstrated that the transgenic rice generated was highly resistant to its target insects.

Materials and Methods

Rice cultivar

Elite rice (*Oryza sativa spp. Japonica*) cultivar Xiushui 134 originated from the Jiaxing Academy of Agricultural Science in Zhejiang Province was used for *Agrobacterium*-mediated transformation. The homozygous transgenic rice line KMD1, containing a synthetic truncated *Cry1Ab* gene under control of maize ubiquitin promoter [16,17], was kindly provided by Dr. Gongyin Ye from Institute of Insect Sciences, ZheJiang University.

Construction of binary vector for rice transformation

pCambia1300 (Cambia, Canberra, Australia) was used for construction of binary vectors for plant transformation. This vector was first modified by substituting the hygromycin-resistant gene expression cassette with a glyphosate-tolerant *5-enolpyruvyl-shikimate-3-phosphate synthase* (*EPSPS*) gene expression cassette. The modified vector was named as pCambia1300–GLY, and was further used to clone an expression cassette for a polycistronic gene encoding Cry1Ab and Cry2Ab linked by a 2A peptide (Fig. 1).

Four polycistronic genes of *Cry1Ab* and *Cry2Ab* linked by the coding sequence of F2A or P2A were generated. They were named as *Cry1Ab-F2A-Cry2Ab* (*Cry1-F-Cry2*), *Cry1Ab-P2A-Cry2Ab* (*Cry1-P-Cry2*), *Cry2Ab-F2A-Cry1Ab* (*Cry2-F-Cry1*), and *Cry2Ab-P2A-Cry1Ab* (*Cry2-P-Cry1*), respectively (Fig. 1). *Cry1-F-Cry2* is a fusion gene, in the order from 5' to 3', of coding sequences of Cry1Ab, F2A, and Cry2Ab (GenBank: KJ716232); *Cry1-P-Cry2* is a fusion gene of Cry1Ab, P2A, and Cry2Ab (GenBank: KJ716233); *Cry2-F-Cry1* is a fusion gene of Cry2Ab, F2A, and Cry1Ab (GenBank: KJ716234); and *Cry2-P-Cry1* is a fusion gene of Cry2Ab, P2A, and Cry1Ab (GenBank: KJ716235).

Maize ubiquitin-1 promoter (pUbi) was obtained by PCR from maize genome with primers pUbi-F (5' attaagcttagcttgcatgccta-cagtg 3', with *Hind*III restriction site underlined) and pUbi-R (5' taaggatccctctagagtcgacctgca 3', with *Bam*HI restriction site underlined). The pUbi fragment digested with *Hind*III and *Bam*HI, the polycistronic gene fragment digested with *Bam*HI and *Kpn*I and the vector 1300-GLY predigested with *Hind*III and *Kpn*I were ligated to generate plasmid p1300-Cry1-F-Cry2, p1300-Cry1-P-Cry2, p1300-Cry2-F-Cry1 and p1300-Cry2-P-Cry1, respectively (Fig. 1).

Agrobacterium-mediated transformation

The T-DNA plasmids were separately transformed into *Agrobacterium tumefaciens* LBA4404 by electroporation. The *Agrobacterium*-mediated transformation of rice was carried out according to Hiei Y. *et al.*, except that 2 mM glyphosate (Sigma-Aldrich, St. Louis, MO, USA) was used for selection [18].

RT-PCR analysis of transgenic rice

Total RNA was extracted from rice leaf with Trizol reagent (Invitrogen, Carlsbad, CA,USA). Concentration of RNA was determined by a spectrophotometer (Thermo Scientific, DE, USA). One μg RNA was treated with Dnase I and used as template to synthesize the first strand of cDNA using cDNA synthesis Kit (Thermo Scientific, DE, USA). Finally the synthesized cDNA was used as PCR template. The primers P1-2-F (5' CAGCGGCAACGAGGTGTACA 3') and P1-2-R (5' TAGGCGTCGCAGATGGTGGT 3') were used to amplify the joint parts in Cry1-F-Cry2 and Cry1-P-Cry2 (Fig. 2**a**). The PCR products were expected to be 321 bp. The primers P2-1-F (5' GCCGCTCGACATCAACGTGA 3') and P2-1-R (5' TCAGGCTGATGTCGATGGGG 3') were used to amplify the joint parts in Cry2-F-Cry1 and Cry2-P-Cry1 (Fig. 2**a**). The PCR products were expected to be 314 bp. The procedure for both PCRs was pre-denaturation at 95°C for 3 min; then 30 cycles of denaturation at 95°C for 40 s, annealing at 50°C for 30 s, and extension at 72°C for 30 s; finally followed with extension at 72°C for 5 min. The PCR products were analyzed by electrophoresis in 1% (w/v) agarose gel.

Western blot analysis

Leaf sample of 0.01 g collected from 1-month-old rice was frozen in liquid nitrogen for 45 s and disrupted with TissueLyserII (Qiagen, Hilden, Germany) at 40 Hz for 45 s. The sample was then homogenized in 200 μL 1×PBS Buffer by shaking vigorously for 30 s and centrifuged at 12000 rpm for 15 min at 4°C. The supernatants were collected as protein samples for western blot analysis. These samples were separated by electrophoresis on 8% polyacrylamide gel and transferred to PVDF membrane (Pall, Ann Arbor, MI, USA). The blotted membrane was blocked in 5% (w/v) skim milk in Tris Buffer Saline with 0.1% Tween-20 (TBST) for 1 h at room temperature. The membrane was then incubated with primary antibody against either Cry1Ab or Cry2Ab (diluted 1:1000 in 1% skim milk/TBST) for 1 h at room temperature with gentle shaking. The polyclonal antisera against Cry1Ab and Cry2Ab were prepared from New Zealand white rabbits immunized with purified recombinant Cry1Ab and Cry2Ab from *Escherichia coli*, respectively. After three times of 5 minute wash in TBST, the membrane was incubated with HRP-conjugated secondary antibody (Promega, Madison, WI, USA) diluted at

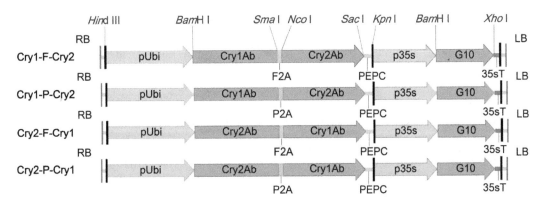

Figure 1. Schematic diagram of T-DNAs for transgenic expression of 2A linked polycistronic genes. Each T-DNA contained two expression cassettes, one for insecticidal polycistronic gene under control of maize ubiquitin-1 promoter (pUbi), the other for glyphosate-tolerant *5-enolpyruvylshikimate-3-phosphate synthase* (*EPSPS*) gene (G10) under control of Cauliflower mosaic virus 35s promoter (p35s). Cry1Ab and Cry2Ab,synthetic Bt inseciticidal gene; F2A and P2A, synthetic DNAs encoding Foot and Mouth Disease Virus 2A and Porcine teschovirus-1 2A, respectively; RB and LB, right border and left border of T-DNA; PEPC, maize *phosphoenolpyruvate carboxylase* gene terminator; 35sT, Cauliflower mosaic virus 35S gene terminator.

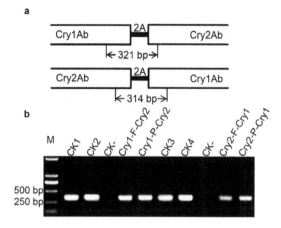

Figure 2. RT-PCR detection of mRNA integrity of the polycistronic genes. a, schematic diagram of the fragments amplified by RT-PCR; **b**, electrophoresis analysis of RT-PCR products. CK1, plasmid containing *Cry1-F-Cry2*; CK2, plasmid containing *Cry1-P-Cry2*; CK3, plasmid containing *Cry2-F-Cry1*; CK4, plasmid containing *Cry2-P-Cry1*; CK-, cDNA prepared from non-transgenic rice; M, DNA marker.

1:5000 with 1% skim milk/TBST for 1 h at room temperature with gentle shaking. After another three times of 5 minute wash in TBST, signals were visualized with DAB substrate (Sigma-Aldrich).

Protein quantification of Cry1Ab and Cry2Ab

Concentrations of Cry1Ab and Cry2Ab in the leaf of transgenic rice at tillering stage were determined by enzyme-linked immunosorbent assay (ELISA) using Envirologix kit AP003 and AP005 (Envirologix, Portland, ME, USA), respectively, according to the manufacturer's recommendation. For each construct, 3 transgenic lines were selected, and the samples prepared with the same method from non-transgenic rice were used to eliminate the basal absorption at 450 nm. After quantification of Cry1Ab and Cry2Ab, the molar ratio of the upstream protein to downstream protein of each transgenic line selected was calculated and analyzed by One-Way ANOVA [19] in SPSS.

Insect bioassay

Insect bioassays were conducted with cotton bollworm (Bt-susceptible and Cry1Ac-resistant), beet armyworm (*Spodoptera exigua*), striped stem borer (*Chilo suppressalis*) and rice leaf roller (*Cnaphalocrocis medinalis*). For beet armyworm and striped stem borer, detached leaf bioassay was carried out [20]. Two leaf blades from the same line were collected at 2-3 cm length at seedling stage and placed in a 70-mm-diameter petri dish lined with a pre-moistened filter paper. The leaf samples in the petri dish were infested with 10 newly hatched neonates. The petri dishes were then sealed with parafilm membrane and placed in dark at 28°C. The result was recorded after 3 days incubation. For cotton bollworm, detached leaf bioassay was carried out except that one leaf blade was placed in a petri dish and infested with one neonate to avoid cannibalism of cotton bollworm. Each line was infested with a total of 10 neonates. Eggs of Bt-susceptible cotton bollworm (CB-S), beet armyworm and striped stem borer were obtained from Genralpest Biotech (Genralpest Biotech, Bejing, China). Eggs of Cry1Ac-resistant cotton bollworm (CB-RR) were kindly provided by Dr. Kongming Wu from Institute of Plant Protection, Chinese Academy of Agricultural Sciences. For rice leaf roller, whole plant assay was carried out. Tillering-stage rice grown in

greenhouse was infested with 10 newly hatched neonates. The result was recorded 2 weeks after infestation. Eggs of rice leaf roller used in the assays were obtained from caged moths collected from rice field. Each assay was repeated for 3 times.

Results

Transgene transcribed as a long intact mRNA

About 30 independent transgenic lines were generated via *Agrobacterium*-mediated transformation for each of the four constructs. To investigate whether the transcript of each polycistronic transgene is a long intact mRNA with sequences encoding both Cry1Ab and Cry2Ab, RT-PCR was used to detect the sequences in the area connecting the two Bt genes, which include sequences from both *Cry1Ab* and *Cry2Ab*, and the 2A peptide coding sequence in between (Fig. 2a). The RT-PCR products with expected sizes were detected clearly in the transgenic lines from all of the four constructs (Fig. 2b), suggesting that each of the polycistronic genes was transcribed into a long intact mRNA as expected.

Co-expression of discrete Cry1Ab and Cry2Ab from polycistronic transgenes

Transgenic lines from each construct were analyzed by western blot analysis using antisera against Cry1Ab and Cry2Ab, respectively. When detected with antiserum against Cry1Ab, a band of approximately 72 kD was detected among different transgenic lines, indicating that the Cry1Ab protein was expressed as a discrete protein rather than a fusion protein (Fig. 3 Upper panel). When detected by antiserum against Cry2Ab, a band of about 68~70 kD was detected among all the transgenic lines (Fig. 3 Lower panel), indicating a discrete Cry2Ab protein was expressed. A possible fusion protein of "Cry-2A-Cry" would be at a size of approximately 147 kD. However, no significant signal of proteins at such size was detected by antiserum against either Cry1Ab or Cry2Ab, suggesting that there was no significant amount of fusion protein expressed in the transgenic rice. The upstream Cry2Ab detected in the Cry2-F-Cry1 and the Cry2-P-Cry1 transgenic lines was slightly larger than the downstream Cry2Ab in the Cry1-F-Cry2 and the Cry1-P-Cry2. This increased size was likely due to the 2A peptide residues attached to the upstream protein during translation, which has been demonstrated in other 2A polycistronic transgenes [21].

Determination of gene expression level and 2A cleavage efficiency

To determine the expression levels of Bt genes with 2A polycistronic transgene and evaluate the cleavage efficiency of 2A peptide in each construct, the amounts of Cry1Ab and Cry2Ab expressed in the transgenic rice leaves were determined by ELISA. Three transgenic lines at tillering stage from each construct were selected. The expression levels of Cry1Ab and Cry2Ab were different among different lines as expected. The concentrations of the soluble Cry1Ab and Cry2Ab in the leaf were in the range of 0.67 to 1.82 µg/g and 0.69 to 2.31 µg/g of leaf fresh weight (LFW), respectively (Fig. 4a).

The molar ratio (R) of the upstream protein to the downstream protein of each construct was calculated to estimate the cleavage efficiency. The R value for polycistronic gene using F2A as 2A cleavage peptide was lower than that using P2A (Fig. 4b), indicating that the cleavage efficiency of F2A peptide was higher than P2A. The R value of polycistronic gene with Cry1Ab at the upstream was lower than that with Cry2Ab at the upstream (Fig. 4b), suggesting that the Cry1Ab at the upstream was likely

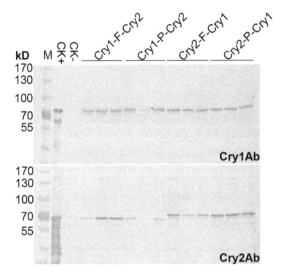

Figure 3. Western bolt analysis of Cry1Ab and Cry2Ab in transgenic rice. Three transgenic lines from each construct were selected for the analysis. Each sample was detected with antiserum against Cry1Ab (upper) and Cry2Ab (bottom) respectively. Cry1Ab or Cry2Ab protein expressed by *E.coli* was used as positive control (CK+). Sample prepared from non-transgenic rice was used as negative control (CK-). M, prestained protein ladder.

more efficient in cleavage than the Cry2Ab at the upstream, regardless of whether F2A or P2A was the cleavage peptide in the synthetic construct. The transgenic lines from the construct of Cry1-F-Cry2 had the R value close to 1, indicating that the two Bt proteins were expressed almost equally (Fig. 4**a**).

Insect-resistant activity of transgenic rice

The transgenic rice lines were assayed for their insecticidal activity using neonates of CB-S, striped stem borer and rice leaf roller. Three transgenic lines from each construct were selected and age-matched non-transgenic rice plants were used as control. Mortalities of all the 3 insect species feeding on transgenic rice were 100% while those on the non-transgenic rice were much lower (Fig. 5). In the assays with CB-S and striped stem borer, the transgenic leaves were only slightly bitten by the insects while non-transgenic controls were severely consumed (Fig. 6**a**). In the assays with rice leaf roller, no rolled leaf was observed in the transgenic plant while several leaves were rolled in the non-transgenic plant (Fig. 6**b**). The bioassays demonstrated that transgenic rice plants generated from the 2A constructs were highly insect-resistant.

The transgenic rice lines were also assayed for their activity against beet armyworm and CB-RR. The KMD1 transgenic rice at the same growing stage was used as control. The mortalities of beet armyworm feeding on the 2A transgenic rice lines were obvious higher than that feeding on KMD1 ($p<0.01$) (Figure 7). The KMD1 showed a rather low activity to CB-RR, while the 2A transgenic rice showed much higher activity ($p<0.01$) (Figure 7).

Discussion

In this study, Cry1Ab and Cry2Ab, two potent Bt crystal proteins with different receptors in the insect midgut [22], were expressed by the polycistronic transgene using 2A peptide. To our best knowledge, this is the first application of 2A peptide for expressing two Bt genes in a transgenic crop. Utilization of two or more Bt genes simultaneously is desirable for management of

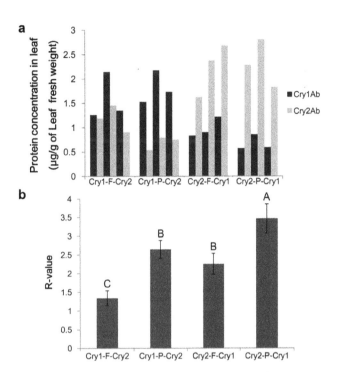

Figure 4. Cry1Ab and Cry2Ab concentrations in different transgenic rice lines and comparison of 2A cleavage efficiency. a, concentrations of Cry1Ab and Cry2Ab in leaves of 3 selected transgenic lines from each construct; **b**, R values of different 2A constructs. R value is the molar ratio of the upstream protein to the downstream protein. Different capital letters on each bar indicate extreme significant differences ($p<0.01$, Fisher's least-significant difference and Duncan's multiple range test).

insect resistance to Bt toxins as well as for broadening insecticidal spectrum of Bt crops. Self-cleaving 2A peptides and 2A-like sequences have also been found in insect virus, and some of them have high self-cleavage activity [23]. While there is no scientific base for any safety concern for using 2A peptide from mammalian virus, it may be better for public perception to use 2A peptides from insect virus for the development of transgenic rice.

Introduction of multiple genes into plant permits complex and sophisticated manipulation of traits in transgenic crops [2]. The number of genes being introduced into crops for genetic engineering is increasing steadily. For instance, the recently released commercial transgenic corn "Genuity SmartStax" (Monsanto, St Louis, MS, USA) has a total of 8 transgenes for insect-resistance and herbicide-tolerance. With more traits under research and development for transgenic improvement, more genes are expected to be utilized for future transgenic crops. Clearly polycistronic strategy using 2A peptide can simplify the process of gene stacking significantly, as it enables us to introduce multiple genes into plants using a single T-DNA with a single promoter [2]. Moreover, polycistronic strategy alleviates the concern of gene silencing induced by the insertion of homologous promoters or multiple T-DNA sequences into recipient genome [2,24]. Additionally, equal expressions of different genes could be achieved by the 2A polycistronic strategy [12,24].

The expression levels of Cry1Ab and Cry2Ab by 2A polycistronic transgene in this study were comparable to traditional monocistronic transgene. The Cry1Ab concentration of transgenic rice KMD1, which is highly resistant to striped stem borer and rice leaf roller [16], was 2.95 µg/g of LFW [25]. The Cry1C concentration in homozygous rice line T1c-19 harboring a

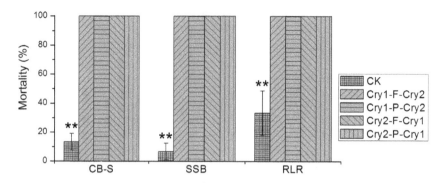

Figure 5. Mortality of Bt-susceptible cotton bollworm, striped stem borer and rice leaf roller feeding on transgenic rice plants. Non-transgenic rice at the same growing stage was used as control (CK). ** on the bar indicate extreme significant differences ($p<0.01$, Kruskal Wallis test). CB-S, Bt-susceptible cotton bollworm; SSB, striped stem borer; RLR, rice leaf roller.

synthetic Cry1C gene under control of maize ubiquitin promoter was 1.38 µg/g of LFW at the heading stage [26]. The Cry2A protein concentrations of the homozygous transgenic rice lines harboring a synthetic Cry2 gene driven by maize ubiquitin promoter was reported at the range from 9.65 to 12.11 µg/g of LFW [27]. In the transgenic *indica* rice line harboring both *Cry1Ac* and *Cry2Ab* gene, the Cry2Ab concentration was reported to be around 1 µg/g LFW [28]. Thus, the expression levels of the Cry1Ab and Cry2Ab in the transgenic rice lines obtained by this study were well within the range of the expression levels of the monocistronic transgene reported previously.

Although cotton bollworm was not a pest of rice, it has been long a good target insect for evaluating Bt toxicity. Moreover, cotton bollworm is a good model insect for the study of insect resistance to Bt toxins because of the availabilities of both sensitive and resistant lines to Cry1Ac. The 2A transgenic lines obtained in this study showed high insect-resistant activity against cotton bollworm, either sensitive or resistant to Cry1Ac, while the KMD1 transgenic rice had little activity toward Cry1Ac-resistant cotton bollworm. This suggested that little or no cross resistance existed between Cry1Ac and Cry2Ab and the 2A peptide based transgenic Bt rice will be useful for the management of Bt resistance developed by insect pests.

The 2A transgenic lines obtained in this study conferred significantly higher activity toward beet armyworm than the KMD1 transgenic rice containing Cry1Ab only [16], suggesting that the Cry2Ab expressed in the 2A transgenic lines indeed contributed to the enhancement of insecticidal activity. A similar phenomenon was also observed in transgenic cotton BollgardII. While Bollgard cotton only had moderate toxic activity to beet armyworm, BollgardII cotton showed greatly improved activity due to the addition of Cry2Ab [29].

The upstream proteins expressed with 2A polycistronic strategy are usually attached with the 2A peptide residues at the C-terminus [30]. The western blot analysis in this study appeared to agree with it. However, due to the limitation of the western blot analysis in determination of protein size, C-terminal amino acid sequencing and mass spectrum are required to confirm. While no adverse impact was observed for the utilization of 2A peptide in gene stacking [12], the Bt toxins attached with a 2A peptide will be considered as different proteins according to the regulation of transgenic crops, and thus additional safety studies will be required for commercialization of transgenic crops using 2A cleaving peptide.

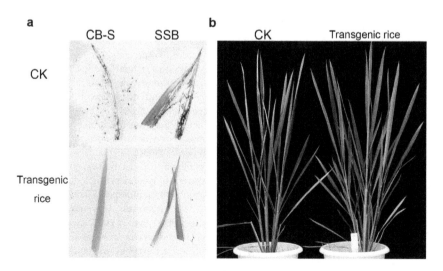

Figure 6. Insect bioassay for transgenic rice with a polycistronic Bt transgene. a, bioassay with Bt-susceptible cotton bollworm (CB-S) and striped stem borer (SSB). **b**, bioassay with rice leaf roller. For each assay, age-matched non-transgenic rice was used as control (CK).

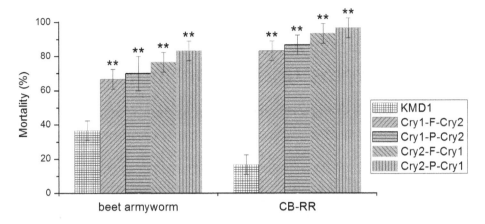

Figure 7. Mortality of beet armyworm and Cry1Ac-resistant contton bollworm feeding on KMD1 and various 2A transgenic rice lines. The transgenic rice KMD1 was used as control. ** on the bar indicate extreme significant difference ($p<0.01$, Fisher's least-significant difference and Duncan's multiple range test). CB-RR, Cry1Ac-resistant cotton bollworm.

Acknowledgments

We thank Kongming Wu and Gemei Liang for the supply of Cry1Ac-resistant cotton bollworm. We thank Gongyin Ye for the supply of KMD1 transgenic rice.

Author Contributions

Conceived and designed the experiments: QcZ MhL JHG ZcS. Performed the experiments: QcZ MhL MmT. Analyzed the data: QcZ MhL MmT. Contributed reagents/materials/analysis tools: QcZ MhL JHG. Wrote the paper: QcZ MhL JHG ZcS.

References

1. Romeis J, Meissle M, Bigler F (2006) Transgenic crops expressing Bacillus thuringiensis toxins and biological control. Nature Biotechnology 24: 63–71.
2. Halpin C (2005) Gene stacking in transgenic plants – the challenge for 21st century plant biotechnology. Plant Biotechnology Journal 3: 141–155.
3. Gassmann AJ, Petzold-Maxwell JL, Keweshan RS, Dunbar MW (2011) Field-Evolved Resistance to Bt Maize by Western Corn Rootworm. PLoS ONE 6.
4. Tabashnik BE, Van Rensburg JBJ, Carriere Y (2009) Field-Evolved Insect Resistance to Bt Crops: Definition, Theory, and Data. Journal of Economic Entomology 102: 2011–2025.
5. Wang P, Zhao J-Z, Rodrigo-Simon A, Kain W, Janmaat AF, et al. (2006) Mechanism of resistance to Bacillus thuringiensis toxin Cry1Ac in a greenhouse population of cabbage looper, Trichoplusia ni. Applied and Environmental Microbiology: 01834–01806.
6. Bates SL, Zhao JZ, Roush RT, Shelton AM (2005) Insect resistance management in GM crops: past, present and future. Nature Biotechnology 23: 57–62.
7. Shelton AM, Tang JD, Roush RT, Metz TD, Earle ED (2000) Field tests on managing resistance to Bt-engineered plants. Nature Biotechnology 18: 339–342.
8. Tabashnik BE, Carriere Y, Dennehy TJ, Morin S, Sisterson MS, et al. (2003) Insect resistance to transgenic Bt crops: Lessons from the laboratory and field. Journal of Economic Entomology 96: 1031–1038.
9. Cao J, Zhao JZ, Tang JD, Shelton AM, Earle ED (2002) Broccoli plants with pyramided cry1Ac and cry1C Bt genes control diamondback moths resistant to Cry1A and Cry1C proteins. Theoretical and Applied Genetics 105: 258–264.
10. Yang Z, Chen H, Tang W, Hua HX, Lin YJ (2011) Development and characterization of transgenic rice expressing two Bacillus thuringiensis genes. Pest Management Science 67: 414–422.
11. Zhao JZ, Cao J, Li YX, Collins HL, Roush RT, et al. (2003) Transgenic plants expressing two Bacillus thuringiensis toxins delay insect resistance evolution. Nature Biotechnology 21: 1493–1497.
12. Szymczak-Workman AL, Vignali KM, Vignali DAA (2012) Design and Construction of 2A Peptide-Linked Multicistronic Vectors. Cold Spring Harbor Protocols 2012: pdb.ip067876.
13. Ha S-H, Liang YS, Jung H, Ahn M-J, Suh S-C, et al. (2010) Application of two bicistronic systems involving 2A and IRES sequences to the biosynthesis of carotenoids in rice endosperm. Plant Biotechnology Journal 8: 928–938.
14. Brown JD, Ryan MD (2010) Ribosome "Skipping": "Stop-Carry On" or "StopGo" Translation Recoding: Expansion of Decoding Rules Enriches Gene Expression. InAtkinsJFGestelandRFeditors: New York:Springerpp101–121.
15. Luke G, Escuin H, De Felipe P, Ryan M (2010) 2A to the fore - research, technology and applications. Biotechnology and Genetic Engineering Reviews 26: 223–260.
16. Shu QY, Ye GY, Cui HR, Cheng XY, Xiang YB, et al. (2000) Transgenic rice plants with a synthetic cry1Ab gene from Bacillus thuringiensis were highly resistant to eight lepidopteran rice pest species. Molecular Breeding 6: 433–439.
17. Ye GY, Shu QY, Yao HW, Cui HR, Cheng XY, et al. (2001) Field evaluation of resistance of transgenic rice containing a synthetic cry1Ab gene from Bacillus thuringiensis Berliner to two stem borers. Journal of Economic Entomology 94: 271–276.
18. Hiei Y, Komari T, Kubo T (1997) Transformation of rice mediated by Agrobacterium tumefaciens. Plant Molecular Biology 35: 205–218.
19. Heiberger R, Neuwirth E (2009) One-Way ANOVA. R Through Excel: New YorkSpringerpp165–191.
20. Chen Y, Tian J-C, Shen Z-C, Peng Y-F, Hu C, et al. (2010) Transgenic Rice Plants Expressing a Fused Protein of Cry1Ab/Vip3H Has Resistance to Rice Stem Borers Under Laboratory and Field Conditions. Journal of Economic Entomology 103: 1444–1453.
21. Halpin C, Cooke SE, Barakate A, Amrani AE, Ryan MD (1999) Self-processing 2A-polyproteins – a system for co-ordinate expression of multiple proteins in transgenic plants. The Plant Journal 17: 453–459.
22. Hernandez-Rodriguez CS, Hernandez-Martinez P, Van Rie J, Escriche B, Ferre J (2013) Shared Midgut Binding Sites for Cry1A. 105, Cry1Aa, Cry1Ab, Cry1Ac and Cry1Fa Proteins from Bacillus thuringiensis in Two Important Corn Pests, Ostrinia nubilalis and Spodoptera frugiperda. Plos One 8.
23. Ryan M (2014) 2A and 2A like sequences. Available: http://www.st-andrews.ac.uk/ryanlab/2A_2Alike.pdf/. Accessed 4 September 2014.
24. Sainsbury F, Benchabane M, Goulet M-C, Michaud D (2012) Multimodal Protein Constructs for Herbivore Insect Control. Toxins 4: 455–475.
25. Tian J-C, Chen Y, Li Z-L, Li K, Chen M, et al. (2012) Transgenic Cry1Ab Rice Does Not Impact Ecological Fitness and Predation of a Generalist Spider. PLoS ONE 7: e35164.
26. Tang W, Chen H, Xu C, Li X, Lin Y, et al. (2006) Development of insect-resistant transgenic indica rice with a synthetic cry1C* gene. Molecular Breeding 18: 1–10.
27. Chen H, Tang W, Xu C, Li X, Lin Y, et al. (2005) Transgenic indica rice plants harboring a synthetic cry2A* gene of Bacillus thuringiensis exhibit enhanced resistance against lepidopteran rice pests. Theoretical and Applied Genetics 111: 1330–1337.
28. Bashir K, Husnain T, Fatima T, Latif Z, Mehdi SA, et al. (2004) Field evaluation and risk assessment of transgenic indica basmati rice. Molecular Breeding 13: 301–312.
29. Perlak FJ, Oppenhuizen M, Gustafson K, Voth R, Sivasupramaniam S, et al. (2001) Development and commercial use of Bollgard cotton in the USA – early promises versus today's reality. The Plant Journal 27: 489–501.
30. Szymczak-Workman AL, Vignali KM, Vignali DAA (2012) Verification of 2A Peptide Cleavage. Cold Spring Harbor Protocols 2012: pdb.prot067892.

15

Soil Chemical Property Changes in Eggplant/Garlic Relay Intercropping Systems under Continuous Cropping

Mengyi Wang, Cuinan Wu, Zhihui Cheng*, Huanwen Meng, Mengru Zhang, Hongjing Zhang

College of Horticulture, Northwest A&F University, Yangling, Shaanxi, China

Abstract

Soil sickness is a critical problem for eggplant (*Solanum melongena* L.) under continuous cropping that affects sustainable eggplant production. Relay intercropping is a significant technique on promoting soil quality, improving eco-environment, and raising output. Field experiments were conducted from September 2010 to November 2012 in northwest China to determine the effects of relay intercropping eggplant with garlic (*Allium sativum* L.) on soil enzyme activities, available nutrient contents, and pH value under a plastic tunnel. Three treatments were in triplicate using randomized block design: eggplant monoculture (CK), eggplant relay intercropping with normal garlic (NG) and eggplant relay intercropping with green garlic (GG). The major results are as follows: (1) the activities of soil invertase, urease, and alkaline phosphatase were generally enhanced in NG and GG treatments; (2) relay intercropping significantly increased the soil available nutrient contents, and they were mostly higher in GG than NG. On April 11, 2011, the eggplant/garlic co-growth stage, the available nitrogen content in GG was 76.30 $mg \cdot kg^{-1}$, significantly higher than 61.95 $mg \cdot kg^{-1}$ in NG. For available potassium on April 17, 2012, they were 398.48 and 387.97 $mg \cdot kg^{-1}$ in NG and GG, both were significantly higher than 314.84 $mg \cdot kg^{-1}$ in CK; (3) the soil pH showed a significantly higher level in NG treatment, but lower in GG treatment compared with CK. For the last samples in 2012, soil pH in NG and GG were 7.70 and 7.46, the highest and lowest one among them; (4) the alkaline phosphatase activity and pH displayed a similar decreasing trend with continuous cropping. These findings indicate that relay intercropping eggplant with garlic could be an ideal farming system to effectively improve soil nutrient content, increase soil fertility, and alleviate soil sickness to some extent. These findings are important in helping to develop sustainable eggplant production.

Editor: Wen-Xiong Lin, Agroecological Institute, China

Funding: This research was supported by a project of the National Natural Science Foundation of China (No. 31171949), the Special Fund for Agro-scientific Research in the Public Interest (No. 200903018), and the University Undergraduates Innovating Experimentation Project (No. 2201110712036). The funders had no role in study design, data collection and analysis, decision to publish, or preparation of the manuscript.

* Email: chengzh2004@163.com

Introduction

Eggplant (*Solanum melongena* L.), because of its rich nutritional value, good taste and easily cultivation, is grown in most parts of the world, among which China is the greatest producer [1]. However, in recent years, particularly under continuous cropping in facilities, many problems have affected the sustainable production of eggplant, such as aggravating plant disease and pests, degraded soil physical and chemical characteristics, and declining production and stress resistance of plants [2,3].

As a simple repetitive agronomic practice, continuous cropping is the practice of growing the same crop year after year in the same field [4]. It makes the soil susceptible to erosion hazard and weed invasion, soilborne pathogens increase, and the survival of certain pathogens enhancing that a certain degree of damage has to be accepted [5]. One of the root causes is that long-term monoculture with a single plant eliminates crop and biological diversity [6]. Consequently, the diversification of crop systems by increasing the number of cultivated species in the same or nearby areas has been proposed to overcome those continuous cropping obstacles. In modern agriculture, crop rotation is the most common for a vast range of crop species worldwide [7]. If properly designed, crop rotation is the most efficient practice to reduce the incidence and severity of soilborne diseases [8]. However, crop rotation is not always practiced because of the difficulty in design of a proper rotation and relatively high risk of losing the lower-value crop. In addition, as the cultivated land is limited and Chinese farmers are accustomed to plant crops of the same species or the same families, it is difficult to carry out crop rotation under protected cultivation in China [9]. Another significant cropping technique - relay intercropping, which is defined as after-crop planting between the rows or plants in later periods of the preceding crops' growth with a shorter symbiotic time [7], is claimed to promote biodiversity and diversify agricultural outcome compared with monocropping in sustainable agriculture. Intercropping, being looked as an efficient and most economical production system, is drawing more and more attention of small growers [10]. The most common advantage of intercropping is to produce a greater yield and diversified production per unit area and time by achieving higher efficient use of the available growth resources that would not be utilized by each crop grown alone. From the view of diversity restoration, intercropping provides high insurance against crop failure and overall provides greater financial stability for farmers, making the system particularly suitable for labor-intensive small farms or greenhouses [11]. Besides, intercropping offers effective weed suppression [12,13] and pest and disease control. Crops grown simultaneously enhance the predators and parasites, which in turn prevent the pest build-up, thus minimizing the use of expensive and dangerous chemical insecticides. Mixed crop species can also delay the diseases onset by reducing the spread of disease carrying spores and by modifying environmental conditions so that they are less favorable to the spread of certain pathogens. Moreover, intercropping is an excellent practice for controlling soil erosion and improving soil fertility [11] and quality [14].

Soil sickness, which means serious decline in soil quality, determines the sustainability and productivity of agroecosystems [15]. The changes of the physical, chemical, and biochemical properties of the soil must be taken into account in assessing changes in soil quality [16]. Soil enzyme activities, as mediators and catalysts of most soil transformation processes, have been proposed as appropriate indicators of the health and sustainability of soil quality and ecosystems [17] due to their sensitivity to ecosystem stress [18], intimate relationship with soil biology, and rapid response to changes in soil management [19]. Invertase, widely exists in the soil, plays an important role in increasing the soluble nutrients in the soil [20]. Urease is of great importance in the soil nitrogen cycle and utilization because it can hydrolyze urea to ammonia, one of the sources of plant nitrogen. Another hydrolase, alkaline phosphatase, can mineralize organic phosphorus (P) to inorganic P [21] for plant absorption. Thus, soil enzymes can provide indications of changes in metabolic capacity and nutrient cycling in management practices [22]. Soil nutrients are important factors affecting the growth and development of plants. Nitrogen (N) is the most important element for plant development; it stimulates shoot growth and produces the rich green color that is characteristic of a healthy plant. P is the second most frequently limiting macronutrient for plant growth [23] and plays a major role in the processes requiring a transfer of energy in plants. Another essential nutrient, potassium (K), is a key factor in plant tolerance to stresses such as cold/hot temperatures, drought, and pests. Soil nutrient contents relate to the soil productivity, and soil enzymatic characteristics can reflect the status of key biochemical reactions that participate in the transformation of soil nutrients [24]. However, the reaction rates of soil enzymes are markedly dependent on pH and the presence or absence of inhibitors [25]. In addition, the availability of mineral elements to plants may also be affected by soil pH. Soil pH may affect plant root growth directly or indirectly by impairment of nutrient relations [26]. In turn, growing roots affect the pH of the rhizosphere during plant growth processes and nutrient uptake [27,28].

In relay intercropping systems, although the increased crop species are expected to overcome the continuous cropping obstacles, the selection of companion crops is still critical. Garlic (*Allium sativum* L.), belonging to the Liliaceae family, is a common vegetable and food spice that is used widely throughout the world. Especially, it is an important economic crop and a good cover crop in vegetable production in China [29]. Garlic products-green garlic, garlic bulbs, and garlic sprouts are all important vegetables favored especially for Asian. It is also commonly used as natural broad-spectrum antibiotic. Khan and Cheng [30] found that garlic root exudates is an effective and environment-friendly management measure against *Phytophthora* blight of pepper and may be used in the organic vegetables production. In addition, it has been reported that the exudates secreted by the rooting system of garlic can produce noteworthy effects on soil structure and ecology [9,31]. Thus, garlic is expected to be an ideal companion crop for relay intercropping with eggplant.

There is an increasing population and decline in arable land in China, so sustainable agriculture has gained more attention owing to its efficient use of resources, balance with the environment, and the ultimate goal of providing human benefit [32]. In recent years, an increasingly number of studies have focused on intercropping of different grain crops [33–35], in addition to intercropping cucumbers [9], peppers [31] or Chinese cabbages [36] with garlic, but eggplant/garlic relay intercropping systems are rarely studied, and intensive study of the soil properties change has been considerably less. For this reason, we have concentrated on comparing the activity of enzymes, content of available nutrients, and pH value in the soil of eggplant/normal garlic or green garlic relay intercropping systems with those in an eggplant monoculture system under continuous cropping to assess if relay intercropping eggplant with normal or green garlic is effective on improving the soil fertility level and maintaining the soil quality, which will help ensure the sustainable long-term development of eggplant cropping.

Materials and Methods

Experimental site

The experiment was conducted from September 2010 to November 2012 under a plastic tunnel at the Horticultural Experimental Station (34°17' N, 108°04' E) of Northwest A&F University, Yangling, Shaanxi Province, China, where it is hot in summer and cold in winter, and the annual mean temperature is 12.9°C, with a frost-free period over 200 days. Under plastic tunnel, the highest temperature can achieve around 50°C, and the lowest is around $-10°C$ (Fig. 1).

The soil used for this experiment was brown loamy, alkaline Orthic Anthrosol (Table 1). The soil pH was 7.8 (1:1 water), and it contained 27.02 g of organic matter, 1.38 g of total N, 0.96 g of total P, and 14.31 g of total K per kilogram dry soil. The ammonium N, available P, and exchangeable K concentrations were 57.17 mg·kg^{-1}, 57.65 mg·kg^{-1}, and 224.01 mg·kg^{-1} and invertase, urease, and alkaline phosphatase activities were 18.12 glucose mg·g^{-1}, 1.99 NH$_3$-N mg·g^{-1} and 0.94 P$_2$O$_5$ mg·g^{-1} respectively in the 0–20 cm soil layer before crop transplanting in March 2010. Cucumber had been planted for approximately ten years before the planting of eggplant in the spring of 2010.

Experimental design

Eggplant as the main crop was relay intercropped with garlic (Fig. 2). A completely randomized design was used consisting of three treatments with three replications: eggplant monoculture (CK) (Fig. 2 A and D), eggplant relay intercropping with normal garlic cv. G110 (NG) (Fig. 2 B and E), and eggplant relay intercropping with green garlic cv. G064 (GG) (Fig. 2 C and F). In NG treatment, normal garlic means garlic bulb, where garlic cloves of uniform size were manually planted into the soil among the eggplant plants in autumn (Fig. 2 B), and expanded garlic bulbs were harvested one by one using shovel in the next spring (Fig. 2 E). In GG treatment, green garlic means green garlic sprouts, where garlic bulbs of uniform size were planted directly by hand in autumn (Fig. 2 C), and the green sprouts were harvested three or four times within the next three months once they were about 20 cm high (Fig. 2 F).

There were two beds per plot of the three treatments. Each bed was 3.5 m long and 1.2 m wide (Fig. 3). There were two rows of eggplants per bed and seven plants per row, and it was 50 cm for plant spacing and 80 cm for row spacing in both the monoculture and relay intercropping treatments. In the relay intercropping treatments, three rows of garlic cloves (20 cm for row spacing and

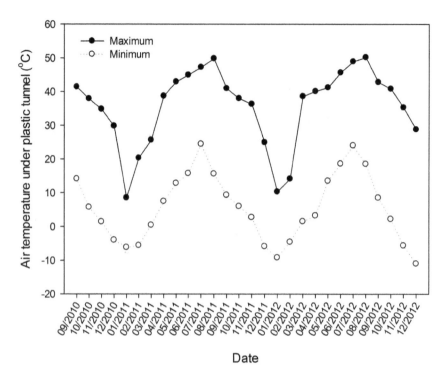

Figure 1. Monthly maximum and minimum air temperature under plastic tunnel in the three experimental years from September 2010 to December 2012. Fig. 1 was drawn using the software program Sigmaplot 12 (Systat Software, Inc.).

6 cm for plant spacing, with 141 cloves for each bed) for NG treatment and four rows of garlic bulbs (10 cm for row spacing and adjacent in each row, with 8.48 kg of bulbs for each bed) for GG treatment were planted in the middle of the bed within the eggplant rows.

Under plastic tunnel, all field operations were performed manually because of the limited land area. Eggplants were manually transplanted on March 19, 2010, March 22, 2011 and March 24, 2012, and uprooted one by one using spade around November 25 in the three years. In 2010, seed cloves and bulbs were planted on September 15, and the three treatments were applied; whereas in 2011 and 2012, the bulbs and seed cloves were planted on August 1/July 20 and September 15, respectively. Every year before eggplant transplantation, the experimental areas were plowed into two shallow furrows and fertilized with 1.5 kg of "PengDiXin" (organic fertilizer, manure substitute; made in Henan Province, China, containing organic matter$\geq$30%, N+ $P_2O+K_2O\geq$4%, humic acid$\geq$20%, trace element$\geq$2%, and organic sylvite$\geq$5%), 0.15 kg of double superphosphate (chemical fertilizer which can be used to improve alkali soil and supply phosphorus and calcium plant needs; total P$\geq$46%, available P$\geq$ 44%) and 0.15 kg of "SaKeFu" fertilizer (NPK compound chemical fertilizer; made in SACF, Hebei Province, China, containing total primary nutrient$\geq$40%) per bed as base fertilizer following local farming convention. In the eggplant-only period and eggplant/garlic co-growth period, the same amount of "JinBa" fertilizer (compound chemical fertilizer which is most often used in local vegetable production; made in Beijing, China, containing humic acid $\geq$3%, trace element $\geq$6%, N+K_2O $\geq$18%, and phosphate and K-solubilizing agent $\geq$5%) was top dressed on each bed according to the instructions. In the garlic-only stage for NG and GG treatments, only water was administered as required. For eggplant, ving tying, pruning, and other farm management

were administered following local convention. No other farm management techniques were needed on garlic (Table 2).

Measurements

Soil samples were collected from the plow layer (0–20 cm) in the plots of each treatment (Table 3). Eight soil cores (40 mm in diameter) were removed in a serpentine pattern from the center of two eggplant rows of each bed resulting in 16 soil cores per plot. Subsequently, all sub-samples taken from a single plot were pooled. The first sampling dates per year were nine days before garlic planting in 2010 (September 6) and five days before eggplant transplantation in 2011 (March 17) and 2012 (March 19). In 2010, the other two soil sampling dates were October 16 (eggplant/garlic co-growth period) and November 26 (before eggplant uprooting). Then, in 2011 and 2012, soil samples were taken on April 11/ April 17 (eggplant/garlic co-growth period), June 20/June 17 (eggplant full bearing period after all garlic harvested), July 25/ July 15 (five days before planting green garlic), August 30/ September 10 (fifteen/five days before planting normal garlic), October 10/October 20 (eggplant/garlic co-growth period), and November 20/November 23 (five days before eggplant uprooted) separately.

Determination of soil enzyme activity, available nutrient content, and pH value

The soil collected from each treatment was put in a well-ventilated place to air-dry then sieved (1 mm) to analyze the activity of enzymes, content of available nutrients, and pH value in the soil. Determinations of all parameters were performed in triplicate, with values reported as means of each treatment.

The activities of invertase, urease, and alkaline phosphatase in soil were assayed on the basis of the release and quantitative determination of the products of glucose, NH_3-N and P_2O_5; soil

Table 1. Basic characteristics of original soil in the experiment.

Soil type	pH value (1soil: 1water)	Organic matter (g·kg⁻¹)	Total nitrogen (g·kg⁻¹)	Total phosphorus (g·kg⁻¹)	Total potassium (g·kg⁻¹)	Ammonium nitrogen (mg·kg⁻¹)	Available phosphorus (mg·kg⁻¹)	Exchangeable potassium (mg·kg⁻¹)	Invertase activity (glucose mg·g⁻¹)	Urease activity (NH$_3$-N mg·g⁻¹)	Alkaline phosphatase activity (P$_2$O$_5$ mg·g⁻¹)	Cultivation history
Brown loamy, alkaline Orthic Anthrosol	7.8	27.02	1.38	0.96	14.31	57.17	57.65	224.01	18.12	1.99	0.94	Cucumber planted for ten years

samples were incubated with 8% sucrose solution, 10% urea solution or 0.5% disodium phenyl phosphate solution, respectively, in suitable buffer solutions for 24 hours at 37°C, and then, spectrophotometric measurements were performed at 508 nm, 578 nm, and 660 nm, respectively [37].

Alkaline hydrolysis diffusion was used to determine the available N content in soil according to the method of Bermner and Shaw [38] with some modifications. The available N in soil was hydrolyzed by 1.0 mol·L^{-1} NaOH solution to NH$_3$, which was absorbed by H$_3$BO$_3$ indicator solution, and then the NH$_3$ absorbed by H$_3$BO$_3$ was titrated with 0.005 mol·L^{-1} (1/2 H$_2$SO$_4$) standard acid. Available P was extracted with 0.5 mol·L^{-1} NaHCO$_3$ solution, and then the level using molybdenum-antimony-D-isoascorbic-acid-colorimetry (MADAC) at 880 nm by the modified method of Olsen and Dean [39]. Available K was extracted with 1 mol·L^{-1} ammonium acetate neutral solution, and then the level using atomic absorption spectrometry according to the method of Pratt [40] with some modifications.

The soil pH value was determined in a soil:water suspension (1:1 ratios) with glass electrodes [41].

Data analyses

Data obtained for each year in this study were analyzed by analysis of variance (ANOVA) using PASW Statistics 18.0 software (IBM, Armonk, New York, USA). The significant differences between the means of the soil enzyme activities, available nutrient contents, and pH values among the monoculture and relay intercropping systems were examined according to Duncan's multiple range test at a P<0.05 level.

Results

Effects of relay intercropping eggplant with normal or green garlic on soil enzyme activities

Soil invertase activity. The activities of soil invertase from 2010 to 2012 are shown in Fig. 4 A. In 2010, the invertase activities of all treatments declined as the weather became cold. Then, in 2011 and 2012, the overall trend of invertase activity for all treatments first rose and then fell over time, although there were slight fluctuations in different periods. In 2011, the second experimental year, the soil invertase activity in NG and GG treatments was higher than that in CK treatment during co-growth periods both on April 11 and October 10. On June 20, during the eggplant full bearing period, the invertase activity in NG was 26.87 glucose mg·g^{-1}, which was significantly higher than 23.84 glucose mg·g^{-1} in CK. At the same time, it was 29.78 glucose mg·g^{-1} in GG which was significantly higher than in NG. However on August 30, when the eggplant was relay intercropped with green garlic, but the normal garlic had not been planted, the invertase activity in GG treatment was significantly lower than that in NG treatments. The peak activity appeared on June 20 for the GG treatment and on July 25 for the CK and NG treatments. However, the maximum value of invertase activity of all three treatments appeared earlier in 2012 (on April 17) than that in 2011. On March 19, 2012, the garlic-only stage for NG and GG treatments, the invertase activity in NG and GG treatments showed a significantly lower level than in CK treatment. Then on April 17, during the co-growth period, the activity level in GG treatment was 47.54 glucose mg·g^{-1}, significantly higher than that of the CK and NG treatments (33.66 and 33.60 glucose mg·g^{-1}). From September 10, 2012, the invertase activity of both NG and GG treatments was higher than that of CK treatment, and the difference even reached a significant level on November 23. In

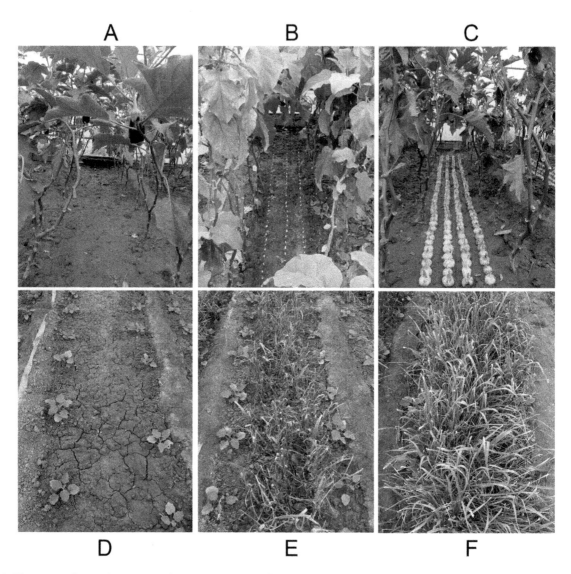

Figure 2. Three experimental treatments in autumn (A–C) and spring (D–F). Fig. 2 was made by graphics software Adobe Photoshop CS6 (Adobe Systems, Inc.). Eggplant monocropping (CK, A and D), eggplant relay intercropping with normal garlic (NG, B and E), and eggplant relay intercropping with green garlic (GG, C and F).

summary, soil invertase activity in relay intercropping treatments was higher than that of the monoculture treatment during the eggplant/garlic co-growth periods, but on other sampling dates, there was no regular routine.

Soil urease activity. The activity of soil urease fluctuated in different periods (Fig. 4 B). In the eggplant/garlic co-growth period on October 16, 2010, the urease activity of all three treatments increased over that on September 6 before the garlic was planted. On November 26, though a slight decline, the values of both the NG and GG treatments were still higher than that of the CK treatment, and the difference between NG and CK was significant.

In 2011, the second year of continuous cropping, the overall trend of urease activity continued increasing until it reached the maximum on August 30. On March 17, when there was only garlic in the field, the urease activity of the NG and GG treatments was 3.30 and 3.33 NH$_3$-N mg·g^{-1}, significantly higher than 2.74 NH$_3$-N mg·g^{-1} in the CK treatment. However, the urease activity of GG was significantly lower than that of CK and NG before the

green garlic were planted on July 25. In the subsequent eggplant/green garlic co-growth period on August 30, the urease activity of GG treatment was no longer lower than CK despite the fact that it was still significantly lower than NG, and the activity of the CK treatment was significantly lower than that of NG as well. Later, the urease activity dropped markedly with the decrease of temperature; yet, it presented an abnormal phenomenon for the CK and GG treatment on November 20 that the activity increased again.

Then in 2012, in the third continuous cropping year, the urease activity of the NG and GG treatments was significantly higher than that of the CK treatment during the eggplant/garlic co-growth period, and this positive effect lasted until the green garlic were planted on July 15. However, there was a sharp decrease for all the three treatments on June 17, which may be the result of the application of adequate fertilizer in time during the eggplant vigorous growth stage. On September 10 and October 20, there was no marked difference among the treatments; but on November 23, the urease activity of the

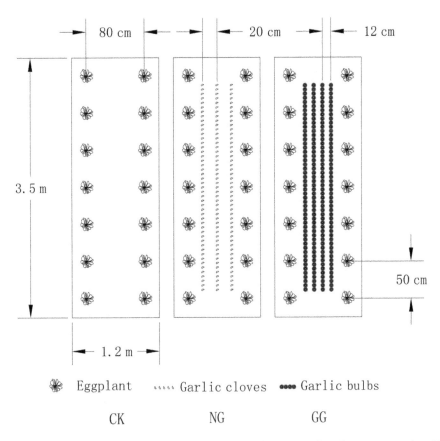

Figure 3. Diagram of three cropping systems in the experiment. Fig. 3 was drawn using the software program AutoCAD 2012 (Autodesk Inc.). CK: eggplant monoculture; NG: eggplant relay intercropping with normal garlic cv. G110; GG: eggplant relay intercropping with green garlic cv. G064.

relay intercropping treatments was again significantly higher than that of CK.

Soil alkaline phosphatase activity.   As shown in Fig. 4 C, the overall change trend of the soil alkaline phosphatase activity was similar to urease activity, but varied among treatments and sampling dates. In addition, the alkaline phosphatase activity in 2012 displayed a general decline compared to that in 2011. For the two relay intercropping treatments, the activity was generally higher than that of the CK treatment during the three years of continuous cropping.

In 2010, the alkaline phosphatase activities decreased slightly in the eggplant/garlic co-growth stage on October 16 compared with that on September 6 before the garlic was planted but rebounded before eggplant uprooting. Then in the second year of eggplant continuous cropping, on March 17, 2011 when there was only garlic in the field, the enzyme activity of the GG treatment was 1.64 P_2O_5 mg·g^{-1}, significantly higher than that of CK (1.15 P_2O_5 mg·g^{-1}). For the rest of 2011, the three treatments had no significant difference on the alkaline phosphatase activity. In the third year of continuous cropping in 2012, the alkaline phosphatase activity of NG treatment was significantly higher than that of CK treatment on March 19. And for the GG treatment, the activity was always higher than CK, even on March 19 (garlic-only stage), June 17 (eggplant-only stage), and October 20 (eggplant/garlic co-growth stage), the difference reached a significant level.

Effects of relay intercropping eggplant with normal or green garlic on soil available nutrients

Soil available N content.   Relay intercropping eggplant with garlic affected the content of the main available nutrients in soil. Fig. 5 A shows that, in 2010, the available N content kept rising during the three sampling dates, but there was no significant difference among the three treatments. Then in 2011, the soil available N content of the NG and GG treatments was almost higher than that of the CK treatment, and most reached significant levels. However on November 20, soil available N content in CK was 110.31 mg·kg^{-1}, significantly higher than 80.79 mg·kg^{-1} in NG and 75.31 mg·kg^{-1} in GG treatments. In the third continuous cropping year of 2012, the soil available N content of the NG or GG treatments was always significant higher than that of the CK treatment in different periods. The results also indicate that the available N content of the GG treatment was higher than that of the NG treatment in many cases. As a general view, the available N content in the soil of the NG and GG treatments was higher than that in the CK treatment, highlighting a positive effect of eggplant/garlic relay intercropping patterns on increasing the soil available N content.

Soil available P content.   As shown in Fig. 5 B, the soil available P content of the NG and GG treatments kept increasing over time in 2010 and increased most rapidly for the GG treatment, which was significantly higher than that of the CK treatment on November 26; but for the CK treatment, the available P content first increased and then slightly decreased. In 2011, for the NG treatment, it was always higher than that of the CK treatment, and the difference was significant on most of the

Table 2. Field operations in the three years.

Year	Date	CK		NG		GG	
		growth stage	field management	growth stage	field management	growth stage	field management
2010	03/19		eggplant transplant		eggplant transplant		eggplant transplant
	03/20–09/14	eggplant only	eggplant management	eggplant only	eggplant management	eggplant only	eggplant management
	09/15				plant garlic cloves		plant garlic bulbs
	09/16–11/27			eggplant/garlic co-growth	eggplant management	eggplant/garlic co-growth	eggplant management
	10/01 10/16 11/02 12/07			eggplant/garlic co-growth	eggplant management	eggplant/garlic co-growth	green garlic harvest
	11/28		eggplant uprooted		eggplant uprooted		eggplant uprooted
	11/29–12/31	blank	without any operation	garlic only	garlic watering	garlic only	garlic watering
	01/01–03/21	blank		garlic only		garlic only	
2011	03/22		eggplant transplant		eggplant transplant		eggplant transplant
	03/23–04/14			eggplant/garlic co-growth	eggplant management	eggplant/garlic co-growth	eggplant management
	04/15				normal garlic (garlic bulbs) harvest		green garlic uprooted
	04/16–07/31			eggplant only	eggplant management	eggplant only	eggplant management
	08/01	eggplant only	eggplant management	eggplant only		eggplant only	plant garlic bulbs
	08/02–09/14			eggplant only	eggplant management	eggplant only	eggplant management
	09/15				plant garlic cloves		eggplant management
	09/16–11/24			eggplant/garlic co-growth	eggplant management	eggplant/garlic co-growth	eggplant management
	09/15 09/30 10/19			eggplant/garlic co-growth	eggplant management	eggplant/garlic co-growth	green garlic harvest
	11/25		eggplant uprooted		eggplant uprooted		eggplant uprooted
	11/26–12/31	blank	without any operation	garlic only	garlic watering	garlic only	garlic watering
	01/01–03/23	blank		garlic only		garlic only	
2012	03/24		eggplant transplant		eggplant transplant		eggplant transplant
	03/25–04/17			eggplant/garlic co-growth	eggplant management	eggplant/garlic co-growth	eggplant management
	04/18				normal garlic (garlic bulbs) harvest		green garlic uprooted
	04/19–07/19	eggplant only	eggplant management	eggplant only	eggplant management	eggplant only	eggplant management
	07/20						plant garlic bulbs
	07/21–09/14	eggplant only	eggplant management	eggplant only	eggplant management	eggplant only	eggplant management
	09/15				plant garlic cloves		plant garlic bulbs
	09/16–11/27			eggplant/garlic co-growth	eggplant management	eggplant/garlic co-growth	eggplant management
	09/11 09/27 10/21			eggplant/garlic co-growth	eggplant management	eggplant/garlic co-growth	harvest green garlic
	11/28		eggplant uprooted		eggplant uprooted		eggplant uprooted

CK: eggplant monoculture; NG: eggplant relay intercropping with normal garlic cv. G110; GG: eggplant relay intercropping with green garlic cv. G064.
Eggplant managements include irrigation, fertilization, pruning, and ving tying. When there was eggplant in the field, water and fertilizer were given only when eggplant needed. During the growth stages, eggplant was pruned to dichotomous branching. The two branches were ving tying when they grew 1 m high between June and July every year. When there was only garlic in the field, garlic managements include only watering.

Table 3. Soil sampling dates and corresponding eggplant/garlic growth stages in the three experimental years.

Sampling dates	2010		2011							
			03/17	04/11	06/20	07/25	08/30	10/10	11/20	
	09/06	10/16	11/26		2012					
				03/19	04/17	06/17	07/15	09/10	10/20	11/23
Stage	nine days before planting normal garlic and green garlic	eggplant/garlic co-growth stage	eggplant/garlic co-growth and before eggplant uproot	five days before eggplant transplant	eggplant/garlic co-growth stage	eggplant full bearing period after all garlic harvested	five days before planting green garlic	fifteen days before planting normal garlic	eggplant/garlic co-growth stage	eggplant/garlic co-growth and five days before eggplant uprooted

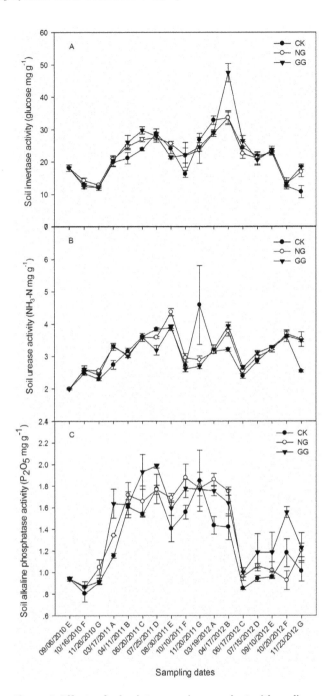

Figure 4. Effects of relay intercropping eggplant with garlic on the activities of invertase (A), urease (B), and alkaline phosphatase (C) in soil from September 2010 to November 2012. Fig. 4 was drawn using the software program Sigmaplot 12 (Systat Software, Inc.). CK: eggplant monoculture; NG: eggplant relay intercropping with normal garlic cv. G110; GG: eggplant relay intercropping with green garlic cv. G064 Error bars represent the standard error of the mean. The capital letters from A to G behind dates in the X-axis represent different periods of crop cycles in the experiment: A represents five days before eggplant transplanted in spring (garlic-only); B represents eggplant/garlic co-growth stage in spring; C represents eggplant-only stage; D represents five days before green garlic planted (eggplant-only); E represents several days before normal garlic planted; F represents eggplant/garlic co-growth stage in autumn; G represents several days before eggplant uprooted (co-growth).

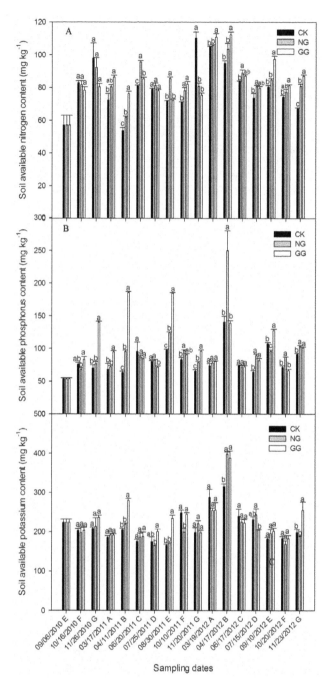

Figure 5. Effects of relay intercropping eggplant with garlic on the contents of available nitrogen (A), available phosphorus (B), and available potassium (C) in soil from September 2010 to November 2012. Fig. 5 was drawn using the software program Sigmaplot 12 (Systat Software, Inc.). CK: eggplant monoculture; NG: eggplant relay intercropping with normal garlic cv. G110; GG: eggplant relay intercropping with green garlic cv. G064 Error bars represent the standard error of the mean. Different letters above the bars indicate significant differences at a P<0.05 level (ANOVA and Duncan's multiple range test), n = 3. The capital letters from A to G behind dates in the X-axis represent different periods of crop cycles in the experiment: A represents five days before eggplant transplanted in spring (garlic-only); B represents eggplant/garlic co-growth stage in spring; C represents eggplant-only stage; D represents five days before green garlic planted (eggplant-only); E represents several days before normal garlic planted; F represents eggplant/garlic co-growth stage in autumn; G represents several days before eggplant uprooted (co-growth).

sampling dates, but for the GG treatment, the P content was significantly lower than that of the CK before the green garlic were planted on July 25. In addition to that, the available P content in GG treatment was significantly higher than that in CK treatment on many stages, including garlic-only stage (March 17, 2011) and eggplant/garlic co-growth stages (April 11, August 30, and November 20, 2011). Besides, the available P content of the GG treatment was significantly higher than that of the NG treatment on most sampling dates. In the third continuous cropping year in 2012, no fixed change rule was observed among the three treatments. Before the eggplant was transplanted on March 19, the available P content of the three treatments was about the same. Then on April 17, the eggplant/garlic co-growth stage, the P content of the NG treatment was 249.73 mg·kg^{-1}, significantly higher than 140.45 mg·kg^{-1} in the CK treatment, but for GG, there was no difference with CK. In the subsequent eggplant-only stage on June 17, when the eggplant grew vigorously, there were no significant differences among the three treatments. On September 10, 2012, when the green garlic had started rooting but the normal garlic had not planted, the available P content of the GG treatment was 127.56 mg·kg−1, significantly higher than 107.24 mg·kg-1 in the CK treatment, but in the NG treatment was 95.73 mg·kg-1, significantly lower than CK. Then on October 20, when the normal garlic grew together with eggplant, the available P content was also significantly higher in the NG treatment than that in the CK treatment. For the last samples in 2012, the available P content in the soil of the NG and GG treatments was significantly higher than that of the CK treatment.

Soil available K content. The soil available K content of the relay intercropping treatments on most sampling dates was higher than that of the CK treatment from September 2010 to November 2012 (Fig. 5 C). In 2010 from September to November, the available K content firstly decreased and then increased for all the three treatments. Then in the following spring, decline was seen on the soil available K content on March 17, 2011. In the eggplant/garlic co-growth period on April 11, the available K content of all three treatments increased again compared with that on March 17, and it was 279.52 mg·kg^{-1} in the GG treatment which was significantly higher than 204.49 mg·kg^{-1} in the CK treatment. On July 25 before the garlic was planted, while during the eggplant vigorous growth period, there was no significant difference between the CK and NG or GG treatment. After the green garlic grew up again on August 30, the available K content of the GG treatment was significantly higher than that of the CK treatment. In contrast, the K content was significantly lower than the CK treatment level in the NG treatment after the normal garlic rooting on October 10.

In 2012, the third year of eggplant continuous cropping, the soil available K content in the NG treatment was significantly higher than the CK treatment only during the eggplant/garlic co-growth stage on April 17. For the GG treatment, the K content was significantly higher than that of the CK treatment at two eggplant/garlic co-growth stages on April 17 and November 23.

Effects of relay intercropping eggplant with normal or green garlic on soil pH

The soil pH value varied from 7.36 to 8.00 during the three experimental years (Fig. 6). In 2010 from the pre-planting garlic on September 6 to pre-uprooting eggplant on November 26, the soil pH first increased and then decreased in CK and NG treatments, but kept decreasing in GG treatment. Then in both 2011 and 2012, the soil pH of all the three treatments was an initial decrease followed by an increase and a decrease, which

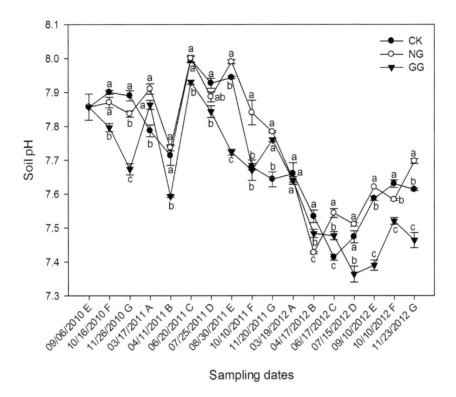

Figure 6. Effects of relay intercropping eggplant with garlic on the soil pH from September 2010 to November 2012. Fig. 6 was drawn using the software program Sigmaplot 12 (Systat Software, Inc.). CK: eggplant monoculture; NG: eggplant relay intercropping with normal garlic cv. G110; GG: eggplant relay intercropping with green garlic cv. G064 Error bars represent the standard error of the mean. Different letters above the bars indicate significant differences at a P<0.05 level (ANOVA and Duncan's multiple range test), n = 3. The capital letters from A to G behind dates in the X-axis represent different periods of crop cycles in the experiment: A represents five days before eggplant transplanted in spring (garlic-only); B represents eggplant/garlic co-growth stage in spring; C represents eggplant-only stage; D represents five days before green garlic planted (eggplant-only); E represents several days before normal garlic planted; F represents eggplant/garlic co-growth stage in autumn; G represents several days before eggplant uprooted (co-growth).

might be largely associated with the temperature under plastic tunnel and the water and fertilizer application situation.

On September 6, 2010, the initial value of soil pH was 7.86. Then on October 16, the preliminary growth stage of the normal garlic and green garlic, the soil pH in the GG treatment dropped to 7.80, which was significantly lower than that of CK, might because of a large amount of root exudates from green garlic or the interaction of eggplant and green garlic. However, for the NG treatment, the soil pH decreased unremarkably might because of the lower quantity of garlic. However, by the late eggplant growth period when the garlic thrived, the soil pH of both the NG and GG treatments was significantly below the soil pH value of the CK treatment. Contrary to the results of 2010, in the spring of 2011, when there was only garlic in the fields on March 17, the soil pH in the NG and GG treatments was significantly higher than that in the CK treatment. However, after the eggplant transplanted, the situation changed again. Except on July 25, the soil pH in the NG treatment was higher than CK, and some of the differences were significant. Especially on June 20, soil pH in NG reached up to 8.00. However, contradictory results were found in the GG treatment except on November 20, and the minimum value even dropped to 7.59 on April 11. These results indicate that the soil pH increased in NG with fewer root exudates but decreased in GG with more root exudates. Then in 2012, the results were similar to those in 2011 that soil pH of NG was always higher and in GG was lower in most cases than CK, and the differences were significant on some sampling dates. In addition, the soil pH levels

were generally lower in 2012 than that in 2011 in the corresponding period under continuous cropping.

Discussion

Conventional continuous monocropping may degrade soil quality and negatively affect soil physical processes, and even crop growth potential and yield. Relay intercropping is believed to reduce these negative aspects by maintaining soil quality and it continues to be an important farming practice in developing countries. Higher species richness may be associated with nutrient cycling characteristics that often can regulate soil fertility [42] and limit nutrient losses [43]. Enzyme assays can indicate the situation in terms of soil quality improvement, functional diversity of critical soil processes, rapid responses to changes in management, and sensitivity to environmental stresses [44–46]. In turn, soil enzymes are also mainly influenced by vegetation species [21] and land management practices [47–49].

Invertase and urease are the most important enzymes in the transformation of carbon and nitrogen in soils [50]. The activity of phosphatase is also positively correlated with the content of soil carbon and nitrogen, and it is also related to the soil pH and organic phosphorus content. Therefore, higher enzyme activities are expected to hold in soil. According to many studies, soil enzyme activities and soil nutrient contents are higher under intercropping systems than under monoculture system [34,51]. Dai [52] found that intercropping of peanut with *A. lancea* effectively increased soil urease and invertase activities. Li [53] also found that urease activities of intercropping sugarcane and soybean were promoted by 89% and 81%, respectively, compared with that of the monoculture models. In addition, Ahmad et al. [31] found that pepper intercropping with green garlic significantly increased the activities of invertase and alkaline phosphatase in soil. Our results of soil enzymes are in agreement with their conclusions. In our work, the invertase activity in relay intercropping systems was always higher than in eggplant monocropping system during the eggplant/garlic co-growth periods in the three experimental years. For urease and alkaline phosphatase activities, the relay intercropping treatments were higher than those of the CK treatment for most sampling dates. These indicate that the garlic relay intercropped with eggplant stimulated the soil invertase, urease, and alkaline phosphatase activities. In the eggplant/garlic relay intercropping systems, the eggplant and garlic secrete different root exudates, and the root exudates of the two crops interact with each other, affecting the microorganism in the soil, thus increase the soil enzyme activities. It is also possible that the garlic root exudates stimulate the soil enzyme activity by directly acting on them. Exceptional cases appear may result from the influence of many complex factors beyond cropping patterns, such as temperature, fertilizer and water management, or the plant growth situation. In 2011 and 2012, the overall trend of invertase activity of all treatments first increased then decreased over time. This trend could be related to the temperature in the plastic tunnel, which was increasing from March to July then decreasing from August to the end of the year (Fig. 1). In addition, continuous monoculture is detrimental to soil enzyme activities. This was obviously demonstrated by the alkaline phosphatase activity, which displayed a general decrease in 2012, the third year of eggplant continuous cropping, compared with that in 2011. The higher activity of soil alkaline phosphatase in relay intercropping treatments compared with the eggplant monoculture treatment alleviated this decline caused by continuous cropping.

Soil enzyme activity can reflect the level of soil fertility [21,54]. Increased enzyme activities promote the transformation of soil nutrients and improve the soil fertility. Conversely, soil enzyme activity is also affected by the soil nutrient contents. N, P, and K are the main nutrients in the soil, and plants absorb nutrients in available nutrient form. Previous research has demonstrated that the utilization efficiency of N, P, and K in a maize/mung bean intercropping system were significantly higher than that of a monoculture system [55]. It was also reported by Li [53] that the effective N and P contents of rhizospheric soil of intercrops sugarcane and soybean were increased by 66% and 311.7%, respectively, compared with those in monoculture systems. These results were similar to our work that the available N, P, and K content in relay intercropping treatments were always higher than that in the CK treatment. Urease activity is strongly indicative of enhanced nitrogen transformation in soil [56,57]. At the initial stage of relay intercropping in 2010, the available N content kept rising, which might be related to the increased urease activity because the urease can promote the replacement of organic nitrogen with available nitrogen. However, the available N content in NG and GG treatments was slightly lower than that in CK, which might because the superiority of relay intercropping was not so obvious at the early stage of relay intercropping, and two crops grown together could absorb more available N than there was only one crop. However, the absorption by garlic did not cause a significant drop in soil available N content and the bad effect is negligible. The results of 2011 and 2012 indicated that the available N content in NG and GG treatments was significantly higher than CK in most stages, but on November 20, 2011, it was significantly lower in NG and GG treatments than CK. This

exactly reversed result might because that, the available N was hardly needed for eggplants growing weakly or even dying before being uprooted, but the normal garlic and green garlic were still thriving at that time and needed more available N.

The higher available N, P, and K content in relay intercropping systems compared with that in monoculture systems demonstrates that the root exudates of normal garlic or green garlic stimulated the nutrition availability in soil. In addition, this may be the result of higher enzyme activity stimulated by garlic root exudates increasing the soil available nutrients [58,59]. Soil enzyme enrichment clearly occurs in response to soil nutrients and vegetation types [21]. This implies that increased enzyme activity is proportionally linked to the improved nutrient cycling and availability. Our study demonstrates that soil enzyme activities and nutrient contents had a similar variation trend in general and relay intercropping eggplant with garlic is better to improve soil fertility. As a result, the external input of N, P, and K chemical fertilizers can be reduced. Furthermore, increased soil fertility leads to good results on crop growth, yield, and land use efficiency. In our study, we also found that the eggplant grew stronger in relay intercropping systems than that in monoculture one, and the eggplant yield and combined output value of per unit area were also slightly higher. Although the eggplant yield declined with the continuous cropping year, relay intercropping could retard the production decrease to ensure the eggplant sustainable production (Data not shown). All these positive results on crop growth and yield could well be related to the higher soil nutrition in relay intercropping systems.

Soil pH is another important property related to soil characteristics and crop growth. Soil pH affects the activity of enzymes and the availability of nutrients [60]. As Acosta-Martínez [61] reported, phosphatase was significantly affected by soil pH, which controlled P availability by the transformation between organic and inorganic P. In other words, the availability of phosphorous in soil depends on the pH. Apparently in our work, the changing patterns of the soil phosphatase activity and pH displayed a similar downward trend in the three years of continuous cropping, which verified that the phosphatase activity was not only harmed by the continuous monoculture but also affected by the decreased soil pH. Results in this study also demonstrated the soil pH in the NG treatment was higher than that in the CK treatment. Increased soil pH led to large increases in nutrient availability [62]. The changes of available N, P, and K content in the CK and NG treatments were the same with soil pH. These results can be explained that the soil urease hydrolyzes urea to form ammonium carbonate, resulting in increased pH [63]. However, for the GG treatment, the soil pH was lower than CK, but the available nutrients contents were still higher than CK. This result was consistent with a study reported that in a wheat/faba bean intercropping system, the rhizosphere pH decreased, but the rhizosphere P availability increased compared with monocropped faba beans and wheat [64]. As is well-known, cropping systems have significant but different effects on soil with time. In relay intercropping systems, the roots of different crops can come into direct or indirect contact, change nutrient conditions and increase interactions, such as competition or mutualism of the two plants [65]. Soil properties based on biological and biochemical activities, especially those involved in energy flow and nutrient cycling, have often been demonstrated to respond to small changes in soil, thus providing sensitive information regarding subtle alterations in soil quality [66]. In eggplant/normal garlic or green garlic relay intercropping systems, both the soil enzyme activities and soil available nutrition content were promoted. However, some results revealed a discrepant change between the two parameters on the same sampling dates. These changes may arise from many intricate aspects of the system that are not yet clear.

Relay intercropping has been shown to motivate higher soil quality and produce more stable yields in a wide range of crop combinations. However, nothing is perfect. We can infer from the results that, the relay intercropping, although being better than monocropping, still exist some problem for improving the sustainability of vegetable production worldwide. Relay intercropping stimulated nutrition more available in soil than monocropping, making crops capture more nutrition, so in the long term, yield advantages of intercropping would have to pay for the higher fertilizer inputs. Besides, mechanization is another problem in intercropping, especially under plastic tunnel with limited land area, intercropping is very labor intensive. However in the developing countries, where manual labor is plentiful and cheap and the work is mainly done by hand in vegetable production, intercropping is still a better cultivation mode.

Conclusion

Conclusions are drawn from the study that the patterns of eggplant relay intercropping with normal garlic or green garlic can increase soil enzyme activities and available nutrient content and change soil pH, thus improve soil quality and ecological

environment. They are expected to help overcome soil sickness and continuous cropping obstacles. It certainly suggests that relay intercropping eggplant with garlic represents a potentially important contribution to meet challenge to sustainable increase the supply of vegetables in China.

Furthermore, it is a reasonable hypothesis that enhanced soil fertility is related to microbial community functions, thus contributing to increased crop growth and yield of the two relay intercropping crops. Clearly, further work is needed to test the microbial community in soil and to elucidate relationships among the soil microorganism, enzyme activity, nutrition, and crop growth and yield. Besides, soil sickness is a result of long term continuous cropping. Longer study periods and larger study plots will help get more convincing results. However, in order to approach local actual vegetable production practice, natural greenhouse environment and field technique used by local farmers need to be kept.

Author Contributions

Conceived and designed the experiments: ZHC MYW. Performed the experiments: MYW CNW MRZ. Analyzed the data: HWM HJZ. Contributed reagents/materials/analysis tools: HWM MRZ. Wrote the paper: MYW.

References

1. FAOSTAT website (2014) Available: http://faostat3.fao.org/faostat-gateway/go/to/compare/Q/QC/E.
2. Kibunja CN (2007) Impact of long-term application of organic and inorganic nutrient sources in a maize-bean rotation to soil nitrogen dynamics and soil microbial populations and activity: University of Nairobi.
3. Kibunja CN, Mwaura FB, Mugendi DN, Gicheru PT, Wamuongo JW, et al. (2012) Strategies for maintenance and improvement of soil productivity under continuous maize and beans cropping system in the sub-humid highlands of kenya: case study of the Long-term trial at Kabete. In: Bationo A, editor. Lessons learned from Long-term Soil Fertility Management Experiments in Africa: Springer. 59–84.
4. Nafziger E (2009) Cropping systems. Illinois Agronomy Handbook.24 ed. University of Illinois at Urbana-Champaign: Cooperative Extension Service. 49–63.
5. Shipton PJ (1975) Take-all decline during cereal monoculture. In: Bruehl GW, editor. Biology and Control of Soil-Borne Plant Pathogens. St. Paul, Minnesota, USA: The American Phytopathological Society. 137–144.
6. Blanco-Canqui H, Lal R (2008) Principles of Soil Conservation and Management. Heidelberg, Germany: Springer. 617 p.
7. Hiddink GA, Termorshuizen AJ, van Bruggen AHC (2010) Mixed Cropping and Suppression of Soilborne Diseases. In: Lichtfouse E, editor. Genetic Engineering, Biofertilisation, Soil Quality and Organic Farming. France: Springer. 119–146.
8. Cook RJ, Veseth RJ (1991) Wheat Health Management. St. Paul, Minnesota, USA: The American Phytopathological Society Press.
9. Xiao X, Cheng Z, Meng H (2012) Intercropping with garlic alleviated continuous cropping obstacle of cucumber in plastic tunnel. Acta Agr Scand B-S P 62: 696–705.
10. Bhatti IH, Ahmad R, Jabbar A, Nazir MS, Mahmood T (2006) Competitive behaviour of component crops in different sesame-legume intercropping systems. Int J Agr Biol 8: 165–167.
11. Lithourgidis AS, Dordas CA, Damalas CA, Vlachostergios DN (2011) Annual intercrops: an alternative pathway for sustainable agriculture. Aust J Crop Sci 5: 396–410.
12. Famaye AO, Iremiren GO, Olubamiwa O, Aigbekaen AE, Fademi OA (2011) Intercropping cocoa with rice and plantain influencing cocoa morphological parameters and weed biomass. J Agric Sci Technol: 745–750.
13. Gomes JKO, Silva PSL, Silva KMB, Rodrigues Filho FF, Santos VG (2007) Effects of weed control through cowpea intercropping on mayze morphology and yield. Planta Daninha 25: 433–441.
14. Latif MA, Mehuys GR, Mackenzie AF, Alli I, Faris MA (1992) Effects of legumes on soil physical quality in a maize crop. Plant Soil 140: 15–23.
15. Sebastiana M, Engracia M, Carlos RJ, Francisco HJ (2007) Chemical and biochemical properties of a clay soil under dryland agriculture system as affected by organic fertilization. Eur J Agron 26: 327–334.
16. Yakovchenko V, Sikora LJ, Kaufman DD (1996) A biologically based indicator of soil quality. Biol Fertil Soils 21: 245–251.
17. Dick RP (1994) Soil enzyme activities as indicators of soil quality. In: Doran JW, editor. Defining Soil Quality for a Sustainable Environment. Madison, WI: Soil Science Society of America Journal. 107–124.
18. Zhang W, Zhang M, An S, Xiong B, Li H, et al. (2012) Ecotoxicological effects of decabromodiphenyl ether and cadmium contamination on soil microbes and enzymes. Ecotoxicol Environ Saf 82: 71–79.
19. Ndiaye EL, Sandeno JM, McGrath D, Dick RP (2000) Integrative biological indicators for detecting change in soil quality. Am J Alternative Agr 15: 26–36.
20. Li H, Shao H, Li W, Bi R, Bai Z (2012) Improving soil enzyme activities and related quality properties of reclaimed soil by applying weathered coal in opencast-mining areas of the Chinese Loess Plateau. Clean - Soil, Air, Water 40: 233–238.
21. Wang B, Xue S, Liu GB, Zhang G, Li G, et al. (2012) Changes in soil nutrient and enzyme activities under different vegetations in the Loess Plateau area, Northwest China. Catena: 186–195.
22. Saha S, Prakash V, Kundu SS, Kumar N, Mina BL (2008) Soil enzymatic activity as affected by long term application of farm yard manure and mineral fertilizer under a rainfed soybean-wheat system in N-W Himalaya. Eur J Soil Biol 44: 309–315.
23. Schachtman DP, Reid RJ, Ayling SM (1998) Phosphorus uptake by plants: from soil to cell. Plant Physiol 116: 447–453.
24. Zhang Y, Chen L, Chen Z, Sun C, Wu Z, et al. (2010) Soil nutrient contents and enzymatic characteristics as affected by 7-year no tillage under maize cropping in a meadow brown soil. R C Suelo Nutr Veg 10: 150–157.
25. Burns RG (1978) Enzyme activity in soil: Some theoretical and practical considerations. In: Burns RG, editor. Soil Enzymes. New York: Academic Press. 295–340.
26. Marschner H (1995) Mineral nutrition of higher plants. Harcourt, Brace, London: Academic Press.
27. Jaillard B, Ruiz L, Arvieu JC (1996) pH mapping in transparent gel using color indicator videodensitometry. Plant Soil 183: 85–93.
28. Ruiz I, Arvieu JC (1990) Measurement of pH gradients in the rhizosphere. Symbiosis 9: 71–75.
29. Han X, Cheng Z, Meng H (2012) Soil properties, nutrient dynamics, and soil enzyme activities associated with garlic stalk decomposition under various conditions. PLoS ONE 7: e50868.
30. Khan MA, Cheng Z (2010) Influence of garlic root exudates on cyto-Morphological alteration of the hyphae of Phytophthora capsici, the cause of phytophthora blight in pepper. Pak J Bot 42: 4353–4361.
31. Ahmad I, Cheng Z, Meng H, Liu T, Wang M, et al. (2013) Effect of pepper-garlic intercropping system on soil microbial and bio-chemical properties. Pak J Bot 45: 695–702.
32. Mousavi SR, Eskandari H (2011) A general overview on intercropping and its advantages in sustainable agriculture. J Appl Environ Biol Sci 1: 482–486.
33. Dahmardeh M, Rigi K (2013) The influence of intercropping maize (Zea mays L.) green gram (Vigna Radiata L.) on the changes of soil temperature, moisture and nitrogen. Int J Ecosyst 3: 13–17.

34. Inal A, Gunes A, Zhang F, Cakmak I (2007) Peanut/maize intercropping induced changes in rhizosphere and nutrient concentrations in shoots. Plant Physiol Biochem 45: 7.

35. Zhang F, Li L (2003) Using competitive and facilitative interactions in intercropping systems enhances crop productivity and nutrient-use efficiency. Plant Soil 248: 305–312.

36. Unlu H, Sari N, Solmaz I (2010) Intercropping effect of different vegetables on yield and some agronomic properties. J Food Agric Environ 8: 723–727.

37. Guan S (1986) Soil enzyme and research method (in Chinese). Beijing, China: Agriculture Press 376 p.

38. Bermner JM, Shaw K (1955) Determination of ammonia and nitrate in soil. J Agr Sci 46: 320–328.

39. Olsen SR, Cole CV, Watanabe FS, Dean LA (1954) Estimation of available phosphorus in soils by extraction with sodium bicarbonate. U S Dept Agr Circular 939: 1–19.

40. Pratt PF (1965) Potassium. In: Black CA, Evans DD, Ensminger LE, White JL, Clark FE et al., editors. Methods of Soil Analysis. Madison, Wisconsin, USA: American Society of Agronomy, Inc. 1025–1027.

41. Barber SA, Chen JH (1990) Using a mechanistic model to evaluate the effect of soil pH on phosphorus uptake. Plant Soil 124: 183–186.

42. Russell AD (2002) Antibiotic and biocide resistance in bacteria: Introduction. J Appl Microbiol 92: 1S–3S.

43. Hauggaard-Nielsen H, Ambus P, Jensen ES (2003) The comparison of nitrogen use and leaching in sole cropped versus intercropped pea and barley. Nutr Cycl Agroecosys 65: 289–300.

44. Dick RP (1997) Soil enzyme activities as integrative indicators of soil health. In: Pankhurst CE DB, Gupta VVSR, editor. Biological Indicators of Soil Health. Wallingford: CAB. 121–156.

45. Nannipieri P, Kandeler E, Ruggiero P (2002) Enzyme activities and microbiological and biochemical processes in soil. In: Burns RG, Dick RP, editors. Enzymes in the Environment. New York: Marcel Dekker, Inc. 1–33.

46. Caldwell BA (2005) Enzyme activities as a component of soil biodiversity: a review. Pedobiologia 49: 637–644.

47. Acosta-Martínez V, Burow G, Zobeck TM, Allen VG (2010) Soil microbial communities and function in alternative systems to continuous cotton. Soil Sci Soc Am J 74: 1181–1192.

48. Carney KM, Matson PA, Bohannan BJM (2004) Diversity and composition of tropical soil nitrifiers across a plant diversity gradient and among land-use types. Ecol lett 7: 684–694.

49. Yao HY, Bowman D, Wei S (2006) Soil microbial community structure and diversity in a turfgrass chronosequence: land-use change versus turfgrass management. Appl Soil Ecol 34: 209–218.

50. Eivazi F, Bayan MR (1996) Effects of long-term prescribed burning on the activity of select soil enzymes in an oak hickory forest. Can J For Res 26: 1799–1804.

51. Ghosh PK, Mohanty M, Bandyopadhyay KK, Painuli DK, Misra AK (2006) Growth, competition, yields advantage and economics in soybean/pigeonpea intercropping system in semi-arid tropics of India: II. Effect of nutrient management. Field Crops Res 96: 90–97.

52. Dai C, Chen Y, Wang X, Li P (2013) Effects of intercropping of peanut with the medicinal plant *Atractylodes lancea* on soil microecology and peanut yield in subtropical China. Agroforest Syst 87: 417–426.

53. Li X, Mu Y, Cheng Y, Liu X, Nian H (2013) Effects of intercropping sugarcane and soybean on growth, rhizosphere soil microbes, nitrogen and phosphorus availability. Acta Physiol Plant 35: 1113–1119.

54. Burke DJ, Weintraub MN, Hewins CR, Kalisz S (2011) Relationship between soil enzyme activities, nutrient cycling and soil fungal communities in a northern hardwood forest. Soil Biol Biochem 43: 795–803.

55. Chowdhury MK, Rosario EL (1994) Comparison of nitrogen, phosphorus and potassium utilization efficiency in maize/mungbean intercropping. J Agr Sci 122: 193–199.

56. Edwards IP, Bürqmann H, Miniaci C, Zeyer J (2006) Variation in microbial community composition and culturability in the rhizosphere of *Leucanthemopsis alpina* (L.) heywood and adjacent bare soil along an alpine chronosequence. Microb Ecol 52: 679–692.

57. Xing S, Chen C, Zhou B, Zhang H, Nang Z, et al. (2010) Soil soluble organic nitrogen and active microbial characteristics under adjacent coniferous and broadleaf plantation forests. J Soils Sediments 10: 748–757.

58. Kroehler CJ, Linkins AE (1988) The root surface phosphatases of *Eriophorum vaginatum*: Effects of temperature, pH, substrate concentration and inorganic phosphorus. Plant Soil 105: 3–10.

59. Johnson D, Leake JR, Read DJ (2005) Liming and nitrogen fertilization affects phosphatase activities, microbial biomass and mycorrhizal colonisation in upland grassland. Plant Soil 271: 157–164.

60. Dick WA, Cheng L, Wang P (2000) Soil acid and alkaline phosphatase activity as pH adjustment indicators. Soil Biol Biochem 32: 1915–1919.

61. Acosta-Martínez V, Upchurch DR, Schubert AM, Porter D, Wheeler T (2004) Early impacts of cotton and peanut cropping systems on selected soil chemical, physical, microbiological and biochemical properties. Biol Fertil Soils 40: 44–54.

62. Bagayoko M, Alvey S, Neumann G, Buerkert A (2000) Root-induced increases in soil pH and nutrient availability to field-grown cereals and legumes on acid sandy soils of Sudano-Sahelian West Africa. Plant Soil 225: 117–127.

63. Zahir ZA, Malik MAR, Arshad M (2001) Soil enzymes research: a review. Online Journal of Biological Sciences: Asian Network for Scientific Information. 299–307.

64. Song YN, Zhang FS, Marschner P, Fan FL, Gao HM, et al. (2007) Effect of intercropping on crop yield and chemical and microbiological properties in rhizosphere of wheat (*Triticum aestivum* L.), maize (*Zea mays* L.), and faba bean (*Vicia faba* L.). Biol Fertil Soils 43: 565–574.

65. Zhang NN (2010) Effets of intercropping and Rhizobium inoculation on yeild and rhizosphere bacterial community of faba bean (*Vicia faba* L.). Biol Fertil Soils 46: 625–639.

66. Pascual JA, Garcia C, Hernandez T, Moreno JL, Ros M (2000) Soil microbial activity as a biomarker of degradation and remediation processes. Soil Biol Biochem 32: 1877–1883.

16

Reduced Levels of Membrane-Bound Alkaline Phosphatase Are Common to Lepidopteran Strains Resistant to Cry Toxins from *Bacillus thuringiensis*

Juan Luis Jurat-Fuentes[1]*, Lohitash Karumbaiah[2], Siva Rama Krishna Jakka[1], Changming Ning[3], Chenxi Liu[3], Kongming Wu[3], Jerreme Jackson[4], Fred Gould[5], Carlos Blanco[6], Maribel Portilla[7], Omaththage Perera[7], Michael Adang[2,8]

1 Department of Entomology and Plant Pathology, University of Tennessee, Knoxville, Tennessee, United States of America, 2 Department of Entomology, University of Georgia, Athens, Georgia, United States of America, 3 State Key Laboratory of Plant Disease and Insect Pests, Institute of Plant Protection, Chinese Academy of Agricultural Science, Beijing, People's Republic of China, 4 Genome Science and Technology Program, University of Tennessee, Knoxville, Tennessee, United States of America, 5 Department of Entomology, North Carolina State University, Raleigh, North Carolina, United States of America, 6 Animal and Plant Health Inspection Service, Biotechnology Regulatory Services, United States Department of Agriculture, Riverdale, Maryland, United States of America, 7 Southern Insect Management Research Unit, Agricultural Research Service, United States Department of Agriculture, Stoneville, Mississippi, United States of America, 8 Department of Biochemistry and Molecular Biology, University of Georgia, Athens, Georgia, United States of America

Abstract

Development of insect resistance is one of the main concerns with the use of transgenic crops expressing Cry toxins from the bacterium *Bacillus thuringiensis*. Identification of biomarkers would assist in the development of sensitive DNA-based methods to monitor evolution of resistance to Bt toxins in natural populations. We report on the proteomic and genomic detection of reduced levels of midgut membrane-bound alkaline phosphatase (mALP) as a common feature in strains of Cry-resistant *Heliothis virescens*, *Helicoverpa armigera* and *Spodoptera frugiperda* when compared to susceptible larvae. Reduced levels of *H. virescens* mALP protein (HvmALP) were detected by two dimensional differential in-gel electrophoresis (2D-DIGE) analysis in Cry-resistant compared to susceptible larvae, further supported by alkaline phosphatase activity assays and Western blotting. Through quantitative real-time polymerase chain reaction (qRT-PCR) we demonstrate that the reduction in HvmALP protein levels in resistant larvae are the result of reduced transcript amounts. Similar reductions in ALP activity and mALP transcript levels were also detected for a Cry1Ac-resistant strain of *H. armigera* and field-derived strains of *S. frugiperda* resistant to Cry1Fa. Considering the unique resistance and cross-resistance phenotypes of the insect strains used in this work, our data suggest that reduced mALP expression should be targeted for development of effective biomarkers for resistance to Cry toxins in lepidopteran pests.

Editor: David Ojcius, University of California Merced, United States of America

Funding: This material is based upon work supported by Cooperative State Research, Education, and Extension Service, U.S. Department of Agriculture, under National Research Initiative Award No. 2004-35607-14936 to M.J.A. and J.L.J.-F., and Biotechnology Risk Assessment Grant Award No. 2008-39211-19577 to J.L.J.-F. and O.P.P. The funders had no role in study design, data collection and analysis, decision to publish, or preparation of the manuscript. Mention of trade names or commercial products in this publication is solely for the purpose of providing specific information and does not imply recommendation or endorsement by the U.S. Department of Agriculture.

Competing Interests: The authors have declared that no competing interests exist.

* E-mail: jurat@utk.edu

Introduction

Cry toxins produced as crystalline inclusions by the bacterium *Bacillus thuringiensis* (Bt) are the most widely used insecticidal trait in transgenic crops for insect control [1]. Due to the wide adoption of Bt transgenic crops, the future efficacy of this technology is threatened by the evolution of resistance by target pests. After more than a decade of commercialization, recent reports support field-evolved resistance to Bt crops in *Helicoverpa zea* [2], *Spodoptera frugiperda* [3], and *Busseola fusca* [4]. Key to the implementation of strategies to delay and manage resistance outbreaks in field environments is the development of efficient methods for early detection. Development of successful DNA-based monitoring methods relies on the identification of biomarker molecules that are specifically and consistently altered in resistant insects.

Optimally, resistance biomarkers should efficiently differentiate susceptible and resistant insects, independent of the resistance mechanism, Bt crop, or Cry toxin involved. However, the multi-step mode of Cry toxin action and the diverse resistance mechanisms described to date [5,6] highlight the difficulty of identifying biomarkers with these ideal characteristics.

Cry toxins target the insect midgut cells to compromise the gut epithelium barrier and facilitate the onset of septicemia [7]. Although the specific mechanism resulting in enterocyte death is still controversial [8], commonly accepted steps in the intoxication process include solubilization of the crystal toxin and activation by the insect gut fluids. Activated toxins are attracted to the brush border membrane of the midgut cells through low affinity binding to glycosylphosphatidylinositol-anchored (GPI-) proteins [9], such as aminopeptidase-N (APN) or membrane-bound alkaline phosphatase (mALP). This initial binding step facilitates subsequent binding of higher affinity to cadherin-like proteins [10], which leads to further processing of the toxin, resulting in formation of a toxin oligomer. Toxin oligomers display high binding affinity towards N-acetylgalactosamine (GalNAc) residues on GPI-anchored proteins [11], resulting in concentration of toxin oligomers on specific membrane regions called lipid rafts, where they insert into the membrane forming a pore that leads to osmotic cell death [12]. Alternatively, binding of toxin monomers to cadherin has been reported to activate intracellular signaling pathways that result in cell death by oncosis [13].

Based on the crucial role of toxin binding to cadherin and the observation that mutations in cadherin genes are linked with resistance to Cry1Ac in *Heliothis virescens* [14], *Helicoverpa armigera* [15], and *Pectinophora gossypiella* [16], DNA-based assays to detect cadherin-gene alterations have been tested for resistance detection [17,18]. However, the existence of alternative resistance mechanisms [19,20,21,22] suggests that, at least in some cases, tests based on detection of cadherin alterations would be inefficient in detecting Bt resistance.

The main goal of the present study was to identify an efficient biomarker for resistance to diverse Cry toxins. Using differential proteomics (2D-DIGE), we detected reduced levels of mALP from *H. virescens* larvae (HvmALP) in three strains displaying diverse Cry resistance phenotypes when compared to susceptible larvae. Quantitative RT-PCR data supported that this reduction in HvmALP levels was due to reduction in amounts of HvmALP transcripts. Reduced levels of HvmALP homologues were also detected for a Cry1Ac-resistant strain of *H. armigera*, and Cry1Fa-resistant strains of *S. frugiperda*, further evidence supporting the potential development of resistance monitoring methods using reduced mALP levels as an efficient Cry resistance biomarker in lepidopteran insects.

Materials and Methods

Insect strains

H. virescens laboratory strains YDK, YHD2-B, CXC, and KCBhyb have been previously described [21,23,24]. Briefly, YDK is an un-selected susceptible strain, while YHD2-B was generated after continuous selection of larvae from the YHD2 strain with Cry1Ac. Both CXC and KCBhyb originated by selecting with Cry2Aa the offspring from backcrosses of moths from Cry1Ac/Cry2Aa resistant strains (CP73-3 and KCB, respectively) to susceptible adults. Both CXC and KCBhyb larvae were resistant to Cry1Ac (200- to 300-fold) and Cry2Aa (more than 250-fold) when compared to YDK larvae [22]. In contrast, YHD2-B larvae are 73,000-fold resistant to Cry1Ac [25], but display only low levels (4 to 25-fold) of cross-resistance to Cry2Aa [23].

The Cry1Ac-susceptible strain of *H. armigera* 96S was originally collected from Xinxiang County (Henan Province, P. R. China) in 1996, and the larvae have since been reared in the laboratory on an artificial diet without exposure to any Bt toxins or chemical insecticides. The *H. armigera* Cry1Ac-resistant strain BtR was derived from 96S by selection with Cry1Ac protoxin incorporated into the diet [26]. After continuous selection for 87 generations, larvae of this strain display more than 2,900-fold resistance to Cry1Ac when compared to 96S larvae.

The 456 and 512 strains of *S. frugiperda* were originated from isofamilies of insects collected in Puerto Rico in 2009 [27]. Two Cry-susceptible strains of *S. frugiperda* were used as reference in our studies. Eggs of one of the strains (Mon) were kindly supplied by Nancy Adams (Monsanto), while eggs of the second strain (Ben) were purchased from Benzon Research (Carlisle, PA).

All insects were reared in the laboratory using artificial diet as previously described [23,26]. Fifth instar larvae from each strain were dissected, and midguts frozen and kept at $-80°C$ until used to prepare BBMV as described below, or placed in RNA*later* (Ambion) overnight at $4°C$ and then stored at $-80°C$.

BBMV purification

BBMV were isolated by the differential centrifugation method of Wolfersberger *et al.* [28] with minor modifications for *H. virescens* and *S. frugiperda* [25]. BBMV proteins were quantified by the method of Bradford [29], using BSA as standard, and kept at $-80°C$ until used. Specific activity of N-aminopeptidase (APN) using leucine-*p*-nitroanilide as substrate was used as a marker for brush border enzyme enrichment in the BBMV preparations. APN activities in the final BBMV preparations from all insect species were enriched 5–8 fold when compared to initial midgut homogenates.

2D Differential In Gel electrophoresis (2D-DIGE) analysis of BBMV proteomes

BBMV proteins to be used in 2DE were extracted and precipitated using the 2D Clean-Up Kit (GE Healthcare). Precipitated proteins were dissolved in solubilization buffer (5 M urea [Plus-One; GE Healthcare], 2 M thiourea [Sigma], 2% CHAPS [Plus-One, GE Healthcare] and Complete^{TM} protease inhibitors cocktail [Roche]). After centrifugation at $15,700 \times g$ for 10 min, solubilized proteins in the supernatant were quantified using the 2D Quant Kit (GE Healthcare) following manufacturer's instructions. BBMV proteins (50 μg per sample) were minimally labeled with Cy3 or Cy5 CyDyes (GE Healthcare) following manufacturer's instructions. Three replicates for each strain from independent BBMV preparations were used. Additionally, samples were also reverse CyDye labeled to account for possible differential labeling effects.

BBMV protein samples (50 μg) were used to rehydrate 11 cm (for Western blots) or 18-cm (DIGE analysis) and pH range 4–7 Immobiline DryStrips (GE Healthcare) overnight in rehydration buffer (solubilization buffer plus 0.002% bromophenol blue, 0.018 M dithiothreitol [DTT], and 0.5% ampholytes). Solutions used to rehydrate 18 cm pH 4–7 Immobiline strips for the DIGE analysis contained three samples labeled with a distinct CyDye (Cy2, Cy3, and Cy5). The Cy3and Cy5 labeling was used for experimental samples (three biological replicates for each strain, each sample was labeled with Cy3 and Cy5 to confirm lack of labeling bias), while the Cy2-labeled sample consisted of equal amounts of all the analyzed BBMV protein samples (50 μg total) and was used as internal reference for comparison of diverse gels. Following rehydration, strips were subjected to isoelectric focusing using a Multiphor II unit following manufacturer's recommendations (GE Healthcare). Temperature was maintained at $20°C$ throughout focusing. Focused strips were equilibrated for 15 min in equilibration buffer (6 M urea [Plus-One; GE Healthcare], 2% SDS, 30% glycerol, 0.05 M Tris [pH 8.8], 0.002% bromophenol blue) containing 1% DTT followed by a second equilibration for 15 minutes in equilibration buffer plus 4% iodoacetamide. For second dimension separation we used the Ettan Dalt*six* system (GE Healthcare) and SDS-8% PAGE gels following manufacturer's instructions. Separated proteins were fixed in 30% ethanol, 7.5% acetic acid overnight at room temperature for DIGE analysis.

Gels were imaged with a Typhoon 9400 scanner (GE Heatlhcare), optimizing the photomultiplier tubes for each laser to achieve the broadest dynamic range. Wavelengths for the filters/lasers were 532 nm/580 nm for Cy3 and 633 nm/670 nm for Cy5. Gel images were analyzed using DeCyder software (GE

Healthcare). Reference maps were obtained for each gel using the Cy2-labeled sample and spot correspondence established to compare protein spot abundance within and between gels. Two-fold differences in spot volume were considered as relevant between samples. Statistical significance was estimated using one-way Analysis Of Variance (ANOVA) in the DeCyder software.

Protein identification using peptide mass fingerprinting (PMF) was done at the University of Georgia Proteomics Facility as described elsewhere [30]. Generated PMF data were used in correlative searching strategies to search the Metazoan subset of NCBI using ProFound (http://prowl.rockefeller.edu/) with a confidence level of 0.1 Da and methionine oxidation as a modification.

Quantification of alkaline phosphatase (ALP) and aminopeptidase (APN) activities

Specific ALP and APN enzymatic activities of BBMV proteins (1 µg) from *H. virescens*, *H. armigera*, and *S. frugiperda* were measured as described elsewhere [31], except that for *H. armigera* BBMV specific ALP activity was determined using a commercial kit (Alkaline phosphatase, Hou-Bio, P. R. China). Enzymatic activities were monitored for 2–5 min. as changes in OD at 405 nm wavelength at room temperature in a microplate reader (BioTek), and the maximum initial velocity (Vmax) calculated using the associated KC4 Data Analysis Software. One enzymatic unit was defined as the amount of enzyme that would hydrolyze 1.0 µmole of substrate to chromogenic product per minute at the specific reaction pH and temperature. Data shown are the mean specific activities from at least three independent BBMV batches from each strain measured in at least three independent experiments. Statistical significance was determined through analysis of variance (ANOVA) using Holm-Sidak or Tukey's multiple pairwise comparison tests (overall significance level = 0.05), using SigmaPlot v.11.0 software (Systat Software Inc., San Jose, CA, USA). Since APN activity data failed an equal variance test, in this case we used ANOVA on ranks (Kruskal-Wallis test, overall significance level = 0.05) to determine statistical significance.

Western blotting

BBMV proteins to be analyzed by one-dimension electrophoresis (1D) were solubilized in sample buffer [32]. Solubilized BBMV proteins (20 µg) were then heat-denatured for 5 min. and loaded on SDS-8% PAGE gels. Following electrophoretic separation BBMV proteins were transferred overnight at 4°C to polyvinylidene difluoride Q (PVDF) membrane filters (Millipore) at 20 V constant voltage. Filters were blocked for one hour in PBST (135 mM NaCl, 2 mM KCl, 10 mM Na_2HPO_4, 1.7 mM KH_2PO_4, pH 7.5, 0.1% Tween-20) plus 3% BSA. After blocking, all filter incubations and washes were done in PBST plus 0.1% BSA. Blocked filters were probed with antisera against *Bombyx mori* mALP (a gift from Dr. Masanobu Itoh, Kyoto Institute of Technology, Japan), *Anopheles gambiae* mALP (a gift from Dr. Gang Hua, University of Georgia, USA), *M. sexta* 120-kDa APN [33], or HevCaLP cadherin [21], to detect HvALP, *S. frugiperda* mALP, APN and cadherin on BBMV. Goat anti-rabbit antibodies conjugated with horseradish peroxidase (HRP) were used as secondary antibodies and blots developed using enhanced chemiluminescence (West pico, Pierce).

Quantitative real-time PCR

Total RNA was extracted from frozen midguts using an RNeasy mini kit (Qiagen) for *H. virescens* or Trizol reagent (Invitrogen) for *H. armigera* samples. Purified RNA was subjected to DNaseI treatment (Takara) to remove any residual DNA according to the manufacturer's instructions. For *H. virescens* samples, total RNA was quantified using RiboGreen® reagent (Molecular Probes) and a fluorescence microplate reader (FLUOstar GALAXY; BMG). The integrity of total RNA was verified by the visualization of a distinct band corresponding to 18S rRNA resolved on a 1.2% formaldehyde agarose gel. First strand cDNA was synthesized in reactions containing 2 µg of total RNA pooled from 5 *H. virescens* midguts using PowerScript™ reverse transcriptase (Clontech), anchored oligo(dT)$_{20}$ primer (Invitrogen) and other reaction components in a 10 µl reaction. For samples from *H. armigera* larvae, total RNA was reverse-transcribed with SuperScript III RNaseH⁻ reverse transcriptase (Invitrogen).

The first strand cDNA was used as a template for qRT-PCR. For *H. virescens* samples, we designed specific primers to amplify 100-bp fragments of an internal region in HvmALP1 (accession no. FJ416470) and HvmALP2 (accession no. FJ416471) isoforms of HvmALP. These HvmALP isoforms were selected because they displayed the highest sequence divergence [34] and the targeted region could be used to differentiate between these two HvmALP isoforms. Forward primer 5′ GAT TTA GGA CGC GAC AGT ATG 3′ and reverse primer 5′ CAG CGG TAA CAT CTG GTC GAA 3′ were used to amplify HvmALP1, while forward primer 5′ GGG ATG TTG ATC TAG ACA ACG T 3′ and reverse primer 5′ CAG CTG TAA CAT CTG GTC GAA T 3′ were used to amplify HvmALP2. As endogenous control, we amplified ribosomal protein S18 (RpS18) using forward 5′ ATG GCA AAC GCA AGG TTA TGT TT 3′ and reverse 5′ TTG TCA AGA TCA ATA TCG GCT TT 3′ primers designed based on the *B. mori* RpS18 sequence (accession no. AY69334).

For *H. armigera* samples, primers were made to amplify a 128 bp conserved region between the HaALP1 (accession no. EU729322) and HaALP2 (accession no. EU729323) isoforms. TaqMan probes (Invitrogen) were labeled at the 5′ end by the reporter dye FAM and at the 3′end by the quencher dye TAMRA. Forward primer 5′ ATA GGC GTA GAC GGC ACG G 3′, reverse primer 5′ CGA GTC GTC GTC ACA ATA CCG 3′, and 5′-FAM CGC CGA GGA GAC TGT CAA GCC GCT T3′-TAMARA were used for HaALP fragment amplification. As endogenous control, we amplified a 184 bp fragment of *H. armigera* actin (accession no. X97615) with forward primer 5′ CAC AGA TCA TGT TCG AGA CGT TCA A 3′, reverse primer 5′- GCC AAG TCC AGA CGC AGG AT-3′ and 5′-FAM CCG CCA TGT ACG TCG CCA TCC AGG 3′-TAMARA.

Quantitative RT-PCR reactions were conducted with three technical replicates for each of three independent biological samples on an ABI 7900HT fast real-time PCR system (Applied Biosystems). PCRs for each template and primer combination were conducted in triplicate and replicated with cDNA prepared from at least three independent biological samples. For *H. virescens* samples reactions (12.5 µl) consisted of a cDNA equivalent of 20 ng of total RNA, 6.25 µl Power® SYBR Green PCR Master Mix (Applied Biosystems) and forward and reverse primers at 0.9 µM concentration. Reactions for *H. armigera* samples (25 µl) consisted of 12.5 µl of *Premix Ex* Taq (2×) (TaKaRa), 0.5 µl of Rox Reference Dye (50×), probe (0.2 µM), primers (0.4 µM), 1 µl of sample cDNA and sterilized ultrapure H_2O (Millipore). Amplification conditions consisted of an initial denaturation at 95°C for 10 min followed by 40 cycles of 95°C for 15 s, 58°C for 30 s and 72°C for 30 s for *H. virescens* samples, while for *H. armigera* samples a single step for annealing and extension was done at 60°C for 60 s. The absolute value of the slope (Ct value Vs Log) for each primer set was <0.1 and all amplification efficiencies when compared to the endogenous control were >99%, indicating a passing validation. Data obtained were analyzed with the relative $2^{-\Delta\Delta C_T}$ quantitation method to calculate transcript abundance [35], using S18 (*H. virescens*) or actin (*H. armigera*) as internal standards. Transcript amounts were standardized to 1 with the sample from susceptible larvae containing the highest transcript levels from the three biological replicate reactions performed. Upon completion of a quantitative PCR run, a dissociation curve analysis was conducted to verify the absence of any nonspecific amplicons. Statistical significance was tested with ANOVA in the SigmaPlot v.11.0 software, using the Holm-Sidak method (overall significance level = 0.05) for multiple comparisons.

Results

Reduced levels of alkaline phosphatase are common to diverse Cry-resistant *H. virescens* strains

In an attempt to identify resistance biomarkers, we used a differential proteomics approach (2D-DIGE) to compare BBMV proteomes from susceptible (YDK) and three resistant *H. virescens* strains (CXC, KCBHyb, and YHD2-B) displaying diverse resistance phenotypes [36]. In these assays we detected four protein spots (Fig. 1) that were significantly down-regulated (3 to 4-fold) in BBMV from all resistant larvae compared to vesicles from susceptible insects (one-way ANOVA, P<0.05). In contrast, no significant differences in the levels of these four protein spots were found among BBMV from the resistant strains. All four protein spots were previously identified as membrane-bound alkaline phosphatase using mass spectrometry and Western blotting [30,37].

To further examine the reduction of alkaline phosphatase in BBMV from resistant compared to susceptible insects, we performed ALP activity assays. Compared to YDK samples, specific ALP activity in all resistant strains was significatively

reduced (P<0.05, Holm-Sidak method) by about 50% (Fig. 2A). In contrast, activity of another brush border membrane enzyme, APN, was not statistically different among the BBMV samples (P = 0.415, Kruskal-Wallis ANOVA on ranks). In agreement with these data, antisera against membrane-bound alkaline phosphatase from *B. mori* detected lower levels of HvmALP protein in all resistant strains compared to YDK samples, while the pattern of BBMV proteins reacting to antisera against APN appeared unchanged among strains, and cadherin (HevCaLP) expression was only reduced in BBMV from YHD2-B larvae (Fig. 2B).

Reduced expression of HvmALP is due to reduced transcript levels

To further investigate the mechanism resulting in reduced HvALP expression levels in BBMV from resistant *H. virescens* larvae, we used qRT-PCR to detect HvmALP transcript levels in midguts from susceptible and resistant *H. virescens* larvae. We targeted two HvmALP isoforms displaying the highest degree of sequence divergence (6%) of all the described HvmALP isoforms [34]: HvmALP1 (FJ416470) and HvmALP2 (FJ416471). In our qRT-PCR assays (Fig. 3), both HvmALP1 and HvmALP2 had reduced transcript levels in all resistant when compared to susceptible larvae. The relative levels of transcript reduction were different for each HvmALP isoform. Thus, the biggest reduction in transcript levels was observed for YHD2-B larvae (7.8 fold for HvmALP1 and 59 fold for HvmALP2), while CXC and KCBhyb levels were similar for HvmALP 1 (about 4 fold) but different in the case of HvmALP2 (9 fold for CXC and 2.3 fold for KCBhyb). In all cases, the reduction in levels of HvmALP transcript in resistant larvae was statistically significant (P<0.05, Holm-Sidak method) when compared to the levels detected in samples from susceptible (YDK) larvae.

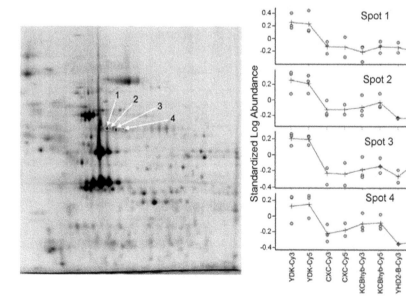

Figure 1. Detection of reduced HvmALP expression in Cry-resistant strains of *H. virescens* using quantitative differential proteomic analysis (2D-DIGE). BBMV proteins from susceptible (YDK) and resistant (YHD2-B, CXC, and KCBhyb) larvae were fluorescently labeled and the corresponding sub-proteome analyzed using 2D-DIGE. A representative gel image is presented with arrows pointing to the four HvmALP spots detected with lower expression levels. The identity of these spots as HvmALP was obtained by peptide mass fingerprinting, de novo sequencing, and Western blotting with specific antisera (data not shown). The standardized log abundance for each spot in all three BBMV samples (labeled with Cy3 or Cy5 from each strain used) is shown. Differences in protein levels between HvmALP spots in BBMV from susceptible and resistant larvae were statistically significant (p<0.001; Student T-test). HvmALP expression levels among BBMV samples from resistant larvae were not significantly different.

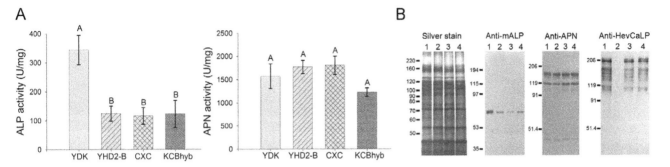

Figure 2. BBMV from Cry1Ac-resistant *H. virescens* larvae display reduced ALP levels. (A) BBMV proteins from *H. virescens* strains as indicated were used in specific ALP or APN activity assays. Different letters indicate statistically significant differences (p<0.05; Holm-Sidak method) among the samples. (B) Western blot analysis of HvmALP, APN, or HevCaLP in BBMV proteins from *H. virescens* strains: Lane 1, YDK; lane 2, YHD2-B; lane 3, CXC; lane 4, KCBhyb.

Detection of reduced expression of mALP in Cry-resistant *H. armigera* and *S. frugiperda* larvae

Considering that efficient biomarkers would allow detection of resistance in diverse insects with unique resistance phenotypes, we were interested in testing whether reduced mALP expression levels were common to alternative Cry-resistant lepidopteran species. As a taxonomically-close relative to *H. virescens*, we tested BBMV from susceptible and Cry1Ac-resistant strains of *H. armigera* (Fig. 4A). Similar to tests with *H. virescens* BBMV, we detected a significant reduction (P<0.05, Tukey test) in ALP activity in BBMV from resistant *H. armigera* compared to susceptible larvae. However, in the case of *H. armigera* we also detected significant differences (P<0.05, Tukey test) in APN activity between BBMV from susceptible and resistant larvae (Fig. 4B). Since in *H. virescens* reduced mALP expression was controlled at the transcriptional level, we quantified HaALP transcript amounts in susceptible and Cry1Ac-resistant *H. armigera* larvae using qRT-PCR. In agreement with our HvmALP data, we detected a 1.6-fold

reduction (P<0.05, Holm-Sidak method) in levels of HaALP transcripts in guts from resistant larvae compared to susceptible insects (Fig. 4C).

To test for reduction of mALP levels in an alternative lepidopteran pest, we used BBMV prepared from susceptible and Cry1Fa-resistant *S. frugiperda* larvae. In activity assays, we detected a significant reduction (P<0.05, Holm-Sidak method) of ALP activity in BBMV from resistant larvae compared to vesicles from susceptible insects (Fig. 5A). Even though we did not detect differences when comparing the two resistant strains (456 and 512) or the two susceptible strains (Mon and Ben), BBMV from resistant larvae presented a 3–4 fold reduction in ALP activity compared to vesicles from susceptible insects. In contrast, we found no significant differences in APN specific activity in BBMV from all four strains (Fig. 5B). Reduction in ALP specific activity correlated with reduced amounts of mALP in BBMV from both resistant strains compared to vesicles from susceptible strains as detected in Western blots (Fig. 5C).

Discussion

We report on the correlation between reduced ALP protein, activity, and mALP expression levels in strains of three species in the Noctuidae Family with diverse resistance phenotypes against Cry toxins. Currently, *H. virescens* is efficiently controlled by transgenic Bt cotton [38], but numerous reports highlight the potential for the development of field resistance to Bt crops in *H. armigera* [39,40,41], while field-evolved resistance to Bt toxins and Bt corn has already been reported for *S. frugiperda* [3,27]. Considering that the *H. virescens* resistant strains in this study present unique alterations in toxin binding [21,22], processing of toxin [24], or midgut regeneration [42,43], our data suggest that reduced mALP expression is a potential biomarker for resistance to diverse Cry toxins and is independent of the resistance mechanism. Furthermore, our data with Cry-resistant Noctuidae suggests that reduced mALP is a common phenomenon in Bt-resistant lepidopteran larvae.

Even though no correlation between ALP activity levels and resistance to insecticides has been reported in the literature to date, reduced ALP activity levels in insects have been reported to occur after intoxication with lectins [44], infection with cytoplasmic polyhedrosis virus (CPV) or *B. thuringiensis* in *B. mori* [45], and microsporidia in *Barathra brassicae* [46]. However, decreased ALP activity appears to be specific to only certain pathologies, since it was not observed after infection of *B. mori* with nuclear polyhedrosis virus (NPV) or *Serratia marcescens* [45]. These

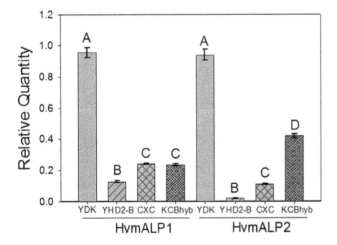

Figure 3. Reduced levels of HvmALP1 and HvmALP2 transcripts in Cry1Ac-resistant *H. virescens* larvae relative to YDK as detected by qRT-PCR. Data shown are the mean transcript quantity relative to the YDK sample with the highest transcript amounts from three independent biological replicates for each HvmALP isoform and strain. All reactions were performed with triplicate technical replicates. Bars denote standard error of the mean; different letters on each bar indicate significant differences (P<0.05; Holm-Sidak method).

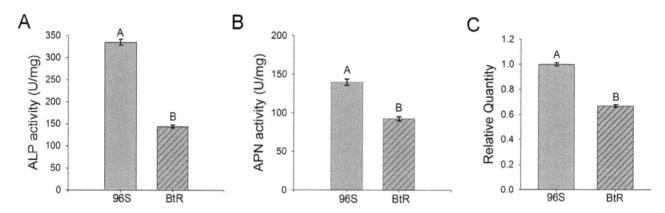

Figure 4. Reduced ALP activity correlates with reduced levels of HaALP transcripts in Cry1Ac-resistant *H. armigera* larvae. BBMV proteins from susceptible (96S) and Cry1Ac-resistant (BtR) *H. armigera* strains were used in specific ALP (A) or APN (B) activity assays. Different letters indicate statistically significant differences (p<0.05; Holm-Sidak method) among the samples. C) Mean relative transcript quantity of HaALP1 and HaALP2 isoforms. Data shown are the mean transcript quantity relative to the 96S sample with the highest transcript amounts from three independent biological replicates for each HaALP isoform and strain. All reactions were performed with triplicate technical replicates. Bars denote standard error of the mean. Different letters indicate significant differences (P<0.05; Holm-Sidak method).

observations suggest that although monitoring methods based on reduction in ALP activity levels would allow detection of resistance to Bt toxins, it may need to be combined with additional biomarkers to assure accurate detection of resistance. Another limitation of using reduced ALP levels in monitoring for resistance to Cry toxins is that, considering that in most cases Bt resistance is recessive, heterozygote larvae present similar levels of HvmALP and ALP activity as susceptible parents [31]. To overcome these limitations, we expect that further characterization of the molecular mechanism involved in reduction of ALP levels in Bt-resistant larvae would result in identification of specific alleles to target in designing DNA probes for real time RT-PCR capable of discriminating heterozygotes and ALP reduction due to Bt resistance or infection by entomopathogens.

Previous reports have suggested that direct interaction between *B. thuringiensis* Cry toxins and lepidopteran midgut ALP results in decreased ALP activity [47,48]. However, our data on the reduction in ALP activity in resistant insects was independent of feeding on Cry-contaminated diet, or on diet composition, as larvae used in our work were reared on diverse diets (including

fresh corn leaf tissue) and were not exposed to Cry toxins before dissection. This observation may suggest that a method based on detection of mALP down-regulation would detect Cry-resistant larvae regardless of larvae feeding on transgenic Bt crops or non-Bt refugia.

Considering that mALP has been proposed as receptor for Cry toxins in *M. sexta* [49], *H. virescens* [30], *H. armigera* [50,51], *Aedes aegypti* [52], *Anopheles gambiae* [53], and *Anthonomus grandis* [54], down-regulation of mALP expression may result in reduced Cry toxin binding to the brush border membrane in resistant insects. However, we detected HvmALP down-regulation in CXC and KCBhyb larvae, while Cry1Ac binding in BBMV from these larvae was not altered when compared to vesicles form susceptible insects [21]. Since Cry1Ac has multiple binding sites in *H. virescens* BBMV [55], and Cry1Ac binding to HvmALP has not been quantified to date, it is possible that changes in Cry1Ac binding due to reduced HvmALP levels are masked by binding to alternative receptors in the BBMV. However, CXC and KCBhyb larvae are cross-resistant to Cry2Aa toxin, which does not share binding sites with Cry1Ac [56].

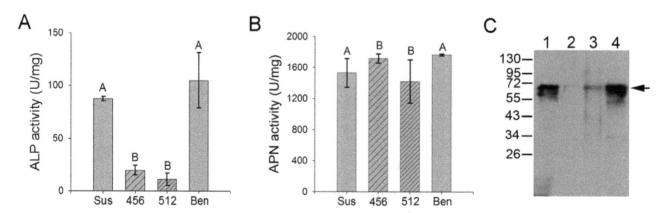

Figure 5. Reduced ALP activity correlates with reduced levels of mALP in BBMV from Cry1Fa-resistant *S. frugiperda* larvae. BBMV proteins from susceptible (Mon and Ben) and Cry1Fa-resistant (456 and 512) strains of *S. frugiperda* were used in specific ALP (A) or APN (B) activity assays. Different letters indicate statistically significant differences (p<0.05; Holm-Sidak method) among the samples. C) Detection of mALP in BBMV from susceptible (Mon and Ben) and resistant (456 and 512) strains of *S. frugiperda* using Western blotting. Lane 1, Mon, lane 2, 456, lane 3, 512, lane 4, Ben. Arrow points to ALP protein band.

Mammalian ALP transcript expression is modulated by members of the mitogen-activated protein kinase (MAPK) family. For example, the p38 kinase-dependent pathway modulates changes in ALP activity levels during development or stress in human osteoblast-like [57,58] and intestinal cells [59]. The homologous p38 pathway in *Caenorhabditis elegans* [60,61], in *Manduca sexta*, and *A. aegypti* [62] controls the gut defense response against Cry intoxication. Considering this information, it is possible that reduced ALP levels in Cry-resistant insects reflect a constitutively expressed enhanced defensive response to intoxication, as previously suggested for *Ephestia kuehniella* [63] or some of the *H. virescens* resistant strains tested in our work [42,43]. Based on reported functions for insect ALP enzymes [64], we expect the gut ALP isoforms down-regulated in the tested Bt-resistant larvae are involved in digestion and gut development. However, down-regulation of these enzymes seems not to affect insect development or survival in the field, as larvae from the 456 and 512 strains of *S. frugiperda* used in this paper were generated from field-collected eggs [27] and do not present affected development when reared on corn leaf tissue compared to susceptible insects (data not shown). Lack of a direct correlation between reduced HvmALP levels and levels of resistance in the tested *H. virescens* strains does not support a direct role for HvmALP in resistance, although multiple resistance mechanisms are present in these insects [21,22,65]. Further research is needed to determine whether correlation between reduced levels of mALP expression and resistance represents a direct decrease in functional Cry toxin receptors and/or compensatory alterations in resistant larvae.

Development of insect resistance is one of the most crucial issues related to increased adoption rates for transgenic Bt crops. As with other insect pest groups, monitoring for heliothine resistance to Bt is currently performed using the F1 or F2 screening test [38,66], which are lengthy and labor intensive. As a monitoring alternative, DNA-based methods would greatly increase sensitivity and speed of detection of resistance alleles, but their development is dependent on the identification of resistance alleles. Cadherins that serve as high affinity Bt toxin receptors have been proposed as optimal targets for the development of DNA-based strategies to detect resistance to Cry toxins in field pest populations [17,67]. However, cadherin resistance alleles have not been detected in field populations of *P. gossypiella* and *H. virescens* [18,68], possibly due to the low frequency of Bt resistance alleles in field populations. In the case of *P. gossypiella* screening, extensive parallel bioassay evidence indicated that cadherin alleles linked to resistance were rare or absent in *P. gossypiella* populations screened by PCR [69]. Recent reports suggest that the existence of multiple cadherin resistance alleles in field populations of *H. armigera* may hinder successful screening when using DNA-based methods to detect specific cadherin resistant alleles [70]. More recently, lack of expression of an ABC transporter protein in *H. virescens* has been demonstrated to result in lack of Cry1Ac binding and resistance to this toxin, supporting a potential role for this protein during Cry1Ac intoxication [65]. However, monitoring for alterations in Cry toxin receptors, as in the case of cadherin or ABC transporter-based monitoring methods, would only be effective in detecting resistance to Cry toxins binding to these receptors. The introduction of pyramided Bt crops represents increased levels of selection for resistance mechanisms to diverse Cry toxins that do not share receptors. In addition, other mechanisms not involving cadherin modifications are associated with resistance phenotypes [5]. Based on our data, reduced ALP expression represents a biomarker that would detect resistant insects independently of the resistance mechanism or cross-resistance phenotype. Therefore, a biomarker not based on individual DNA sequence but rather diagnostic of a phenotype associated with resistance to Cry toxins, such as the one identified in this study, would be desirable in the development of monitoring methods. Future work aimed at developing effective and sensitive Bt-resistance monitoring methods will include characterization of the molecular mechanism resulting in ALP down-regulation, testing the range of mALP expression in natural field populations, and optimizing biochemical detection of ALP activity associated with resistance alleles in field populations.

Acknowledgments

The authors would like to thank Dr. Bruce Tabashnik (University of Arizona) and Dr. Brenda Oppert (USDA-ARS Stored Product Insect Research Unit) for insightful comments to earlier versions of this manuscript.

Author Contributions

Conceived and designed the experiments: JLJ-F LK KW MA. Performed the experiments: JLJ-F LK SRKJ CN CL JJ. Analyzed the data: JLJ-F LK SRKJ CN CL KW MA. Contributed reagents/materials/analysis tools: FG CB MP OP MA. Wrote the paper: JLJ-F. Discussed the results, commented, and edited the manuscript: JLJ-F LK SRKJ CN CL KW JJ FG CB MP OP MA.

References

1. James C (2009) Global status of commercialized Biotech/GM crops: 2009. ISAAA briefs No 41 ISAAA, Ithaca, NY.

2. Tabashnik BE, Gassmann AJ, Crowder DW, Carriere Y (2008) Insect resistance to Bt crops: evidence versus theory. Nature Biotechnol 26: 199–202.

3. Storer NP, Babcock JM, Schlenz M, Meade T, Thompson GD, et al. (2010) Discovery and characterization of field resistance to Bt maize: *Spodoptera frugiperda* (Lepidoptera: Noctuidae) in Puerto Rico. J Econ Entomol 103: 1031–1038.

4. Van Rensburg JBJ (2007) First report of field resistance by stem borer, *Busseola fusca* (Fuller) to Bt-transgenic maize. S African J Plant Soil 24: 147–151.

5. Ferre J, Van Rie J (2002) Biochemistry and genetics of insect resistance to *Bacillus thuringiensis*. Annu Rev Entomol 47: 501–533.

6. Heckel DG, Gahan LJ, Baxter SW, Zhao JZ, Shelton AM, et al. (2007) The diversity of Bt resistance genes in species of Lepidoptera. J Invertbr Pathol 95: 192–197.

7. Raymond B, Johnston PR, Nielsen-LeRoux C, Lereclus D, Crickmore N (2010) *Bacillus thuringiensis*: an impotent pathogen? Trends Microbiol 18: 189–194.

8. Soberon M, Gill SS, Bravo A (2009) Signaling versus punching hole: How do *Bacillus thuringiensis* toxins kill insect midgut cells? Cell Mol Life Sci 66: 1337–1349.

9. Arenas I, Bravo A, Soberon M, Gomez I (2010) Role of alkaline phosphatase from *Manduca sexta* in the mechanism of action of *Bacillus thuringiensis* Cry1Ab toxin. J Biol chem 285: 12497–12503.

10. Bravo A, Gomez I, Conde J, Munoz-Garay C, Sanchez J, et al. (2004) Oligomerization triggers binding of a *Bacillus thuringiensis* Cry1Ab pore-forming toxin to aminopeptidase N receptor leading to insertion into membrane microdomains. Biochim Biophys Acta 1667: 38–46.

11. Pardo-Lopez L, Gomez I, Rausell C, Sanchez J, Soberon M, et al. (2006) Structural changes of the Cry1Ac oligomeric pre-pore from *Bacillus thuringiensis* induced by N-acetylgalactosamine facilitates toxin membrane insertion. Biochemistry 45: 10329–10336.

12. Zhuang M, Oltean DI, Gomez I, Pullikuth AK, Soberon M, et al. (2002) *Heliothis virescens* and *Manduca sexta* lipid rafts are involved in Cry1A toxin binding to the midgut epithelium and subsequent pore formation. J Biol Chem 277: 13863–13872.

13. Zhang X, Candas M, Griko NB, Taussig R, Bulla LA, Jr. (2006) A mechanism of cell death involving an adenylyl cyclase/PKA signaling pathway is induced by the Cry1Ab toxin of *Bacillus thuringiensis*. Proc Natl Acad Sci USA 103: 9897–9902.

14. Gahan LJ, Gould F, Heckel DG (2001) Identification of a gene associated with Bt resistance in *Heliothis virescens*. Science 293: 857–860.

15. Xu X, Yu L, Wu Y (2005) Disruption of a cadherin gene associated with resistance to Cry1Ac δ-endotoxin of *Bacillus thuringiensis* in *Helicoverpa armigera*. Appl Environ Microbiol 71: 948–954.

16. Morin S, Biggs RW, Sisterson MS, Shriver L, Ellers-Kirk C, et al. (2003) Three cadherin alleles associated with resistance to *Bacillus thuringiensis* in pink bollworm. Proc Natl Acad Sci USA 100: 5004–5009.

17. Morin S, Henderson S, Fabrick JA, Carriere Y, Dennehy TJ, et al. (2004) DNA-based detection of Bt resistance alleles in pink bollworm. Insect Biochem Mol Biol 34: 1225–1233.

18. Gahan LJ, Gould F, Lopez JD, Jr., Micinski S, Heckel DG (2007) A polymerase chain reaction screen of field populations of Heliothis virescens for a retrotransposon insertion conferring resistance to Bacillus thuringiensis toxin. J Econ Entomol 100: 187–194.

19. Gahan LJ, Ma YT, Coble ML, Gould F, Moar WJ, et al. (2005) Genetic basis of resistance to Cry1Ac and Cry2Aa in Heliothis virescens (Lepidoptera: Noctuidae). J Econ Entomol 98: 1357–1368.

20. Baxter SW, Zhao JZ, Gahan LJ, Shelton AM, Tabashnik BE, et al. (2005) Novel genetic basis of field-evolved resistance to Bt toxins in Plutella xylostella. Insect Molec Biol 14: 327–334.

21. Jurat-Fuentes JL, Gahan LJ, Gould FL, Heckel DG, Adang MJ (2004) The HevCaLP protein mediates binding specificity of the Cry1A class of Bacillus thuringiensis toxins in Heliothis virescens. Biochemistry 43: 14299–14305.

22. Jurat-Fuentes JL, Gould FL, Adang MJ (2003) Dual resistance to Bacillus thuringiensis Cry1Ac and Cry2Aa toxins in Heliothis virescens suggests multiple mechanisms of resistance. Appl Environ Microbiol 69: 5898–5906.

23. Gould F, Anderson A, Reynolds A, Bumgarner L, Moar W (1995) Selection and genetic analysis of a Heliothis virescens (Lepidoptera: noctuidae) strain with high levels of resistance to Bacillus thuringiensis toxins. J Econ Entomol 88: 1545–1559.

24. Karumbaiah L, Oppert B, Jurat-Fuentes JL, Adang MJ (2007) Analysis of midgut proteinases from Bacillus thuringiensis-susceptible and -resistant Heliothis virescens (Lepidoptera: Noctuidae). Comp Biochem Physiol Part B 146: 139–146.

25. Jurat-Fuentes JL, Gould FL, Adang MJ (2002) Altered Glycosylation of 63- and 68-kilodalton microvillar proteins in Heliothis virescens correlates with reduced Cry1 toxin binding, decreased pore formation, and increased resistance to Bacillus thuringiensis Cry1 toxins. Appl Environ Microbiol 68: 5711–5717.

26. Liang GM, Wu KM, Yu HK, Li KK, Feng X, et al. (2008) Changes of inheritance mode and fitness in Helicoverpa armigera (Hubner) (Lepidoptera: Noctuidae) along with its resistance evolution to Cry1Ac toxin. J Invertbr Pathol 97: 142–149.

27. Blanco CA, Portilla M, Jurat-Fuentes JL, Sanchez JF, Viteri D, et al. (2010) Susceptibility of isofamilies of Spodoptera frugiperda (Lepidoptera: Noctuidae) to Cry1Ac and Cry1Fa proteins of Bacillus thuringiensis. Southwest Entomol 35: 409–415.

28. Wolfersberger MG, Luthy P, Maurer A, Parenti P, Sacchi VF, et al. (1987) Preparation and partial characterization of amino acid transporting brush border membrane vesicles from the larval midgut of the cabbage butterfly (Pieris brassicae). Comp Biochem Physiol 86A: 301–308.

29. Bradford M (1976) A rapid and sensitive method for the quantitation of microgram quantities of protein utilizing the principle of protein-dye binding. Anal Biochem 72: 248–254.

30. Krishnamoorthy M, Jurat-Fuentes JL, McNall RJ, Andacht T, Adang MJ (2007) Identification of novel Cry1Ac binding proteins in midgut membranes from Heliothis virescens using proteomic analyses. Insect Biochem Molec Biol 37: 189–201.

31. Jurat-Fuentes JL, Adang MJ (2004) Characterization of a Cry1Ac-receptor alkaline phosphatase in susceptible and resistant Heliothis virescens larvae. Eur J Biochem 271: 3127–3135.

32. Laemmli UK (1970) Cleavage of structural proteins during the assembly of the head of bacteriophage T4. Nature 227: 680–685.

33. Luo K, Sangadala S, Masson L, Mazza A, Brousseau R, et al. (1997) The Heliothis virescens 170 kDa aminopeptidase functions as "receptor A" by mediating specific Bacillus thuringiensis Cry1A delta-endotoxin binding and pore formation. Insect Biochem Molec Biol 27: 735–743.

34. Perera OP, Willis JD, Adang MJ, Jurat-Fuentes JL (2009) Cloning and characterization of the Cry1Ac-binding alkaline phosphatase (HvALP) from Heliothis virescens. Insect Biochem Mol Biol 39: 294–302.

35. Livak KJ, Schmittgen TD (2001) Analysis of relative gene expression data using real-time quantitative PCR and the 2(-Delta Delta C(T)) method. Methods 25: 402–408.

36. Jurat-Fuentes JL, Adang MJ (2006) Cry toxin mode of action in susceptible and resistant Heliothis virescens larvae. J Invertbr Pathol 92: 166–171.

37. Jurat-Fuentes JL, Adang MJ (2007) A proteomic approach to study Cry1Ac binding proteins and their alterations in resistant Heliothis virescens larvae. J Invertbr Pathol 95: 187–191.

38. Blanco CA, Andow DA, Abel CA, Sumerford DV, Hernandez G, et al. (2009) Bacillus thuringiensis Cry1Ac resistance frequency in tobacco budworm (Lepidoptera: Noctuidae). J Econ Entomol 102: 381–387.

39. Lu MG, Rui CH, Zhao JZ, Jian GL, Fan XL, et al. (2004) Selection and heritability of resistance to Bacillus thuringiensis subsp kurstaki and transgenic cotton in Helicoverpa armigera (Lepidoptera: Noctuidae). Pest Manag Sci 60: 887–893.

40. Meng F, Shen J, Zhou W, Cen H (2004) Long-term selection for resistance to transgenic cotton expressing Bacillus thuringiensis toxin in Helicoverpa armigera (Hubner) (Lepidoptera: Noctuidae). Pest Manag Sci 60: 167–172.

41. Caccia S, Hernandez-Rodriguez CS, Mahon RJ, Downes S, James W, et al. (2010) Binding site alteration is responsible for field-isolated resistance to Bacillus thuringiensis Cry2A insecticidal proteins in two Helicoverpa species. PLoS ONE 5: e9975.

42. Forcada C, Alcacer E, Garcerá MD, Tato A, Martinez R (1999) Resistance to Bacillus thuringiensis Cry1Ac toxin in three strains of Heliothis virescens: proteolytic and SEM study of the larval midgut. Arch Insect Biochem Physiol 42: 51–63.

43. Martinez-Ramirez AC, Gould F, Ferre J (1999) Histopathological effects and growth reduction in a susceptible and a resistant strain of Heliothis virescens (Lepidoptera : Noctuidae) caused by sublethal doses of pure Cry1A crystal proteins from Bacillus thuringiensis. Biocont Sci Tech 9: 239–246.

44. Kaur M, Singh K, Rup PJ, Kamboj SS, Singh J (2009) Anti-insect potential of lectins from Arisaema species towards Bactrocera cucurbitae. J Environ Biol 30: 1019–1023.

45. Miao Y-G (2002) Studies on the activity of the alkaline phosphatase in the midgut of infected silkworm, Bombyx mori L. J Appl Ent 126: 138–142.

46. Kucera M, Weiser J (1974) Alkaline-phosphatase in last larval instar of Barathra brassicae (Lepidoptera) infected by Nosema Plodiae. Acta Entomol Bohem 71: 289–293.

47. Sangadala S, Walters FS, English LH, Adang MJ (1994) A mixture of Manduca sexta aminopeptidase and phosphatase enhances Bacillus thuringiensis insecticidal CryIA(c) toxin binding and ^{86}Rb$^{(+)}$-K$^+$ efflux in vitro. J Biol Chem 269: 10088–10092.

48. English L, Readdy TL (1989) Delta endotoxin inhibits a phosphatase in midgut epithelial membranes of Heliothis virescens. Insect Biochem 19: 145–152.

49. McNall RJ, Adang MJ (2003) Identification of novel Bacillus thuringiensis Cry1Ac binding proteins in Manduca sexta midgut through proteomic analysis. Insect Biochem Molec Biol 33: 999–1010.

50. Sarkar A, Hess D, Mondal HA, Banerjee S, Sharma HC, et al. (2009) Homodimeric alkaline phosphatase located at Helicoverpa armigera midgut, a putative receptor of Cry1Ac contains alpha-GalNAc in terminal glycan structure as interactive epitope. J Proteome Res 8: 1838–1848.

51. Ning C, Wu K, Liu C, Gao Y, Jurat-Fuentes JL, et al. (2010) Characterization of a Cry1Ac toxin-binding alkaline phosphatase in the midgut from Helicoverpa armigera (Hubner) larvae. J Insect Physiol 56: 666–672.

52. Fernandez LE, Aimanova KG, Gill SS, Bravo A, Soberon M (2006) A GPI-anchored alkaline phosphatase is a functional midgut receptor of Cry11Aa toxin in Aedes aegypti larvae. Biochem J 394: 77–84.

53. Hua G, Zhang R, Bayyareddy K, Adang MJ (2009) Anopheles gambiae alkaline phosphatase is a functional receptor of Bacillus thuringiensis jegathesan Cry11Ba toxin. Biochemistry 48: 9785–9793.

54. Martins ES, Monnerat RG, Queiroz PR, Dumas VF, Braz SV, et al. (2010) Midgut GPI-anchored proteins with alkaline phosphatase activity from the cotton boll weevil (Anthonomus grandis) are putative receptors for the Cry1B protein of Bacillus thuringiensis. Insect Biochem Molec Biol 40: 138–145.

55. Jurat-Fuentes JL, Adang MJ (2001) Importance of Cry1 delta-endotoxin domain II loops for binding specificity in Heliothis virescens (L.). Appl Environ Microbiol 67: 323–329.

56. Hernandez-Rodriguez CS, Van Vliet A, Bautsoens N, Van Rie J, Ferre J (2008) Specific binding of Bacillus thuringiensis Cry2A insecticidal proteins to a common site in the midgut of Helicoverpa species. Appl Environ Microbiol 74: 7654–7659.

57. Suzuki A, Palmer G, Bonjour JP, Caverzasio J (1999) Regulation of alkaline phosphatase activity by p38 MAP kinase in response to activation of Gi protein-coupled receptors by epinephrine in osteoblast-like cells. Endocrinology 140: 3177–3182.

58. Suzuki A, Guicheux J, Palmer G, Miura Y, Oiso Y, et al. (2002) Evidence for a role of p38 MAP kinase in expression of alkaline phosphatase during osteoblastic cell differentiation. Bone 30: 91–98.

59. Ding Q, Wang Q, Evers BM (2001) Alterations of MAPK activities associated with intestinal cell differentiation. Biochem Biophys Res Comm 284: 282–288.

60. Bischof LJ, Kao CY, Los FC, Gonzalez MR, Shen Z, et al. (2008) Activation of the unfolded protein response is required for defenses against bacterial pore-forming toxin in vivo. PLoS Pathog 4: e1000176.

61. Huffman DL, Abrami L, Sasik R, Corbeil J, van der Goot FG, et al. (2004) Mitogen-activated protein kinase pathways defend against bacterial pore-forming toxins. Proc Natl Acad Sci USA 101: 10995–11000.

62. Cancino-Rodezno A, Alexander C, Villasenor R, Pacheco S, Porta H, et al. (2010) The mitogen-activated protein kinase p38 is involved in insect defense against Cry toxins from Bacillus thuringiensis. Insect Biochem Mol Biol 40: 58–63.

63. Rahman MM, Roberts HL, Sarjan M, Asgari S, Schmidt O (2004) Induction and transmission of Bacillus thuringiensis tolerance in the flour moth Ephestia kuehniella. Proc Natl Acad Sci USA 101: 2696–2699.

64. Eguchi M (1995) Alkaline phosphatase isozymes in insects and comparison with mammalian enzyme. Comp Biochem Physiol 111B: 151–162.

65. Gahan LJ, Pauchet Y, Vogel H, Heckel DG (2010) An ABC transporter mutation is correlated with insect resistance to Bacillus thuringiensis Cry1Ac toxin. PLoS Genet 6: e1001248.

66. Xu Z, Liu F, Chen J, Huang F, Andow DA, et al. (2009) Using an F$_2$ screen to monitor frequency of resistance alleles to Bt cotton in field populations of Helicoverpa armigera (Hübner) (Lepidoptera: Noctuidae). Pest Manag Sci 65: 391–397.

67. Tabashnik BE, Biggs RW, Higginson DM, Henderson S, Unnithan DC, et al. (2005) Association between resistance to Bt cotton and cadherin genotype in pink bollworm. J Econ Entomol 98: 635–644.

68. Tabashnik BE, Fabrick JA, Henderson S, Biggs RW, Yafuso CM, et al. (2006) DNA screening reveals pink bollworm resistance to Bt cotton remains rare after a decade of exposure. J Econ Entomol 99: 1525–1530.

69. Tabashnik BE, Biggs RW, Fabrick JA, Gassmann AJ, Dennehy TJ, et al. (2006) High-level resistance to Bacillus thuringiensis toxin cry1Ac and cadherin genotype in pink bollworm. J Econ Entomol 99: 2125–2131.

70. Zhao J, Jin L, Yang Y, Wu Y (2010) Diverse cadherin mutations conferring resistance to Bacillus thuringiensis toxin Cry1Ac in Helicoverpa armigera. Insect Biochem Mol Biol 40: 113–118.

17

Movement of Soil-Applied Imidacloprid and Thiamethoxam into Nectar and Pollen of Squash (*Cucurbita pepo*)

Kimberly A. Stoner[1]*, **Brian D. Eitzer**[2]

1 Department of Entomology, The Connecticut Agricultural Experiment Station, New Haven, Connecticut, United States of America, 2 Department of Analytical Chemistry, The Connecticut Agricultural Experiment Station, New Haven, Connecticut, United States of America

Abstract

There has been recent interest in the threat to bees posed by the use of systemic insecticides. One concern is that systemic insecticides may translocate from the soil into pollen and nectar of plants, where they would be ingested by pollinators. This paper reports on the movement of two such systemic neonicotinoid insecticides, imidacloprid and thiamethoxam, into the pollen and nectar of flowers of squash (*Cucurbita pepo* cultivars "Multipik," "Sunray" and "Bush Delicata") when applied to soil by two methods: (1) sprayed into soil before seeding, or (2) applied through drip irrigation in a single treatment after transplant. All insecticide treatments were within labeled rates for these compounds. Pollen and nectar samples were analyzed using a standard extraction method widely used for pesticides (QuEChERS) and liquid chromatography mass spectrometric analysis. The concentrations found in nectar, 10±3 ppb (mean ± s.d) for imidacloprid and 11±6 ppb for thiamethoxam, are higher than concentrations of neonicotinoid insecticides in nectar of canola and sunflower grown from treated seed, and similar to those found in a recent study of neonicotinoids applied to pumpkins at transplant and through drip irrigation. The concentrations in pollen, 14±8 ppb for imidacloprid and 12±9 ppb for thiamethoxam, are higher than those found for seed treatments in most studies, but at the low end of the range found in the pumpkin study. Our concentrations fall into the range being investigated for sublethal effects on honey bees and bumble bees.

Editor: Subba Reddy Palli, U. Kentucky, United States of America

Funding: This study was supported by a grant from Project Apis m. (URL: www.ProjectApism.org). The funders had no role in study design, data collection and analysis, decision to publish, or preparation of the manuscript.

Competing Interests: The authors have declared that no competing interests exist.

* E-mail: Kimberly.Stoner@ct.gov

Introduction

The long-term security of insect pollination for food crops is a major concern in the U.S. [1] and around the world [2,3]. Beekeepers have suffered major losses of honey bee (*Apis mellifera*) colonies annually for the last four years in the U.S. [4], and in parts of Europe [5]. In addition, formerly common species of bumble bees (*Bombus* spp.) have undergone major losses in range in North America [6] and Europe [7]. Many potential factors could be involved in these global declines of managed and wild pollinating insects. For honey bees, losses of managed populations have been attributed to the worldwide movement of parasitic mites, viruses, and the pathogen *Nosema ceranae*; loss of genetic diversity; loss of bee forage; and global trade and economic changes; as well as changes in pesticide use [1,4,5]. For bumble bees, losses of species diversity have been attributed to changes in land use with reduced season-long bee forage and nesting habitats, spread of pathogens (*Nosema bombi* and *Crithidia bombi*) from commercial bumble bee colonies to wild populations, and fragmented populations with low genetic diversity, with changes in pesticides use cited as a possible additional factor [1,6,7].

Although honey bees are exposed to a wide range of pesticides – including those applied to the hive by beekeepers as well as those in the environment [8] – a class of systemic insecticides known as neonicotinoids has come under particular scrutiny as a result of

heavy mortality of honey bee colonies associated with seed treatment of sunflower and corn with imidacloprid in France [9] and seed treatment of corn with clothianidin in Germany [5]. Neonicotinoids include imidacloprid, thiamethoxam, clothianidin, acetamiprid, thiacloprid, nitenpyram, and dinotefuran, and as a group comprise 24% of the global insecticide market [10]. Imidacloprid is the largest selling insecticide in the world, with sales of $1091 million in 2009 and registered for 140 crop uses in over 120 countries [10]. Thiamethoxam is the second largest selling neonicotinoid with sales of $627 million in 2009 and registered for 115 crop uses in at least 65 countries [10]. Neonicotinoids applied to the seed are taken up by the roots and travel through the entire plant to the flowers [10,11]. Previous field studies measuring the concentration of neonicotinoids in canola, corn or sunflowers, where the seed was treated with the insecticide before sowing, found mean concentrations from 2 to 3.9 ppb in pollen [12–14] and from 2.2 ppb to 3.0 ppb in nectar [12,13]. Two studies using radiolabeled imidacloprid applied to sunflower seed under more controlled conditions found concentrations of 3.9 ppb in pollen and 1.9 ppb in nectar [15] and a concentration of 13±13 ppb (mean ± sd) in pollen [11].

Neonicotinoids are applied to plants in other ways besides direct treatment of the seed [10]. They are applied by foliar spray treatment, by trunk injection in trees, as granules to potting mix or soil, and as liquid sprayed directly to soil or applied through drip irrigation [10,16]. Little research has been done to quantify the exposure of bees and other pollinators to these pesticides applied in other ways besides seed treatment [16]. Concurrent with our own research, a similar study comparing methods of application of neonicotinoids to pumpkins and measuring concentrations of parent compounds in nectar and pollen was conducted in Maryland [17].

The goal of our project was to quantify movement of two neonicotinoid insecticides into the pollen and nectar of plants when applied directly to the soil, either by direct spray to the soil just before seeding or through drip irrigation. Although we did not quantify bee exposure to these insecticides, knowledge of the neonicotinoid concentrations in the matrices consumed by bees can be compared to those found to have sublethal effects on bees in the scientific literature.

We chose squash (*Cucurbita pepo*) for study because it is routinely treated in the U.S. for control of striped cucumber beetles with systemic insecticides through soil application of neonicotinoids by direct spray to the seed furrow or through irrigation [18]; the flowers are large and both pollen and nectar can be collected in quantities suitable for analysis [19]; and insect pollination is required for fruit set [20,21]. The major pollinators of squash in the eastern U.S. are squash bees, *Peponapis pruinosa*, a specialist feeding its larvae exclusively on pollen from the genus *Cucurbita* [22], bumble bees (*Bombus impatiens*), and honey bees [20–23].

Materials and Methods

Planting and Insecticide Application

In 2009, yellow summer squash, *Cucurbita pepo* L. cv. "Multipik," was grown on black plastic mulch in rows on 1.5 m centers with seed holes spaced at 0.9 m. For the direct-seeded treatments, three seeds were planted per hole. For the transplanted treatments, three seeds per cell were started in the greenhouse before transplanting, and one cell was transplanted per seed hole. Fertilizer (NPK 10-10-10) was applied at a rate of 90 kg/ha of nitrogen, and lime was applied as recommended based on soil tests. The field was laid out in a randomized complete block design with three blocks and five treatments: 1) untreated control; 2) imidacloprid (at 358 g [AI]/ha; Admire Pro®, Bayer Crop Science, Research Triangle Park, NC) applied by surface spray to the soil in the planting hole (11 cm diameter) and immediately incorporated into soil with hand tools, one day before seeding; 3) thiamethoxam (at 140 g [AI]/ha; Platinum®, Syngenta Crop Protection, Greensboro, NC) applied to the soil in the planting hole as above; 4) imidacloprid applied at the same rate per ha as #2 using a Venturi injector through drip irrigation to the entire row five days after transplanting; and 5) Platinum® applied at the same rate per ha as #3 using a Venturi injector through drip irrigation to the entire row five days after transplanting. The chronology of planting, pesticide applications and sampling for 2009 and 2010 are presented in Table S1 of the supplementary material.

In 2010, in a different field where neonicotinoid insecticides had not previously been used, three blocks were planted with yellow summer squash "Sunray F1" and a fourth block was planted with winter squash, *Cucurbita pepo* L. cv. "Bush Delicata." All five treatments were applied as in 2009, but the rates were different: imidacloprid was applied at 411 g [AI]/ha and thiamethoxam was applied at 143 g[AI]/ha. In both years, the rates of imidacloprid and thiamethoxam applied were within the range of labeled rates (281–420 g[AI]/ha for imidacloprid as Admire Pro® and 89–193 g[AI]/ha for thiamethoxam as Platinum®).

Rainfall was very different during the two growing seasons of the study. In 2009, there were 19.6 cm of rain in June and 16.6 cm in July. In 2010, there were 9.1 cm of rain in June and 9.5 cm. in July. In 2009, no irrigation was used other than the irrigation to apply the insecticides through the drip lines. In 2010, one additional irrigation of the entire field was applied through the drip lines on 8 July.

Sample Collection

Plant samples were collected over a longer period in 2010 than in 2009 (Table S1) because there was a greater spread among flowering times of the different types of squash (summer and winter), treatments, and even among blocks within a treatment. As female flowers appeared in each plot, they were collected with a clean razor blade, the petals and stigmata were removed, and the remaining bases of the flowers, where the nectaries are located, were saved for chemical analysis. Collection continued in each plot until a 50-ml centrifuge tube was packed full or until all available female flowers from the center row of the plot were collected. Similarly, as male flower buds appeared, the fully developed flower buds were opened before anthesis and the synandria (cone-like male flower structures made of fused anthers) were collected for later chemical analysis.

In 2009, whole-plant samples were taken by randomly selecting a single seed hole from the center row of each plot and collecting all squash plants growing from that hole (generally three plants, but some seed holes had only one or two plants, if not all seed germinated). The total weight of all plant material from that seed hole was recorded.

Nectar was collected with an Eppendorf pipette from female flowers that had been enclosed the previous afternoon in a pollinating bag (Lawson #217, Lawson Pollinating Bags, Northfield, IL). Nectar collection continued as long as female flowers were available in order to get as much nectar as possible for analysis. The nectar from all three blocks was pooled in 2009, and nectar from the three blocks of summer squash was pooled in 2010 in order to have enough nectar for reliable chemical analysis. Nectar from the winter squash in 2010 was collected later and analyzed separately.

Pollen was collected by hand-collecting open male flowers that had been enclosed the previous afternoon in a pollinating bag (as above). Flowers were collected from 6 until 10 am into a large plastic bag, which was then taken back to the laboratory where pollen was scraped by hand, using a thin plastic sheet, from the synandrium of each flower. The plastic bags of flowers were stored for up to one week at 4 C. After pollen was collected, it was stored at −18 C until analysis. In 2009, the pollen was pooled across all three blocks in order to have enough for analysis, but a second sample was taken a week later, also pooled across blocks. In 2010, we collected enough pollen to analyze the blocks separately.

Each plot consisted of 3 rows, and all samples were taken from the center row in order to avoid edge effects. All plant material collected was kept in a cooler on an ice pack during the day of collection and then stored at −18 C until analysis, except for the male flowers for pollen analysis, which were handled as described above.

Chemical Analysis

Extraction. All samples were extracted using a modified version of the QuEChERS (for Quick, Easy, Cheap, Effective, Rugged and Safe) protocol [24]. In brief, vegetative samples (1–5 g pollen/synandria, 5 g female flower base, 15 g whole chopped plant) were combined with water to a final volume of 15 mL. To this sample was added 100 ng of isotopically labeled (d-4)

Table 1. Neonicotinoid insecticide residues observed in 2009 in various tissues of summer squash after application either to the seed hole just before planting (Soil) or to the transplanted plant through drip irrigation (Drip).

Tissue	Imidacloprid (ppb ± SD)		Thiamethoxam (ppb ± SD)	
	Soil	Drip	Soil	Drip
Whole Plant	47±37	218±52	154±44	362±22
Female Flower Bases	10±5	31±17	10±2	22±5
Synandria	15±5	46±4	19±6	31±4

imidacloprid (Cambridge Isotope Laboratories) as an internal standard. The samples were combined with 15 mL of acetonitrile, 6 g magnesium sulfate and 1.5 g sodium acetate. After shaking and centrifuging, 10 mL of the supernatant was combined with 1.5 g magnesium sulfate, 0.5 g PSA, 0.5 g C-18 silica and 2 mL toluene. The samples were shaken and centrifuged and 6 mL of the supernatant was concentrated to 1 mL for instrumental analysis.

Analysis. Extracts were analyzed with liquid chromatography/mass spectrometry/mass spectrometry (LC/MS/MS). In 2009, the LC system was an Agilent 1100 LC; 6 μL of the extract was injected onto a Zorbax SB-C18, 2.1×150 mm, 5 micron column. The column is gradient eluted at 0.25 mL per minute from 12.5% methanol in water to 100% methanol. Both solvents have 0.1% formic acid added. In 2010 the LC system was an Agilent 1200 Rapid Resolution system with a Zorbax SB-C18 Rapid Resolution HT 2.1×50 mm, 1.8 micron column using a 3 ul injection with the gradient going from 5% methanol in water to 100% methanol at 0.45 mL/min. In both years, the LC was coupled to a Thermo-LTQ, a linear ion trap mass spectrometer. The system is operated in the positive ion electrospray mode, with a unique scan function for each compound allowing for MS/MS monitoring. Metabolites of imidacloprid (5-hydroxy imidacloprid; imidacloprid urea) and thiamethoxam (clothianadin) were also monitored. The specific parent and product ions monitored for each compound are listed in Table S2 of the supplementary material. Using these extraction and analysis conditions in spiked control samples the compounds averaged 95±18% recovery with detection limits ranging from 0.5 to 2 PPB depending on matrix and the amount of sample available.

Statistical Analysis

Effects of application method were analyzed for each year and for each of the pesticides and metabolites using an analysis of variance, including blocks in the model [25]. Results for nectar were not analyzed statistically because samples were pooled over blocks in order to have enough material for chemical analysis.

Results

Both imidacloprid and thiamethoxam were detected in all parts of the squash. Data for whole plants and flower parts for each year are presented in Tables 1 and 2, and the data for nectar and pollen are summarized over both years in Table 3. As expected, higher concentrations were observed in the whole plants than in the flower parts, pollen or nectar. Two metabolites of imidacloprid (5-OH imidacloprid and imidacloprid urea) and one metabolite of thiamethoxam (clothianidin) were also detected in whole plant samples. In 2009, when the two application methods were compared in the cultivar "Multipik," the concentrations of imidacloprid and the two metabolites and the concentration of thiamethoxam and the metabolite clothianidin were significantly higher in whole plant tissue in the drip irrigation treatment than in the soil treatment (df for all tests $= 1,2$; imidacloprid: $F = 58.386$, $p = 0.017$; 5-OH imidacloprid: $F = 27.106$, $p = 0.035$; imidacloprid urea: $F = 30.439$, $p = 0.031$; thiamethoxam: $F = 79.6$, $p = 0.012$; clothianidin: $F = 23.253$, $p = 0.040$). Also in 2009, the concentration of imidacloprid was significantly higher in the synandria (df $= 1,2$; $F = 411.857$; $p = 0.002$) and thiamethoxam was significantly higher in the base of female flowers (df $= 1,2$; $F = 26.518$, $p = 0.036$) in the drip than in the soil treatment. No other comparisons in 2009 between application methods were significantly different.

In 2010, the whole plant tissue was not monitored as the focus was movement of the pesticides into flower parts and then into pollen and nectar. The data for the 2010 yellow summer squash cultivar "Sunray" are presented in Table 2. There were no significant differences between the application methods during this year (for imidacloprid in female flower parts: df $= 1,2$, $F = 4.646$, $p = 0.164$; synandria: df $= 1,2$; $F = 1.240$, $p = 0.381$; pollen, df $= 1,3$, $F = 82.561$, $p = 0.116$; for thiamethoxam in female flower parts: df $= 1,2$; $F = 5.128$, $p = 0.152$; synandria: df $= 1,2$, $F = 2.469$, $p = 0.257$; pollen, df $= 1,3$, $F = 0.586$, $p = 0.500$). Although the data did not rise to the level of significance, the trend in 2010 was for the residues to be higher in the soil treatment than in the drip irrigation treatment for imidacloprid in the female flower bases

Table 2. Neonicotinoid insecticide residues observed in 2010 in various tissues of summer squash after application either to the seed hole just before planting (Soil) or to the transplanted plant through drip irrigation (Drip).

Tissue	Imidacloprid (ppb ± SD)		Thiamethoxam (ppb ± SD)	
	Soil	Drip	Soil	Drip
Female Flower Bases	28±10	15±2	26±12	13±3
Synandria	9±1	11±3	29±22	14±6

and thiamethoxam in both the synandria and female flower bases – this trend was the reverse of the 2009 data.

Table 3 presents a summary of the concentrations found in nectar and pollen across years, treatments, and varieties, including the winter squash variety "Bush Delicata." There were no significant differences in pesticide concentration in pollen with treatment in either year. All samples from treated plants across all three cultivars sampled over the two year period had concentrations of imidacloprid and thiamethoxam in both nectar and pollen greater than 4 ppb. Residues ranged between 5 and 35 ppb for pollen in 12 samples for each insecticide and 5 and 20 ppb for nectar in 6 samples for each insecticide. Averaging across the varieties and years gives overall mean pesticide concentrations in these matrices after insecticide use at labeled rates. In pollen, 14±8 ppb of imidacloprid and 12±9 ppb of thiamethoxam were detected, while in nectar 10±3 ppb of imidacloprid and 11±6 ppb of thiamethoxam were detected.

Discussion

In assessing the potential hazard of neonicotinoid insecticides to pollinators, two kinds of data are required: 1) levels of exposure and 2) effects of exposure at those levels on the biology of the pollinators. Past risk assessments have based their assumptions about levels of exposure on concentrations of neonicotinoids found in nectar and pollen of crops treated as seeds because those were the only data available at the time. Rortais et al. [26] and Halm et al. [27] used 3.4 ppb of imidacloprid for pollen and 1.9 ppb for nectar as maximum levels, and Cresswell [28] considered 0.7–10 ppb to be the field-realistic range of concentration of imidacloprid in nectar. Cresswell [28] noted that "more studies of the amounts of neonicotinoids in nectar and pollen are needed to establish the field-realistic range because the available data is meager."

Our results partially confirm those of Dively and Kamel [17] in expanding the range of concentration of neonicotinoids found in nectar and pollen in the field. Of the treatments Dively and Kamel used, their "transplant-drip" treatment is the most similar to our treatments, and had similar levels of concentration of neonicotinoids in nectar. Our levels of concentration of neonicotinoids in pollen were similar to the levels they found in 2010, although the levels they found in 2009 were 6–7X higher than ours. They also had higher levels of metabolites in both nectar and pollen than we found (Data not shown here). They also were able to test a wider range of metabolites.

Table 3. Summary of neonicotinoid measurements in pollen and nectar of squash, combining all treatments, years, and varieties.

	Imidacloprid (ppb)		Thiamethoxam (ppb)	
	Pollen	Nectar	Pollen	Nectar
Mean concentration (± SD)	14±8	10±3	12±9	11±6
Number of samples	12	6	12	6
Minimum concentration	6	5	5	5
Maximum concentration	28	14	35	20

The differences in concentrations between application methods we observed in both male and female flower parts in 2009 were not repeated in 2010, perhaps due to differences in weather or crop varieties. Dively and Kamel [17] also had no consistent significant differences when comparing application in transplant water, through drip irrigation, and by foliar spray, although they had significantly lower concentrations in nectar and pollen when imidacloprid was applied in a drench to bedding plants before transplant and when thiamethoxam was applied as a seed treatment.

One reason for the higher concentration of neonicotinoids in nectar and pollen with soil or drip application compared to crops treated as seeds may be because the labeled rates of neonicotinoid applied per unit area are higher for the application methods we used. The highest rate we found for a seed treatment with imidacloprid, for corn in Northern Europe - 95 g AI/ha, [29], was one-third the lowest labeled rate for soil application of imidacloprid on squash, 281 g AI/ha [30] and 27% of the lowest rate of imidacloprid used in this experiment (358 g AI/ha). The seed treatment tested by Dively and Kamel [17], not yet available to us in Connecticut, uses thiamethoxam at 0.75 mg AI per seed. At recommended seeding rates for pumpkin, that would be 13 g AI/ha or 9% of the rate used here.

What would be the effects of the concentrations measured here on the exposure of honey bees and other bees? The concentrations of imidacloprid and thiamethoxam found in nectar are particularly important because honey bees consume far more sugar (as nectar or processed into honey) than pollen over their lifespan. Each worker bee during the summer, going through all the stages of development and a succession of house bee and foraging tasks, consumes 736–1575 mg. of sugar, while each worker bee surviving over the winter consumes an additional 792 mg of sugar maintaining the temperature of the hive [26]. The estimated pollen consumed per bee (stored as bee bread, and processed by nurse bees into glandular secretions for feeding to bee larvae) is only 70.4 mg [26]. Since *C. pepo* nectar is 28–42% sugars by weight [19], each worker bee would consume a minimum equivalent of 1750 mg of nectar over a summer lifespan. The extent to which imidacloprid and thiamethoxam are broken down when pollen and nectar are processed and stored as bee bread and honey is unknown.

A number of studies have been conducted on the sublethal effects of imidacloprid on honey bees. Cresswell [28] did a meta-analysis of 13 studies feeding imidacloprid to honey bee colonies in sugar water (50% sucrose) and modeled the reduction in honey bee colony performance that would be predicted at sublethal doses that have been found in field studies, including the range of doses found here. In addition, recent studies have found interactions of sublethal concentrations of neonicotinoids with honey bee immune systems and with the pathogen *Nosema ceranae* causing increased mortality of honey bees at concentrations of 0.7 and 7 ppb in sugar water [31] or 5 ppb in pollen [32].

There is much less information available on sublethal effects of pesticides on other species of bees. Whitehorn et al. [33] found that *Bombus terrestris* (a European species of bumble bee) had an 85% reduction in queen production over the season when fed imidacloprid at concentrations of 0.7 ppb in sugar water and 6 ppb in pollen for two weeks before being placed in the field.

Both honey bees and bumble bees are generalist feeders on a very wide range of other pollen and nectar sources in addition to *Cucurbita*, so their actual feeding exposure to neonicotinoids would depend on the range of alternative food sources available in addition to treated crop plants. However, squash bees are specialists on *Cucurbita*, feeding their larvae exclusively on *Cucurbita* pollen [22], and also build their nests in soil, often directly beneath squash and pumpkin vines [21], so they could have much more exposure to the soil-applied insecticides used on these crops.

There is much research still to be done on modes of exposure of bees to pesticides [14,34], and effects of pesticides on bees [16]. Very little research has been done on fruit and vegetable crops like squash, which are frequently treated with insecticides, and which are entirely dependent on pollination by bees in order to set fruit and produce a yield.

Supporting Information

Table S1 Chronology of planting, treatments and sampling.

Table S2 MS/MS transitions monitor.

Acknowledgments

We were assisted by Morgan Lowry and Tracy Zarrillo in the field, and by Joe Hawthorne in the laboratory. We also appreciate the assistance of Richard Cecarelli, our farm manager, and his crew.

Author Contributions

Conceived and designed the experiments: KAS. Performed the experiments: KAS BDE. Analyzed the data: KAS. Contributed reagents/materials/analysis tools: KAS BDE. Wrote the paper: KAS BDE.

References

1. National Research Council (2007) Status of pollinators in North America. Washington, DC: National Academies Press. 307 p.
2. Aizen MA, Garibaldi LA, Cunningham SA, Klein AM (2008) Long-term global trends in crop yield and production reveal no current pollination shortage but increasing pollinator dependency. Curr Biol 18: 1572–1575. doi:10.1016/j.cub.2008.08.066.
3. Potts SG, Biesmeijer JC, Kremen C, Neumann P, Schweiger O, et al. (2010) Global pollinator declines: trends, impacts and drivers. Trends Ecol Evol 25: 345–353.
4. Van Engelsdorp D, Hayes J, Underwood RM, Caron D, Pettis J (2011) A survey of managed honey bee colony losses in the USA, fall 2009 to winter 2010. J Apic Res 50(1): 1–10. doi: 10.3896/IBRA.1.50.1.01.
5. Van Engelsdorp D, Meixner MD (2010) A historical review of managed honey bee populations in Europe and the United States and factors that may affect them. J Invertebr Pathol 103: S80–S95.
6. Cameron SA, Lozier JD, Strange JP, Koch JB, Cordes N, et al. (2011) Patterns of widespread decline in North American bumble bees. Proc Natl Acad Sci U S A 108: 662–667.
7. Goulson D, Lye GC, Darvill B (2008) Decline and conservation of bumble bees. Annu Rev Entomol 53: 191–208.
8. Mullin CA, Frazier M, Frazier JL, Ashcraft S, Simonds R, et al. (2010) High levels of miticides and agrochemicals in North American apiaries: Implications for honey bee health. PLoS One 5(3): e9754. doi: 10.1371/journal.pone.0009754.
9. Maxim L, van der Sluis JP (2010) Expert explanations of honeybee losses in areas of extensive agriculture in France: Gaucho® compared with other supposed causal factors. Environ Res Lett 5: 140006. doi:10.1088/1748-9326/5/1/014006.
10. Jeschke P, Nauen R, Schindler M, Elbert A (2011) Overview of the status and global strategy for neonicotinoids. J Agric Food Chem 59: 2897–2908.
11. Laurent FM, Rathahao E (2003) Distribution of [^{14}C] imidacloprid in sunflowers (*Helianthus annuus* L.) following seed treatment. J Agric Food Chem 51: 8005–8010.
12. Cutler GC, Scott-Dupree CD (2007) Exposure to clothianidin seed-treated canola has no long-term impact on honey bees. J Econ Entomol 100: 765–772.
13. Bonmatin JM, Marchand PA, Charvet R, Moineau I, Bengsch ER, et al. (2005) Quantification of imidacloprid uptake in maize crops. J Agric Food Chem 53: 5336–5341.
14. Krupke CH, Hunt GJ, Eitzer BD, Andino G, Given K (2012) Multiple routes of pesticide exposure for honey bees living near agricultural fields. PLoS One 7(1): e29268. doi:10.1371/journal.pone.0029268.
15. Schmuck R, Schöning R, Stork A, Schramel O (2001) Risk posed to honeybees (*Apis mellifera* L, Hymenoptera) by an imidacloprid seed dressing of sunflowers. Pest Manag Sci 57: 225–238.
16. Fischer D, Moriarty T (2011) Pesticide risk assessment for pollinators: Summary of a SETAC Pellston workshop. Pensacola, FL. Society of Environmental Toxicology and Chemistry (SETAC). Available: http://www.setac.org/node/265. Accessed 2012 May 21.
17. Dively GP, Kamel A (2012) Insecticide residues in pollen and nectar of a cucurbit crop and their potential exposure to pollinators. J Agric Food Chem. 60: 4449–4456.

18. Macintyre AJK, Scott-Dupree CD, Tolman JH, Harris CR (2001) Evaluation of application methods for the chemical control of striped cucumber beetle (Coleoptera: Chrysomelidae) attacking seedling cucurbits. J Veget Crop Prod 7: 83–95.
19. Nepi M, Guarinieri M, Paccini E (2001) Nectar secretion, reabsorption, and sugar composition in male and female flowers of *Cucurbita pepo*. Int J Plant Sci 162: 353–358.
20. Shuler R, Roulston TH, Farris GE (2005) Farming practices influence wild pollinator populations on squash and pumpkin. J Econ Entomol 98: 790–795.
21. Julier HE, Roulston TH (2009) Wild bee abundance and pollination service in cultivated pumpkins: Farm management, nesting behavior and landscape effects. J Econ Entomol 102: 563–573.
22. Willis DS, Kevan PG (1995) Foraging dynamics of *Peponapis pruinosa* (Hymenoptera: Anthophoridae) on pumpkin (*Cucurbita pepo*) in southern Ontario. Can Entomol 127: 167–175.
23. Artz DR, Nault BA (2011) Performance of *Apis mellifera, Bombus impatiens,* and *Peponapis pruinosa* (Hymenoptera: Apidae) as pollinators of pumpkin. J Econ Entomol 104: 1153–1161.
24. Lehotay SJ (2005) Quick, easy, cheap, effective, rugged and safe (QuEChERS) approach for determining pesticide residues. In: Martinez Vidal JL, Garrido Frenich A, editors. *Methods in Biotechnology,* Vol. 19, Pesticide Protocols. Totowa, NJ: Humana Press. 239–261.
25. SYSTAT (2009) Systat Software, Inc. Chicago, IL.
26. Rortais A, Arnold G, Halm MP, Touffet-Briens F (2005) Modes of honeybees exposure to systemic insecticides: Estimated amounts of contaminated pollen and nectar consumed by different categories of bees. Apidologie 36: 71–83.
27. Halm MP, Rortais A, Arnold G, Tasei JN, Rault S (2006) New risk assessment approach for systemic insecticides: The case of honey bees and imidacloprid (Gaucho). Environ Sci Technol 40: 2448–2454.
28. Cresswell JE (2011) A meta-analysis of experiments testing the effects of a neonicotinoid insecticide (imidacloprid) on honey bees. Ecotoxicology 20: 149–157.
29. Schnier HF, Wenig G, Laubert F, Simon V, Schmuck R (2003) Honey bee safety of imidacloprid corn seed treatment. Bull Insectology 56: 73–75.
30. CDMS Agrochemical Database. Admire® 2 Flowable Insecticide. Pesticide label. Available: http://www.cdms.net/LabelsMsds/LMDefault.aspx?pd=7797&t= . Accessed 2012 May 21.
31. Alaux C, Brunet JL, Dussaubat C, Mondet F, Tchamitchan S, et al (2009) Interactions between *Nosema* microspores and a neonicotinoid weaken honeybees (*Apis mellifera*). Environ Microbiol 12. doi:10.1111/j.1462-2920.2009.02123.x.
32. Pettis JS, vanEngelsdorp D, Johnson J, Dively G (2012) Pesticide exposure in honey bees results in increased levels of the gut pathogen *Nosema*. Naturwissenschaften 99: 153–158. doi: 10.1007/s00114-011-0881-1.
33. Whitehorn PR, O'Connor S, Wackers FL, Goulson D (2012) Neonicotinoid residue reduces colony growth and queen production. Science 336: 351–352. doi: 10.1126/science.1215025.
34. Tapparo A, Girio C, Marzaro M, Marton D, Solda L, et al. (2011) Rapid analysis of neonicotinoid insecticides in guttation drops of corn seedlings obtained from coated seeds. J Environ Monit 13: 1564–1568.

18

Effects of Dominance and Diversity on Productivity along Ellenberg's Experimental Water Table Gradients

Andy Hector[1,2]*, Stefanie von Felten[1], Yann Hautier[1¤], Maja Weilenmann[1], Helge Bruelheide[3]

1 Institute of Evolutionary Biology and Environmental Studies, University of Zurich, Zurich, Switzerland, 2 Microsoft Research, Cambridge, United Kingdom, 3 Institute of Biology/Geobotany and Botanical Garden, Martin Luther University Halle-Wittenberg, Halle (Saale), Germany

Abstract

Heinz Ellenberg's historically important work on changes in the abundances of a community of grass species growing along experimental gradients of water table depth has played an important role in helping to identify the hydrological niches of plant species in wet meadows. We present a previously unpublished complete version of Ellenberg's dataset from the 1950s together with the results of a series of modern statistical analyses testing for hypothesized overyielding of aboveground net primary production as a consequence of resource-based niche differentiation. Interactions of species with water table depth and soil type in the results of our analyses are qualitatively consistent with earlier interpretations of evidence for differences in the fundamental and realized niches of species. *Arrhenatherum elatius* tended to dominate communities and this effect was generally positively related to increasing water table depth. There was little overyielding of aboveground net primary production during the two repeats of the experiment conducted in successive single growing seasons. Examination of how the effects of biodiversity on ecosystem processes vary across environmental gradients is an underutilized approach – particularly where the gradient is thought to be an axis of niche differentiation as is the case with water availability. Furthermore, advances in ecology and statistics during the 60 years since Ellenberg's classic experiment was performed suggest that it may be worth repeating over a longer duration and with modern experimental design and methodologies.

Editor: Adam Siepielski, University of San Diego, United States of America

Funding: The authors have no support or funding to report.

Competing Interests: Andy Hector was employed visiting researcher in computational ecology and environmental science at MSR (Microsoft Research) Cambridge.

* E-mail: andrew.hector@uzh.ch

¤ Current address: Department of Ecology, Evolution and Behavior, University of Minnesota, Saint Paul, Minnesota, United States of America

Introduction

There is a long tradition in ecology of investigating interactions by growing species alone and in competition with others. In this article we present a previously unpublished complete dataset from a classic example of this type of experiment: Heinz Ellenberg's "Hohenheim groundwater table experiment" on the effects of water table depth on communities of grassland plant species grown in monoculture and mixture [1,2]. In plant ecology this monoculture *vs.* mixture approach has been used to investigate competition between species and the consequences of species interactions for ecosystem primary productivity [3–8]. Our paper therefore also presents the results of a contemporary analysis of overyielding and the effects of diversity on productivity using Ellenberg's data and additive partitioning methods [6,9].

Heinz Ellenberg is well known for having introduced the concept of indicator values, based on the occurrence of species along gradients in nutrient and water supply, pH and climate and other environmental variables [10]. What is less well known is that his concept of fundamental and realized niches predated the frequently cited paper of Hutchinson [11] and was derived from grass communities grown along experimental gradients of depth to the water table [1,2]. Ellenberg created the experimental gradients using a concrete tank with sides that gradually increased in height from one end to the other (Fig. 1). The tank was filled with soil that also varied in depth along its length by following the height of the

walls. Water flowed through the tank from an inlet at the deep end to a spill way and outlet at the shallow end to produce a gradient of depth to the water table. As a supplement to our paper we present a newly discovered complete version of Ellenberg's data from his Hohenheim experiment [1,2].

Although Ellenberg's Hohenheim experiments date from the early 1950s they more recently played an important role in defining the hydrological niches of plants in wet meadows in S.W. England. Silvertown and colleagues [12] did this using data on species occurrence in the Somerset Levels in relation to the depth of the water table (estimated using bore hole measurements). Randomisation tests of the relative abundance of species from Ellenberg's published Hohenheim data showed that the fundamental niches of species, as measured in monoculture, overlapped far more than expected by chance and more than the realized niches of plants grown together in mixed communities. The first result suggests that, when grown alone, species tend to favour the same conditions (have similar fundamental niches), while the second result suggests that competition drives species to have different realized niches [12].

The aim of Ellenberg's experiments was to create a gradient in soil moisture - a resource-based potential niche axis. Coexistence of species through resource partitioning is generally expected to result in overyielding of community productivity. We recognized that Ellenberg's species abundance data in monoculture and

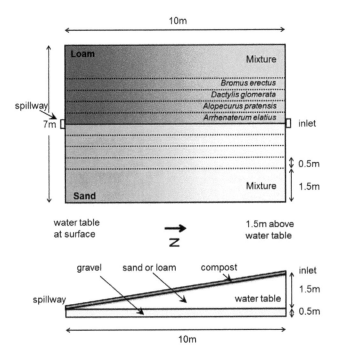

Figure 1. Schematic diagram of Ellenberg's water table depth gradient experiment. The diagram shows the concrete tank from above (top) and the side (below) following Ellenberg (1953, 1954) and Schulze and Beck (2002). Note that the compost layer is on top of the sand, not underneath it as shown in Schulze and Beck (2002). The concrete tank was divided into two halves filled to varying depth with either loam or sand to generate increasing distance to the water table. Each half was divided into strips sown either as monocultures of individual species (4 in 1953, as shown here, 6 in 1952, see Figure S1) or a mixtures of all species combined.

mixture could also be used to test for overyielding of aboveground biomass production along the gradient in depth to the water table. At first it appeared that this would not be possible since Ellenberg's [1,2] published data does not include matched values for the aboveground biomass of species in both monoculture and mixtures as required by most tests of overyielding. However, on investigation, we were able to retrieve the raw data from Ellenberg's handwritten notes that were needed to complete the dataset and to calculate the measures of complementarity necessary to test for the overyielding hypothesized to result if species were differentiated with respect to a resource-based niche, in this case the gradient of depth to the water table.

Materials and Methods

Data

The complete dataset comprises measurements taken from two similar experiments conducted in 1952 [1] and 1953 [2]. The 1952 experiment involved six species (*Poa palustris, Festuca pratensis, Alopecurus, pratensis, Dactylis glomerata, Arrhenatherum elatius, Bromus erectus*; Fig. S1) while the 1953 work involved a subset of four (Fig. 1). In both years species were sown to achieve approximately equal seedling density.

The experiments were carried out in one large concrete tank of 7 m × 10 m, divided into two halves of 3.5 m × 10 m with one side filled with silty loam, the other side with sand ([13], Fig. 1). At the shallow end the tank was 50 cm deep and at its deep end it was 2 m deep. The gradient in water table depth was created by filling

the tanks to create an inclined soil surface, ascending from 50 cm to 200 cm above the bottom of the concrete container. Water was passed through the tank so that it was approximately at ground level at the shallow end of the tank (in 1953 the target was to have the water table exactly at the soil surface while in 1952 the soil was slightly below the level of the tank causing 5 cm of standing water above the soil surface) and 140 cm below the soil surface at the deep end of the tank (the water table was apparently raised by 5–10 cm at the deep end of the tank relative to the shallow presumably due to capillary forces). The two years also differed in the accuracy with which the water table was maintained: in 1952 fluctuations of several decimeters occurred, whereas by 1953 the water table was kept constant through a more continuous inflow and outflow of water.

The experiments were carried out on strips running along the water gradient. The width of the strips differed between 1952 and 1953 because of the different number of species grown in the two years and the fixed 7 m width of the concrete tank. Knowledge of the size of these experimental units is essential for translating the relative yield values reported by Ellenberg [1,2] into biomass per unit area, which had not been published by Ellenberg (Supporting Information S1). The strip width given for 1953 [2] is 0.5 m for the four species when grown alone and 1.5 m for the mixtures of all four species combined (correctly giving a width for each half of the tank of 3.5 m). In 1952, a further two species were grown, bringing the total to six (Fig. S1). The width of the mixture strip was reduced to1.2 m leaving 2.3 m in each half for the monocultures [1]. Assuming each of the six monocultures is grown on a strip of equal width produces a value of 0.38 m. These dimensions were checked against historic photographs of the experiment (Supporting Information S2) and were confirmed with certainty because they correctly reproduce all individual data values given in Ellenberg's papers and notes.

The strips were separated belowground by panels to exclude root interference from adjacent strips (the thickness of these internal walls is not given in Ellenberg [1,2] and Lieth and Ellenberg [13] and since they would reduce the width of each strip by a negligible amount in any case our calculations and the measurements we report (e.g. Fig. 1) do not take them into account). The gradient in water depth was divided in 10 (in 1952) or 11 (in 1953) sections, in which biomass was harvested, either from a species growing alone (monocultures) or in the mixed communities produced by combining the four species grown in 1952 or the six species the following year. Three samples along the gradient were oven-dried, and using these samples, fresh weight was related to dry weight. In 1952, some species in monocultures did not reach complete coverage in some sections due to poor germination (even though the median cover values were high see Fig. S2 and Table S1). Ellenberg predicted what the fresh weight of each species would have been assuming complete 100% cover (e.g. if a species achieved only 50% cover its biomass value would be doubled). Such cover values were not provided in the manuscript for 1953 suggesting that full cover was approximately attained for all species, which is corroborated by photographs of that year (Supporting Information S2). Thus, the values for 1952 were cover-adjusted, while those from 1953 were almost certainly not. For comparability with the earlier works of Ellenberg and colleagues and Silvertown et al., we analyse the data in the same form as in these earlier works, but an open access version of the full dataset is provided as supplementary material to allow further analyses of this historically important data (Supporting Information S3, S4 and S5).

In total the newly assembled complete dataset comprises 416 values (6 species × 10 levels of water table depth in 1952, +4

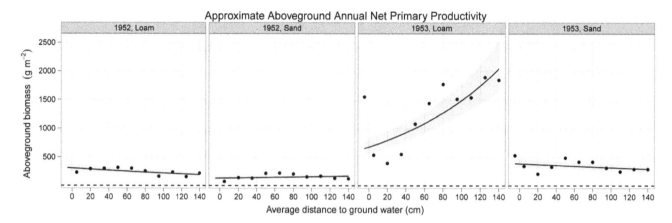

Figure 2. Results of the mixed-effects model analysis of the effect of the water table depth on total aboveground biomass production on sand and loam soils in 1952 and 1953. Note that there is little evidence for limitation of production by water: no drought was imposed. On the contrary, in 1953 there may have been some water logging on the loam since conditions were more productive as depth to the water table increased. The lines are slopes from the mixed-effects model with their 95% confidence intervals (shaded).

species×11 levels of water table depth in 1953, ×2 soil types, ×2 levels of plant diversity).

Measures of overyielding

As explained in the introduction, Ellenberg's data is thought to demonstrate evidence for hydrological niches and resource-based niches of this type are thought to result in overyielding. Our analyses therefore tests whether these hydrological niches along a resource gradient result in overyielding of productivity and how this relationship was affected by differences in soil type and year. For this reason, and since the number of species varied between 4 and 6 in the two different years of the experiment, we concentrated on analyses of community- and ecosystem-level responses, namely additive partitioning of biodiversity effects [6,9] and relative yields and relative yield totals [5]. The additive partitioning of biodiversity effects produces absolute measures while relative yields are, as the name suggests, a relative measure. For brevity, the main text presents complementarity and selection effects [6]. Dominance, trait-dependent and trait-independent complementarity effects from the tripartite additive partitioning [9,14,15] are presented as supplementary material but don't appear to bring any further insights (Fig. S3 and S4). The selection effect examines the covariation between changes in the deviations from expected relative yields of species in mixture and their productivity in monoculture: positive selection effects occur when, on average, species with higher than average monoculture biomass increase their relative abundance in communities while negative selection effects occur when species with lower than average monoculture biomass increase their relative abundance in mixture. Complementarity effects occur when the total biomass of a mixture is more (or less) than expected based on the monoculture yields, producing overyielding (or underyielding). Positive complementarity effects (or overyielding) occur when increases in the biomass of some species are not exactly compensated by decreases in the biomasses of others. This could occur due to complementary resource use, a reduction of natural enemy effects in mixtures, some form of facilitation or some combination of these effects. Negative complementarity effects are consistent with some type of interference competition (although the biological details remain unknown). The relative yield of a species is its biomass in mixture expressed as a proportion of its yield in monoculture. The relative yield total (RYT) of a mixture is the sum of the individual relative yields. Values of RYT greater than one indicate overyielding (equivalent to positive complementarity effect values). Further explanation and comparison of the overyielding methods can be found in [15]. Details of the overyielding calculations are given in the supplementary material (Supporting Information S6).

Analysis

We modelled changes in these response variables across the gradients in depth to the water table using regressions that included linear and quadratic terms only (depth and depth squared) because of the limited size of the dataset and because while it seemed plausible that maximum or minimum values of the responses could occur at intermediate water table depths, we had no *a priori* biological basis for allowing more complex relationships in monoculture.

Competitive displacement of species from the middle of the gradient to both higher and lower water table depths could produce bimodal relationships in mixtures [16] but exploratory generalized additive models for the aboveground biomass yields of the species in mixture provided no support for a higher degree of complexity than the quadratic relationship (Fig. S5). We therefore limited ourselves to the simpler quadratic models for consistency across monocultures and mixtures and due to risk of over-fitting complex higher-order polynomials to this limited dataset without supporting evidence.

We then tested whether the resulting curves varied with soil type and year using linear mixed-effects models containing linear and quadratic terms for water table depth and including a random factor ("gradient") with four levels, one for each soil type in each year, to reflect the grouping of the data into two years and the splitting of the concrete tank into two halves. For the biodiversity effects (trait-dependent and trait-independent complementarity, selection and dominance effects), the variance was not constant, so we transformed the values by taking the square root of the absolute values and restoring the original positive or negative sign [6]. The analyses were performed in R 2.12.1 [17] using the lmer function from the lme4 library [18] for both the species- and community-level data (necessary due to crossed random effects for the species-level data where species repeat in the different year and soil type combinations). For the species-level data, we considered fixed effects of water table depth, soil type and species while treating the halves of the concrete tank, the strips and the species compositions

Individual Species Aboveground Biomass

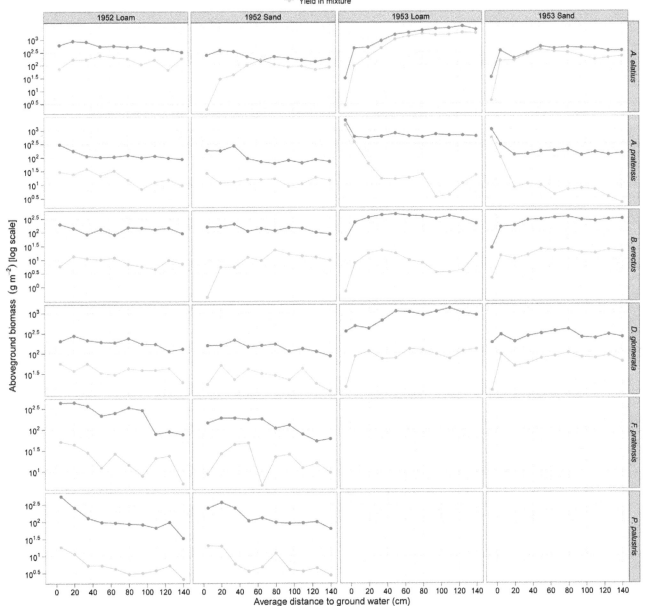

Figure 3. The complete dataset showing the yields of six different species in monoculture and mixture across the water table depth gradient on two soil types in two years. Note the absence of two species in 1953.

(6 monocultures plus 2 mixtures) as random effects. For the community level data, we considered fixed effects of water table depth, soil type and year while treating the four gradients (produced by the two halves of the concrete tank in each of the two years) as random effects. Following examples given by the software authors [19], our analysis took a model simplification approach to identifying the single most parsimonious model from the series of nested models (i.e. the one with the lowest BIC, favouring simpler models when pairs of nested models were tied with BICs within 2 units per difference in the number of parameters). We used the BIC since it is one of the more widely used information criteria and there is less risk of over-fitting

unnecessarily complex models to small datasets than with the AIC. In the supplementary material we also provide graphs of the full quadratic models for each variable.

Results

The aim of Ellenberg's experiments was to create wetter and drier growing conditions by varying depth of the water table on two soils of different water-holding capacity in order to look for evidence for what he coined the terms physiological and ecological behaviour and what we would now call fundamental and realized hydrological niches, respectively (Fig. 1). Interestingly, little

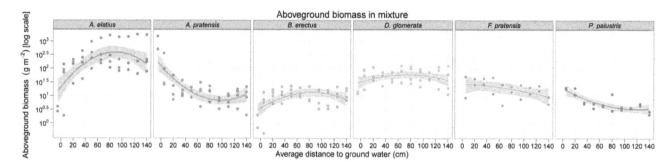

Figure 4. Results of the mixed-effects model analysis of the yields of individual species across the experimental water table depth gradient. Undetectable interactions with soil type have been removed from the model while the results are essentially averaged across the two years by including year as a random factor. The curves are slopes from the mixed-effects model with their 95% confidence intervals (shaded). Note that *F. pratensis* and *P. palustris* were present only in 1952.

previous assessment has been made of the success of this manipulation. Counter to our expectations there is very little evidence for any effect of drought on total aboveground net biomass production (Fig. 2). In fact, the most pronounced effect was of increasing biomass with greater depth to the water table on the loam in 1953, suggesting not drought but rather water-logging on this soil type in this wet year (Fig. 2; Table S2).

The newly assembled complete dataset of the mass of the different species grown in monoculture and mixture across the water table depth gradients on the two soil types in the two years is shown in Figure 3 and Figure S6. Ellenberg [1,2] interpreted subsets of these data as demonstrating differences in the fundamental and realized niches of species, a conclusion supported by the randomisation tests of species relative (percent) abundances reported by Silvertown et al. [12]. The results of our analysis of the full data on the yields of the species in monoculture are also consistent with differences in fundamental niches via the presence of many interactions between species identity and the environmental variables (Table S3), while the same analysis of relative yields (Fig. S7; Table S4) of individual species in mixture is similarly consistent with niche differences. Model simplification for the aboveground biomass yields of the species in mixture selected a model with different quadratic regressions for each species but which omitted interactions with soil type (Fig. 4; Table S3).

Arrhenatherum elatius showed a positive convex relationship with increasing depth to the water table as did, to a lesser extent, *Bromus erectus and Festuca pratensis*. In contrast, *Alopecurus pratensis* showed a negative concave relationship, as did *Poa palustris* (but with less curvature). *Dactylis glomerata* yields peaked at the middle of the gradient but were fairly insensitive to the water table depth.

Relative yields of individual species showed a similar but slightly more complex response with species having different curvatures (quadratic relationships) of their responses to depth to the water table and with soil type (but no 3-way interaction between soil, species and water table; Fig. S8; Table S4). Comparing observed with expected relative yields (where expected relative yields are simply 1/S where S is the number of species in the mixture: 1/6 in 1952 and 1/4 in 1953) showed that *Arrhenatherum elatius* was the species with the highest gains relative to expected values across most of the water table gradient in both years and on both soil types (Fig. S7). The performance of *Arrhenatherum* and *Alopecurus* appears to be somewhat negatively related since where *Arrhenatherum* performed better than expected across most of the gradient, *Alopecurus* under-performed. However, *Arrhenatherum* did only as well as, or worse than expected when the water table was high, conditions where *Alopecurus pratensis* did as well or better than expected. In contrast, *Dactylis glomerata* was fairly unresponsive: it did about as well as expected under all conditions. *Bromus erectus*

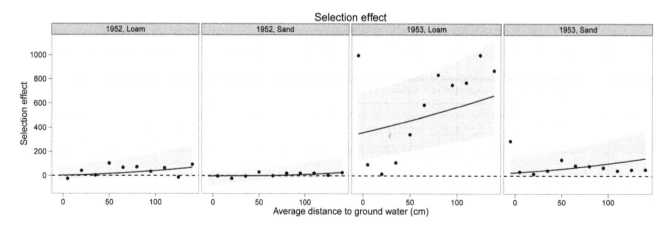

Figure 5. Results of the analysis of the effect of the water table depth on the strength and direction of the Selection Effect on sand and loam soils in 1952 and 1953. The curves are back-transformed slopes from the mixed-effects model with their 95% confidence intervals (shaded). The positive Selection Effect on loam in 1953 is driven by increasing dominance by *Arrhenatherum elatius* as depth to the water table increased.

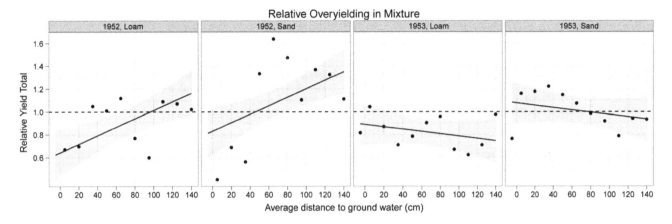

Figure 6. Results of the analysis of the effect of the water table depth on relative overyielding (RYT) on sand and loam soils in 1952 and 1953. The curves are back-transformed slopes from the mixed-effects model with their 95% confidence intervals (shaded). The dotted lines show the zero sum null expectation where species perform the same in mixtures as in monoculture and increases in some species populations are exactly offset by declines in others. Complementarity effects produced similar results but the data are not so well behaved (Figure S9).

generally performed worse than expected across the entire gradient.

Selection effects were strongest on loam in 1953 where they were increasingly positive as depth to the water table increased (Fig. 5). This was largely due to the dominance of *Arrhenatherum elatius*, a species with a relatively high monoculture biomass (Fig. 3), as described above.

There was little evidence for complementarity (Fig. S9): confidence intervals for average relative yield totals contained zero in most cases with some overyielding at greater water table depths in 1952 and underyielding on loam in 1953 when *Arrhenatherum* strongly dominated and drove a positive selection effect (Fig. S8). Since RYTs and complementarity effects are linearly related [6] they show similar patterns (Fig. 6) but the complementarity effect data were statistically less well behaved.

Discussion

Ellenberg's experimental manipulation of water table depth has been highly influential work, both in its original publication and through later use, notably by Silvertown and colleagues to support their demonstration of hydrological niches in wet meadow communities [12]. The results of our analysis seem consistent with this earlier work: the many species-by-environment interactions in the analysis of both the monoculture and mixture data are consistent with differences in the fundamental and realized niches of these grass species. However, we found limited evidence of overyielding.

The limited evidence for positive complementarity effects in Ellenberg's data is in some ways surprising given that the same information has been used as evidence for hydrological niche differences and that resource-based niches are thought to lead to overyielding. We can think of three possible explanations for their absence. First, although previous analyses by Ellenberg and Silvertown and colleagues are consistent with species coexistence they do not conclusively demonstrate that stabilizing differences between species are strong enough to overcome any differences in fitness to ensure stable long-term coexistence. Rather, it could be that in the long-term there would have been competitive exclusion with different species dominating at the points on the water table gradient where they were competitively superior. Second, theory predicts that there are mechanism for the stable coexistence of

species without positive effects on community productivity [20] and it is possible that hydrological niches are one such example (although this seems unlikely given that that the published example of a coexistence mechanism that does not lead to overyielding is the competition-colonisation trade off which is not a resource-based niche like the hydrological gradient examined here). Third, the lack of complementarity effects may simply reflect the short duration of the two single year experiments and insufficient time for overyielding to fully emerge. A meta-analysis of 47 biodiversity experiments [3] found that complementarity effects were initially weak and increased over time and were still linearly increasing even after several years. Testing these alternative hypotheses would require repeating Ellenberg's experiments (with modern improvements in design and methodology) and running them for a longer duration, ideally beyond the range of recent studies [3] (5–10 years or more). Finally, while overyielding is consistent with many forms of stable coexistence through niche differences (but see [20]) it may be an imperfect measure that may also reflect other processes [21,22].

We were also surprised by another aspect of the data. We had the impression that the water table depth gradients were likely to impose drought, at least on some species and particularly on the freely draining sand. However, there is little evidence for a drought effect. On the contrary, on the loam in 1953 aboveground biomass production increased with depth to the water table to very high levels, suggesting if anything the opposite: limitation of production by water logging as the water table approached the soil surface that could be due to the particularly wet summer of 1953 exacerbated by the low oxygen supply on loam compared to sand (Fig. S10). Whether such high productivities would be repeatable is not clear and is another incentive to repeat Ellenberg's experiments with modern improvements. In addition to improved replication and randomisation, these future studies should incorporate measurements of actual water table depths and soil water potential and could then attempt to impose drought at the drier end of the gradient.

Ellenberg's data show evidence for strong positive selection effects on the more fertile loam soil during 1953. Positive selection effects occur when species with higher-than-average monoculture aboveground biomasses dominate mixtures: in this case *Arrhenatherum* mainly drove the effect. Furthermore, this effect was generally positively related to increasing depth to the water table.

While we cannot know the biological mechanisms underlying the positive selection effects, one simple explanation is that species with high aboveground biomass in monoculture also had high belowground biomass enabling them to better access the deeper water table. Both *Festuca* and *Poa* showed declining biomasses with increasing depth to the water table, as did *Alopecurus* in 1953, although only when in competition with other species in the mixture (particularly *Arrhenatherum*).

Not surprisingly, developments in experimental design and statistical analysis since the 1950s mean that Ellenberg's experiments could be improved in several ways if repeated. Lack of true replication in Ellenberg's experiments means that we cannot know how repeatable the yields at a given point on the gradient of water table depth on a particular soil and in a particular year are, and an increase in replication would therefore be an essential feature of a repeat study. The experiments also examine only one set of species grown alone and when combined together. Replicate species pools and mixtures of intermediate diversity - particularly species pairs - would also be useful for generalizing the results and for investigating species interactions in greater detail. The spatial arrangement of the strips also has the (wider) mixed community strips always at the edge of the concrete tanks (Fig. 1). Replicate monocultures and mixtures could be randomised to avoid potential edge effects or, if replication were low, they could be systematically arranged so that equal proportions of the monoculture and mixture strips were at the edge. We hope this article will stimulate these updated repeats of Ellenberg's experiments in order to further the experimental investigation of the hydrological niche and its effects of productivity and ecosystem functioning.

Supporting Information

Supporting Information S1   Ellenberg's adjustment for incomplete cover.

Supporting Information S2   Historical photographs of the Hohenheim-Experiment.

Supporting Information S3   Metadata.

Supporting Information S4   Ellenberg's full data set.

Supporting Information S5   Ellenberg's climate data set from the weather station in Hohenheim.

Supporting Information S6   Additive partitioning of biodiversity effects.

Figure S1   Schematic diagram of Ellenberg's water table depth gradient experiment in 1952.

Figure S2   Means biomass for the monocultures of the 6 species used in the 1952 run of the experiment. Green symbols show Ellenberg's data as adjusted for cover levels below 100% and red symbols show the same mean values allowing for the observed level of cover (i.e. undoing Ellenberg's changes).

Figure S3   Results of the analysis of the effect of the water table depth on the strength and direction of the trait-dependent complementarity effect on sand and

loam soils in 1952 and 1953. The lines are slopes from the mixed-effects model with their 95% confidence intervals (shaded).

Figure S4   Results of the analysis of the effect of the water table depth on the strength and direction of dominance effect on sand and loam soils in 1952 and 1953. The lines are slopes from the mixed-effects model with their 95% confidence intervals (shaded).

Figure S5   Estimated smoothing curves for the yields of individual species across the experimental water table depth gradient. The solid line is the smoother and the dotted lines are 95% point-wise confidence bands. Note that the smoother is centred around zero.

Figure S6   Dry biomass. Aboveground biomass of species in monoculture (blue) and mixture (green) on the two soil types in the two experimental years together with the total community aboveground biomass (black). Note that the black symbols within each column of panels are the same and are the sum of the green symbols.

Figure S7   Relative yields of six different species across the water table depth gradient on two soil types in two years. Note the absence of two species in 1953.

Figure S8   Results of the mixed-effects model analysis of the relative yields of individual species across the experimental water table depth gradient. The curves are back-transformed slopes from the mixed-effects model with their 95% confidence intervals (shaded).

Figure S9   Results of the analysis of the effect of the water table depth on the strength and direction of the trait-independent complementarity effect on sand and loam soils in 1952 and 1953. The curves are back-transformed slopes from the mixed-effects model with their 95% confidence intervals (shaded). The negative Complementarity Effect on loam in 1953 is mainly driven by increasing dominance by Arrhenatherum elatius as depth to the water table increased.

Figure S10   The climate data from the weather station in Hohenheim clearly showing that 1953 had a very wet summer.

Table S1   Available cover data of the monocultures in 1952 with median cover, cover corrected biomass mean (se) in g and the uncorrected biomass mean (se) in g within species and by soil type.

Table S2   Analysis of total aboveground biomass across the water table depth gradient on the two soil types in the two years. Hereafter, mixed-effects models are given in R syntax so that a response is analysed as a function of ($\sim$) fixed effects with random effects given in parentheses, as follows: Response$\sim$fixed explanatory variables+(random effects).

Table S3   Analysis of aboveground biomass yields of the species in mixture across the water table depth gradient on the two soil types.

Table S4 Analysis of relative yields of the species in mixture across the water table depth gradient on the two soil types.

Acknowledgments

We are very grateful to Dr. Hans Heller who, as a long-term associate of Prof. Heinz Ellenberg, maintained the archive of Ellenberg's manuscripts and slides and invested many hours to retrieve the original documents. In addition, he provided essential explanations on the design of Ellenberg's experiment in Hohenheim and pointed out errors in drafts of our manuscript. We also thank: Mike Austin for providing Peter Auckland's CSIRO translation of two of Ellenberg's publications [23]; Eva Vojtech for help with the early stages of data processing and Rheinhard Furrer for advice on the complexity of the analysis appropriate to such limited data (any remaining inadequacies of the analysis are our own).

Author Contributions

Conceived and designed the experiments: AH HB. Analyzed the data: AH YH. Contributed reagents/materials/analysis tools: MW HB SVF. Wrote the paper: AH YH. Additional input analyzing the data: SVF MW HB. Additional input writing the manuscript: SVF MW HB.

References

1. Ellenberg H (1953) Physiologisches und oekologisches Verhalten derselben Pflanzenarten. Berichte der Deutschen Botanischen Gesellschaft 65: 350–361.
2. Ellenberg H (1954) Über einige Fortschritte der kausalen Vegetationskunde. Plant Ecology 5/6: 199–211.
3. Cardinale BJ, Wright JP, Cadotte MW, Carroll IT, Hector A, et al. (2007) Impacts of plant diversity on biomass production increase through time due to species complementarity: A meta-analysis of 44 experiments. Proceedings of the National Academy of the USA 104: 18123–18128.
4. De Wit CT (1960) On competition. Verslagen Landbouwkundige Onderzoekingen 66: 1–82.
5. Harper JL (1977) Population Biology of Plants. London: Academic Press.
6. Loreau M, Hector A (2001) Partitioning selection and complementarity in biodiversity experiments. Nature 412: 72–76 [erratum: 413 548].
7. Trenbath BR (1974) Biomass productivity of mixtures. Advances in Agronomy 26: 177–210.
8. Vandermeer J (1989) The ecology of intercropping. Cambridge: Cambridge University Press. 1–22 p.
9. Fox JW (2005) Interpreting the 'selection effect' of biodiversity on ecosystem function. Ecology Letters 8: 846–856.
10. Mueller-Dombios D, Ellenberg H (1974) Phenological aspects in plant communities. Aims and Methods in Vegetation Ecology. New York: Wiley International Edition. pp. 150–152.
11. Hutchinson GE (1957) Concluding remarks. Cold Spring Harbor Symposium. Quantitative Biology 22: 515–427.
12. Silvertown J, Dodd ME, Gowing DJG, Mountford JO (1999) Hydrologically defined niches reveal a basis for species richness in plant communities. Nature 400: 61.
13. Lieth H, Ellenberg H (1958) Konkurrenz und Zuwanderung von Wiesenpflanzen. Ein Beitrag zum Problem der Entwicklung neu angelegten Grünlands. Zeitschrift für Acker- und Pflanzenbau 106: 205–223.
14. Fox JW, Harpole S (2008) Revealing how species loss affects ecosystem function: the trait-based Price equation partition. Ecology: 1139–1143.
15. Hector A, Bell T, Connolly J, Finn J, Fox J, et al. (2009) The analysis of biodiversity experiments: From pattern toward mechanism. In: Naeem S, Bunker DE, Hector A, Loreau M, Perrings C, editors. Biodiversity, Ecosystem Functioning, and Human Wellbeing: An Ecological and Economic Perspective. Oxford: Oxford University Press. pp. 94–104.
16. Austin MP, Belbin L, Meyers JA, Doherty MD, Luoto M (2006) Evaluation of statistical models for predicting plant species distributions: role of artificial data and theory. Ecological Modelling 199: 197–216.
17. R Development Core Team (2010) A language and environment for statistical computing. R.2.12.1 ed. Vienna, Austria: R Foundation for Statistical Computing. Available: http://www.r-project.org/. Accessed 2012 Jul 31.
18. Bates D, Maechler M, Bolker B (2005) lme4: Linear mixed-effects models using S4 classes. Vienna, Austria: R Foundation for Statistical Computing. Available: http://www.r-project.org/. Accessed 2012 Jul 31.
19. Pinheiro JC, Bates DM (2000) Mixed-effects models in S ans S-plus. New York: Springer.
20. Moore J, Mouquet N, Loreau M, Lawton JH (2001) Coexistence, saturation and invasion resistance in simulated plant assemblages. Oikos 94: 303–314.
21. Carroll IT, Cardinale BJ, Nisbet RM (2011) Niche and fitness differences relate the maintenance of diversity to ecosystem function. Ecology 92: 1157–1165.
22. Loreau M, Sapijanskas J, Isbell F, Hector A (2011) Niche and fitness differences relate the maintenance of diversity to ecosystem function: Comment. Ecology 92: 1157–1165.
23. Auckland P (1978) Translations of two of Heinz Ellenberg's papers. Technical Memorandum 78: 1–34. CSIRO, Division of Land Use Research, Canberra.

Intercropping Competition between Apple Trees and Crops in Agroforestry Systems on the Loess Plateau of China

Lubo Gao[1], Huasen Xu[1], Huaxing Bi[1,2]*, Weimin Xi[3], Biao Bao[1], Xiaoyan Wang[1], Chao Bi[1], Yifang Chang[1]

1 College of Water and Soil Conservation, Beijing Forestry University, Beijing, P.R. China, 2 Key Laboratory of Soil and Water Conservation, Ministry of Education, Beijing, P.R. China, 3 Department of Biological and Health Sciences, Texas A&M University-Kingsville, Kingsville, Texas, United States of America

Abstract

Agroforestry has been widely practiced in the Loess Plateau region of China because of its prominent effects in reducing soil and water losses, improving land-use efficiency and increasing economic returns. However, the agroforestry practices may lead to competition between crops and trees for underground soil moisture and nutrients, and the trees on the canopy layer may also lead to shortage of light for crops. In order to minimize interspecific competition and maximize the benefits of tree-based intercropping systems, we studied photosynthesis, growth and yield of soybean (*Glycine max* L. Merr.) and peanut (*Arachis hypogaea* L.) by measuring photosynthetically active radiation, net photosynthetic rate, soil moisture and soil nutrients in a plantation of apple (*Malus pumila* M.) at a spacing of 4 m × 5 m on the Loess Plateau of China. The results showed that for both intercropping systems in the study region, soil moisture was the primary factor affecting the crop yields followed by light. Deficiency of the soil nutrients also had a significant impact on crop yields. Compared with soybean, peanut was more suitable for intercropping with apple trees to obtain economic benefits in the region. We concluded that apple-soybean and apple-peanut intercropping systems can be practical and beneficial in the region. However, the distance between crops and tree rows should be adjusted to minimize interspecies competition. Agronomic measures such as regular canopy pruning, root barriers, additional irrigation and fertilization also should be applied in the intercropping systems.

Editor: Randall P. Niedz, United States Department of Agriculture, United States of America

Funding: This paper was supported by National Scientific and Technology Program of China (No. 2011BAD38B02) and CFERN & GENE Award Funds on Ecological paper. The funders had no role in study design, data collection and analysis, decision to publish, or preparation of the manuscript.

Competing Interests: The authors have declared that no competing interests exist.

* E-mail: bhx@bjfu.edu.cn

Introduction

The Loess Plateau is the birthplace of China's primitive agriculture. However, because of unsound land use and destruction of forests, the Loess Plateau has suffered serious soil erosion. At the same time, rapid population growth has also brought greater pressure to the environment in the region. The ensuing ecological and environmental problems have slowed down the economic development and living standards of local people. These problems lead to further deterioration of ecological environment, forming a vicious cycle. The local government is facing dual pressures from both economy and ecology.

Agroforestry systems have been considered as an effective practice to alleviate the conflicts between the rapidly growing population and the limited arable land resources [1,2]. In recent years, agroforestry management has been widely applied in the Loess Plateau region for reducing soil erosion and water loss, restoring ecological balance, raising land utilization rate and increasing economic benefits [3,4]. However, in most agroforestry systems, competition for light, moisture and nutrients exists at the interface between trees and crops which can cause a reduction of crop yield [5]. It is a major constraint that has affected stability of the structure and the function of the agricultural ecosystems. The competition between woody tree species and understory crop species not only exists aboveground (competition for light) but also comes from belowground (competition for soil moisture and nutrients), leading to lower crop yield. According to Friday and Fownes, the competition between trees and crops is overwhelmingly for light which is the main reason for the reduction of maize in alley cropping system in Hawii, USA [6]. Similar results were reported by Peng et al. in loess area of Weibei in Shaanxi Province, China [7]. Elsewhere in southern Australia, studies showed that reduced crop yields are associated with the competition for water in windbreak and alley systems [8,9]. Kowalchuk and Jong found that, especially in drought years, competition for water is the principal factor affecting the yield of spring wheat intercropped with shelterbelts in Western Saskatchewan [10]. In some related studies, the results indicated that competition for nutrients does not exist in intercropping systems [11–13]. However, others reported that as one of the main reasons leading to the reduction of crop yield, the competition for soil nutrients does exist in the interface of trees and crops and has a negative impact [14,15]. It is very important to explore the competitive mechanism in intercropping systems, in order to provide optimum management strategies and technologies for managing intercropping system with high-yield, high-efficiency and stabilization.

Table 1. Characteristics of apple trees intercropped with soybean and peanut in the experimental sites in July 2011.

Measurement	Intercropped with soybean	Intercropped with peanut
Tree height (m)	2.4	2.5
DBH (cm)	4.1	4.2
Depth of live crown (m)	1.7	1.8
Mean radius of crown (m)	1.3	1.2

Apple-crop intercropping system is one of the most commonly applied agroforestry systems in the Loess Plateau region owing to its good ecological, social and economic benefits. However, only few studies focused on this intercropping system in the area. In order to explore the biological reasons of the competition in typical intercropping systems and to provide effective management techniques, we report on a study of two apple-crop intercropping systems (apple-soybean, apple-peanut) on the Loess Plateau region in the western portion of Shanxi Province. The objectives of our research were (1) to analyze the interspecies competition relationship between trees and crops; (2) to find the limiting factors in the development of intercropping systems in this area; (3) to offer possible solutions to minimize the interspecies competitions and maximize resource utilization; (4) to enrich the related study and to improve the management of the intercropping systems in this region.

Materials and Methods

Study site

The study site was located in the Baidong Village, Jixian County, Shanxi Province, China (36°06′ N, 110°35′ E, 1025 m a.s.l.). The area is a typical hill and gully region of the Loess Plateau. The annual mean rainfall is about 575 mm, and the mean annual temperature is 10°C (1991–2010). The precipitation is unevenly distributed seasonally, with an average rainfall of 463 mm from June to August (1991–2010), which contributed about 80% of annual precipitation. The parent material of the soil is loess, and the soil properties are uniform. The bulk density, pH, total porosity, $CaCO_3$ content, cation exchange capacity, organic C, total N and available P of the top soil layer (100 cm) were 1.32 Mg•m^{-3}, 8.24, 50.16%, 18.35%, 18.43 cmol•kg^{-1}, 6.27 g•kg^{-1}, 0.39 g•kg^{-1} and 4.39 mg•kg^{-1}, respectively. The main intercropping tree species are Apple (*Malus pumila* M.), Apricot (*Prunus armeniaca* L.), Pear (*Pyrus bretschneideri* R.), Chinese arborvitae (*Platycladus orientalis* (L.) and Franco) and Black locust (*Robinia pseudoacacia* L.).

Ethics Statement

No specific permits were required for the described field studies. The sampling locations were not privately-owned or protected in any way and the field studies did not involve endangered or protected species.

Treatments and Crop Cultivation

Two typical intercropping systems of apple-soybean and apple-peanut were chosen for this study during the crop growing season of 2011 and 2012. The apple trees were planted in an East-West orientation in 2007. The characteristics of the apple trees intercropped with soybeans and peanuts in July 2011 are listed in Table 1. There were four treatments in this study: apple-soybean intercropping treatment (AS), soybean monoculture served as control (CS), apple-peanut intercropping treatment (AP) and peanut monoculture served as control (CP). Each treatment had three replicates. Each replicate of intercropping treatment (AS and AP) was an 8 × 10 m plot that included 12 trees planted in three rows with 4 m between trees and 5 m between rows. Each replicate of control treatment (CS and CP) was the same size of 8 × 10 m. For all treatments, the crops were planted at a spacing of 0.4 m with in rows and 0.5 m between rows and received the same agricultural management practices. Soybean and peanut were grown 0.3 m from an adjacent tree row in the intercropping systems. All plots received 147 kg N ha^{-1}, 30 kg P ha^{-1} and 30 kg K ha^{-1} as basal fertilizer and no additional fertilizer or irrigation in the rest of the year.

Measurements of Plant Photosynthesis, Soil Moisture and Nutrients

For the sampling of plant photosynthesis, soil moisture and soil nutrients, six sampling locations at distances of 0.5 m, 1.5 m and 2.5 m, respectively, from both side of tree row were identified as sampling points in each intercropping plot (Figure 1). The sampling points were further divided into three equal groups and denoted as F0.5, F1.5 and F2.5 based on the distance (0.5 m, 1.5 m and 2.5 m) from the tree row. Measurement parameters of F0.5, F1.5 and F2.5 were used to represent the major locations of 0.5 m, 1.5 m and 2.5 m away from apple tree row. For each control plot, five selected points were established with an S-shaped sampling method.

Photosynthetically active radiation (PAR) and net photosynthetic rate (NPR) of crops were performed by two portable Li-6400 photosynthesis systems which had a 6 cm^2 clamp-on leaf chamber connected to the main engine (Li-6400, Li-Cor Inc., Lincoln, NE, USA) under ambient humidity, temperature and irradiance. One fully expanded leaf from the upper part of the crop canopy in each sampling point was selected and measured five times with 2 h intervals during daytime (0900–1700 h). During each measurement period, all sampling points of intercropping treatment and control treatment were visited. These treatments were measured in mid-August 2011 and again in late August 2012, the typical phenological phases of peanut and soybean. For all measurements, the flow velocity was set at 500 μmol•s^{-1} and the airstream entering the chambers was kept at the growth CO_2 concentration (370 μmol•mol^{-1}) by a computer-controlled CO_2 injector system supplied with Li-6400. PAR and CO_2/H_2O exchanged by the leaf were measured concurrently with the quantum sensor and the infrared gas analyzer on LI-6400. The data were recorded and calculated automatically with the software in the photosynthesis system.

For soil moisture, the samples were taken at different phenological phases of soybean and peanut: 8 July, 23 August, and 23 September in 2011; and 4 July, 11 August, and 22 September in 2012. A drill was used to remove the soil from 0–100 cm in 20 cm intervals in soil profile. The soil moisture content

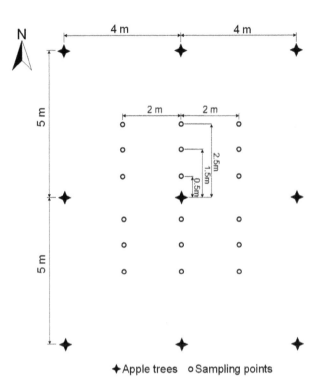

Figure 1. Sampling points of plant photosynthesis, soil moisture and nutrients in the intercropping study sites.

was determined gravimetrically in each layer. The mean soil moisture content of the five layers in all sampling time (2011 and 2012) was calculated and used as the final value of the sampling point.

For the sampling of soil nutrients, the soil samples were taken on 23 August, 2011 and 11 August, 2012, the typical phenological phases of peanut and soybean. The soil samples were collected from a depth of 0–100 cm in soil profile with a drill. Organic matter content was determined by H_2SO_4-$K_2Cr_2O_7$ pyrogenation. Total N was determined using the Kjeldahl method, with a KDY-9830 N Analyzer. Available P was determined by Olsen sodium-bicarbonate extraction. Available K was determined by flame photometer.

Measurements of Crop Growth and Yields

For the sampling of crop growth, we used the same sample locations as for soil moisture. A single crop plant was sampled at each sampling point on 24 August 2011 and 12 August 2012. A total of 69 soybean plants and 69 peanut plants were harvested during each measurement period. In the lab, plant height, hundred leaf dry weight and total above-ground biomass of all plants were measured and recorded.

At the end of the growing season, in each intercropping plot, peanuts and soybeans were harvested from both sides of the tree row in two rectangular areas. The rectangular area was 4.0 m long and 2.7 m wide. As the convenience of the study, the two rectangular areas were divided into three groups: (1) the area of 0.3–1.0 m away from tree row; (2) the area of 1.0–2.0 m away from tree row; and (3) the area of 2.0–3.0 m away from tree row. The yields of the three groups were used to represent the crop yield of F0.5, F1.5 and F2.5, respectively. In the control plots, 2 m × 2 m quadrates of soybean and peanut were harvested to get the grain production. The peanuts and soybeans were dried at 70 °C

and then weighed to obtain an average dry weight. Yield values were reported on a per hectare basis.

Data analysis

All parameters (PAR, NPR, soil moisture, soil nutrients content, crop growth and yields) measured for control treatments and three major locations (F0.5, F1.5 and F2.5) of intercropping treatments were described in terms of mean values followed by respective standard deviations. Simple regression analysis was used to examine the relationships between the data of PAR, NPR, soil moisture and the distance from the tree row. Differences among groups for each crop (soybean or peanut) were determined by one-way ANOVA, and the results of the multiple comparisons were performed with least significant difference (LSD) test at $P<0.05$. NPR, total above-ground biomass and yield values of soybean and peanut had a correlated analysis with environmental parameters to decide the effect of apple trees competition on crop growth and productivity via bivariate correlation (Pearson) analysis at $P<0.05$ and $P<0.01$. All the analyses were performed by using the software IBM SPSS Statistics 20.0 for Windows.

Results

Light Interception and Plant Photosynthesis

For both crops, diurnal variation of photosynthetically active radiation (PAR) in the intercropping systems and the monoculture configurations (control treatments) showed a single peak curve with time (Figure 2). The peak of PAR appeared at 13:00 pm and the minimum value appeared in 17:00 pm. Because of reflectance, absorbance and transmittance by the apple tree canopy, the PAR of crops in the intercropping systems were lower than that in the monoculture configurations during the same period. On the horizontal distribution, the general trend was that the closer the crops to the tree rows, the lower the PAR received. The same tendency was found in diurnal variation of net photosynthetic rate (Figure 3).

The daily mean values of PAR showed a clear linear relationship with distance from the apple tree row in both intercropping treatments (Figure 4A). The trend lines of PAR (Y, $\mu mol \cdot s^{-1} \cdot m^{-2}$) and distance from trees rows (X, m) were $Y = 78.5 \times +865.3$ ($R^2 = 0.999$) in apple-soybean intercropping treatment (AS) and $Y = 82.5 \times +881.9$ ($R^2 = 0.873$) in apple-peanut intercropping treatment (AP). The slopes of both regression lines suggested that the PAR in AP treatment had a higher growth than that in AS treatment as the distance from the tree increased. As shown in Figure 4A, PAR reaching the upper parts of the crop canopy in AP treatment also had higher values than that in AS treatment at the same distance away from the tree row. It indicated that peanut canopy could obtain more solar radiation in AP treatment than soybean canopy in AS treatment. At confidence level of 95%, the control treatment PAR mean fell within the confidence intervals of F2.5 in the corresponding intercropping systems. Compared with the corresponding control treatment, PAR at F0.5 and F1.5 showed a reduction of 17.9% and 10.4% in AS treatment, respectively, 17.8% and 5.4% in AP treatment. Similar linear relationships were also obtained through regression analysis of the relationship between NPR and distance from the apple tree row (Figure 4B). The trend lines of NPR (Y, $\mu mol \cdot s^{-1} \cdot m^{-2}$) and distance from trees rows (X, m) were $Y = 1.025 \times +12.003$ ($R^2 = 0.902$) in AS treatment, and $Y = 0.940 \times +10.983$ ($R^2 = 0.951$) in AP treatment. The NPR in AS treatment had higher values and growth than that in AP treatment as the distance from the tree increased which was different from the measurement of PAR. The control treatments

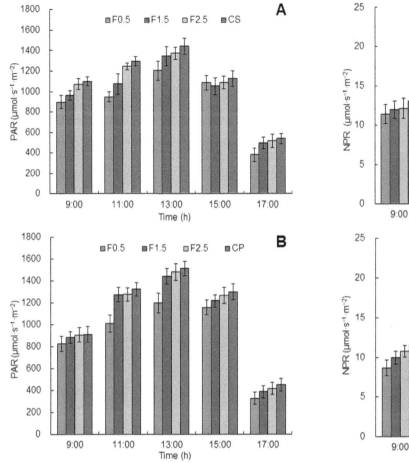

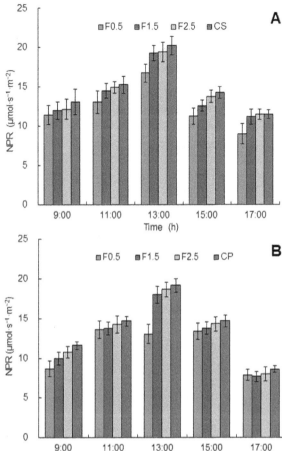

Figure 2. Diurnal variation of photosynthetically active radiation (PAR) for the intercropping systems and its control (A. apple–soybean and B. apple–peanut). F0.5, F1.5 and F2.5 were used to represent the sampling points which had different distance (0.5 m, 1.5 m and 2.5 m) from the tree row. Error bars indicate standard deviation.

Figure 3. Diurnal variation of net photosynthetic rate (NPR) for the intercropping systems and its control (A. apple–soybean and B. apple–peanut). F0.5, F1.5 and F2.5 were used to represent the sampling points which had different distance (0.5 m, 1.5 m and 2.5 m) from the tree row. Error bars indicate standard deviation.

mean fell within the confidence intervals of F2.5 in the corresponding intercropping systems at confidence level of 95%.

Spatial Distribution of Soil Moisture

Although the soil moisture content in the whole soil profile (0 to 100 cm in depth) in AS was different from AP, the trend of spatial distributions of soil moisture was similar (Figure 5). Soil moisture content in AS was related to distance from the apple tree row and showed a clear linear relationship ($Y = 0.465 \times +11.602$, $R^2 = 0.999$), and AP showed the same trend ($Y = 0.590 \times +11.002$, $R^2 = 0.900$). Compared with AP, AS had higher values at the same distance away from the tree row. However, with increasing distance from the tree row, soil moisture in AP had a higher growth than that in AS. The lowest soil moisture content was 11.83% in AS and 11.41% in AP, showed a decrease of 10.31% and 11.14% when compared with the corresponding control treatments. The soil moisture at F2.5 in both intercropping systems also had slightly lower values than that in monoculture configurations, however no difference was observed at significance level of 5%, since the control treatments mean fell within the confidence intervals of F2.5 in the corresponding intercropping

systems (confidence level 95%). Otherwise, the average soil moisture content in AP was lower than that in AS.

Spatial Distribution of Soil Nutrients

The soil nutrients content in the 0 to 100 cm interval was calculated (Table 2). It represented that organic matter, total N, available P and available K in AS had different degrees of reduction when compared with CS, and showed significant differences ($P<0.05$). Similar results were found between AP and CP, except that no significant difference was observed for total N and available K at the location of F2.5. The average content of organic matter, total N, available P and available K in AS decreased by 30.77%, 63.24%, 56.08% and 27.83% when compared with CS–the monoculture configuration. For AP and CP, the decreased percentages were 18.32%, 21.05%, 36.27% and 7.49% respectively. In addition, except available K, soil nutrients content in AP was higher than that in AS at the same spatial location. With the increasing distance from tree row, the distribution trend of soil nutrients was different from that of PAR, NPR or soil moisture in the same intercropping condition. The lowest content of organic matter, total N, available P and available K in AS were present at the location of F1.5. The similar

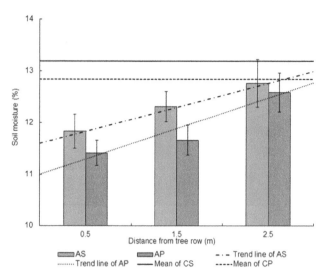

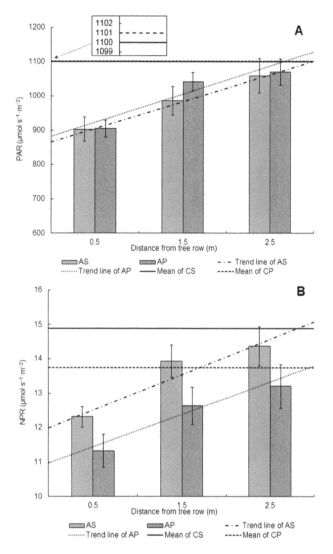

Figure 4. Daily mean of photosynthetically active radiation (PAR) and net photosynthetic rate (NPR) for the intercropping systems and its control (A. PAR and B. NPR). Vertical lines indicate confidence interval at 95% level.

Figure 5. Soil moisture of 0 to 100 cm depth for the intercropping systems and its control. Vertical lines indicate confidence interval at 95% level.

CP. No differences were observed between the locations of F2.5 in both intercropping systems and the corresponding monoculture configuration ($P<0.05$).

Within plot differences in these parameters were significantly correlated (Table 4). NPR was highly correlated with PAR, and soil moisture. The total above-ground biomass of soybean was highly correlated with PAR, soil moisture and available P. The total above-ground biomass of peanut was correlated with PAR, soil moisture, total N and available P. Yield of soybean was highly correlated with PAR, soil moisture, available P and total N, with a trend of soil moisture >PAR>available P>total N. For peanut, the trend was soil moisture > PAR > total N > available P. It showed that, for both of the intercropping system in the study region, the primary factor affecting the yield is soil moisture, and the secondary factor is photosynthetically active radiation, and soil nutrient also have an impact on crop yield in some depth.

Discussion

Agroforestry system has been studied for a long time and has been widely used in the agricultural production practices in China [3,16,17]. However, there has been little research done on the agroforestry system in the Loess Plateau region. The main intercropping models which have been studied are always walnut-wheat and apple-wheat [18–20]. The types of fruit trees intercropping with economic crops such as soybean and peanut has not been well studied. In fact, compared with wheat, soybean and peanut could bring more economic income to farmers. At the same time, these two crops could be rotated with wheat in order to improving land-use efficiency, and re-establishing the economic viability of the Loess Plateau.

Our study observed a clearly positive linear relationship between distance from the apple tree rows and the daily mean values of PAR and NPR in the intercropping systems. For both apple-crop intercropping systems, the shading of the 4–5 years old apple trees had a significant negative effect on the crops in the range of 1.5 m away from the tree rows and further caused the reduction of crop yield. In other researches of temperate agroforestry systems, the similar results were reported by Reynolds et al. [21] about maize and soybean intercropped with poplar and

result was found in AP, except that the lowest value of available K was the location of F0.5.

Crop Growth and Yields

Plant height, hundred leaf dry weight and total above-ground biomass in both intercropping systems had lower values when compared with the monoculture configuration ($P<0.05$; Table 3). The locations of F0.5 and F1.5 in all these parameters showed significant differences with corresponding monoculture configuration ($P<0.05$); however, there were no difference observed in the location of F2.5.

The yield of soybean in AS was significantly related to the distance from the row of apple trees ($Y=0.180\times+1.400$, $R^2 = 0.991$), and the yield of peanut in AP showed the same trend ($Y=0.095\times+1.554$, $R^2 = 0.900$) which showed that yield of soybean had greater impacted by distance from the tree row. The yields at F0.5 and F1.5 in AS were lower than that in CS ($P<0.05$), with a reduction of 22.45% and 11.95%, and in AP the yields of reduction were 13.31% and 11.03% when compared with

Table 2. Soil nutrients for the intercropping systems and control configurations.

Measurement	AS			CS	AP			CP
	F0.5	F1.5	F2.5		F0.5	F1.5	F2.5	
Organic matter (g·kg⁻¹)	4.63±0.27a	3.40±0.29b	4.93±0.27a	6.24±0.31c	6.26±0.36a	5.21±0.26b	6.37±0.36a	7.28±0.27c
Total N (g·kg⁻¹)	0.28±0.05a	0.22±0.04b	0.25±0.04ab	0.68±0.03c	0.31±0.06a	0.24±0.05b	0.35±0.07ac	0.38±0.04c
Available P (mg·kg⁻¹)	2.82±0.56a	2.50±0.54a	3.93±0.31b	7.02±0.22c	5.06±0.61a	3.44±0.44b	4.75±0.42a	6.93±0.33c
Available K (mg·kg⁻¹)	97.33±1.77a	84.02±2.09b	95.43±2.97a	127.84±2.75c	87.45±2.81a	90.73±2.45a	98.11±2.42b	99.55±2.19b

Data were given as the means ± SD.
Different lowercase letters within a row of each crop indicate significant differences (LSD, $P<0.05$).

silver maple in Canada and Peng et al. [22] about mungbean and pepper intercropped with walnut and plum in Weibei area, China. For total above-ground biomass and yield of both crops, PAR of soybean had higher correlations than that of peanut, which indicated that soybean is more adversely impacted by tree shading. Within tree-based intercropping systems, many factors such as tree species, tree height, crown shape, tree row orientation and distance between tree rows can influence tree shading of adjoining agricultural crops. Light reduction would depend on the extent and duration of the shade of trees [21]. Regular pruning of fruit trees could reduce light competition within the intercropping system, improving crop yields.

In semiarid and arid regions, it is still a focus of studies whether intercropping system has an overall negative or positive effect on soil moisture [23]. In some related studies, it was considered that the trees can improve soil moisture holistic conditions in intercropping systems [17,24]. In other studies, the opposite results were reported [11,12,25,26]. However, little research has been carried out in this aspect on the Loess Plateau. Our research confirmed that the competition of water between trees and crops do exist, and showed adverse effects in the study site. A clear linear relationship was observed between the distance from the tree row and soil moisture in both of the intercropping systems. The closer to the tree row, the more intense the competition. The lowest soil moisture content in apple-soybean intercropping system and apple-peanut intercropping system showed a reduction of 10.31% and 11.14%, respectively. Only considering competition of water, the mainly affected region of the apple trees was 1.5 m away from the tree rows under the current tree age.

Another key factor of crop growth is soil nutrients in the intercropping systems. Elsewhere, Thomas et al. [27] and Thevathasan et al. [13] have reported that competition for nutrients in intercropping systems does not exist. In our study, it identified that there were competition for soil nutrients between trees and crops in the intercropping systems. The average content of organic matter, total N, available P and available K showed different degrees of reduction in both of the apple-crops intercropping systems than that of the corresponding control treatments. In particular, total N and available P had higher reduction rate than organic matter and available K, and had significantly correlation with yield of crops. As leguminous plants, soybean and peanut could fix nitrogen from the air via a symbiotic relationship with rhizobium bacteria and increase the mineral soil nitrogen content [28,29]. However, the nitrogen coming from biologically fixed N_2 of symbiosis could not meet all the demand of crops growth, and any gaps between N supply by N_2 fixation and crop N demand must be met by N uptake from soil [30]. The deficiency of light and water in the intercropping systems reduced the physiological activity of the crop, and then affected the N fixation capacity, resulting more intense competition for nitrogen between trees and crops. Compared with soybean and peanut, the growth of other non-nitrogen-fixing crop species (i.e. wheat, maize and millet) would be more severely affected because of nitrogen deficiency in the intercropping systems. Different from understory light distribution and soil moisture, soil nutrients had a different variation pattern in both intercropping systems. The main reasons for this phenomenon might be: (1) the crops close to tree row were seriously affected by tree shading, soil moisture stress and human activities, resulting in low physiological activity and low absorption

Table 3. Crop growth, biomass and yield for soybean and peanut intercropped with apple trees and its control.

Measurement	AS			CS	AP			CP
	F0.5	F1.5	F2.5		F0.5	F1.5	F2.5	
Crop height (cm)	43.7±2.8a	44.4±2.3a	47.7±3.2b	50.2±2.6b	19.1±0.7a	21.9±1.0b	22.6±0.9bc	23.5±0.8c
Hundred leaf dry weight (g)	14.12±0.81a	14.14±0.51a	14.95±0.38b	15.37±0.57b	4.39±0.23a	5.79±0.22b	5.90±0.31bc	6.14±0.28c
Total above-ground biomass (g)	54.55±4.66a	66.06±3.81b	76.96±4.30c	79.76±4.22c	50.10±2.62a	51.57±2.45a	79.16±3.28b	79.65±2.46b
Yield (t/ha)	1.48±0.06a	1.69±0.04b	1.84±0.06c	1.91±0.04c	1.62±0.04a	1.66±0.04a	1.81±0.05b	1.86±0.04b

Data were given as the means ± SD.
Different lowercase letters within a row of each crop indicate significant differences (LSD, $P<0.05$).

Table 4. Correlations of soybean and peanut net photosynthetic rate, biomass, and yield with environmental or physiological parameters measured in Jixian, China.

Independent variable	NPR ($\mu mol \cdot s^{-1} \cdot m^{-2}$)	Total above-ground biomass (g)	Yield (kg/ha)
Soybean			
PAR ($\mu mol \cdot s^{-1} \cdot m^{-2}$)	0.973**	0.996**	0.952**
Soil moisture (%)	0.953**	0.977**	0.957**
Organic matter (g·kg^{-1})	0.441	0.575	0.566
Total N (g·kg^{-1})	0.537	0.555	0.601*
Available P (mg·kg^{-1})	0.697*	0.750**	0.763**
Available K (mg·kg^{-1})	0.469	0.538	0.565
Peanut			
PAR ($\mu mol \cdot s^{-1} \cdot m^{-2}$)	0.986**	0.773**	0.816**
Soil moisture (%)	0.926**	0.965**	0.959**
Organic matter (g·kg^{-1})	0.450	0.562	0.583
Total N (g·kg^{-1})	0.513	0.843**	0.770**
Available P (mg·kg^{-1})	0.424	0.628*	0.646*
Available K (mg·kg^{-1})	0.479	0.531	0.446

*Significant at 5% level.
**Significant at 1% level.

of soil nutrients; (2) the decomposition of tree litter leaded to high nutrients content in the area near the tree row; (3) the overlapping of apple tree roots and crop roots resulting in lower nutrient content at F1.5; (4) the tree roots reduced with the increase of the distance from the tree, therefore, the soil nutrients had relatively high content F2.5. Therefore, in the area of 1.5 m away from tree row, strengthen the application of fertilizer (especially nitrogen and phosphorus) would be helpful to alleviate interspecific competition for soil nutrients.

In the apple-crop intercropping systems, the competition of light, water and nutrients resulted in a greater negative impact on crop growth and yields. For the two apple-crop intercropping systems in our study, the primary factor affecting the crop yield was soil moisture, and the secondary factor was light, and deficiency of the soil nutrient also had a negative impact on crop yields. In the same study area, Yun et al. reported a similar research with a different conclusion: the light is the primary limiting factor leading to reduction of crops, followed by soil moisture [31]. In their research, the apple trees had greater crown width, canopy density and root depth due to elder age (9-year-old) and smaller tree spacing (3 m×4 m). Affected by the impact of canopy structure, the obvious microclimate effect inhibited evapotranspiration of soil moisture to some extent [32], in the same time, the low transmittance led to more intense light stress to crops. Furthermore, the effect of hydraulic lift by tree roots also alleviated the interspecific competition for soil water [24]. Combined with these reasons, different results were found. For different intercropping patterns and tree ages, the intensity of competition for resources would be different in the intercropping system. Therefore, a long-term observation should be carried out in this region to obtain more details about the mechanism of interspecific competition in the intercropping systems. In our study, under the current tree age and growth conditions, the influence scope of the apple trees was 1.5 m away from the tree rows. Compared with the corresponding monoculture configuration, the yield of peanut in the intercropping system had a lower reduction than that of soybean. With comprehensive consideration, peanut is more suitable for intercropping with apple trees in this region.

As we have demonstrated in this study, soil moisture, light and soil nutrients were the limiting factors of crop yield. In order to obtain more production, appropriate management measures were needed to minimize competition between trees and crops. Namirembe [33] and Friday [6] have suggested that the competition for light between trees and crops could be alleviated by pruning of trees crown and increasing the intercropping distance. In general, the aboveground competition could be intuitively observed and managed. However, the competition belowground is invisible and easily ignored by farmers or managers. To avoid these yield losses, root barrier in the intercropping interface is considered to be a useful agricultural management practice according to some related studies [12,34,35]. Combined with their research achievements and our experiment result, we offered several specific recommendations to reduce the competition exist in apple-crop intercropping systems: (1) the selection of crop varieties which is more suitable for apple-crop intercropping systems; (2) appropriate distance increase between the crops and apple tree rows; (3) regular pruning of fruit trees, in order to increase canopy light transmittance rate; (4) additional fertilization and irrigation in the key phenological phase of the crops; (5) differences of irrigation and fertilization based on the distance from the apple trees. Management measures such as plastic film and straw mulching have been widely used in agricultural production. Whether these measures have overall positive effects on intercropping system would be one of the focus of our future research work in this region.

Conclusions

As an effective method to increase the efficiency of land use and economic returns, tree-based intercropping systems are particularly important on Loess Plateau. We concluded that the competitions exist both above-ground and below-ground between apple trees and crops. The competition for soil moisture is the primary limiting factor for the crop productivity in this region. Furthermore, the tree shading and the competition for soil nutrients in the interface of trees and crops also have a negative impact on the understory crops. However, it could be minimized by better agricultural technology and management measures.

In summary, our study suggests that there is great potential for intercropping systems in the Loess Plateau. Therefore, in order to relieve the shortage of arable land and promote the sustainable development of natural resources, the intercropping systems would continue to be the hot spot for future research. Canopy structure, roots distribution of trees, the application of different agronomic measures and the role they play in the competition process in the intercropping systems will be the focus of our future research.

Acknowledgments

We are grateful for the support from the Shanxi Jixian Forest Ecosystem Research Station. We also would like to thank the three anonymous reviewers and the editors for their helpful comments.

Author Contributions

Conceived and designed the experiments: LG HX HB BB. Performed the experiments: LG HX HB BB. Analyzed the data: LG HX WX BB XW CB YC. Contributed reagents/materials/analysis tools: XW CB YC. Wrote the paper: LG HB WX.

References

1. Burel F (1996) Hedgerows and their role in agricultural landscapes. Critical Reviews in Plant Sciences 15: 169–190.
2. Gene Garrett HE, Buck L (1997) Agroforestry practice and policy in the United States of America. Forest Ecology and Management 91: 5–15.
3. Li W, Lai S (1994) Agroforestry in China. Beijing: Chinese Science Press. pp. 14–18.
4. Zhu Q, Zhu J (2003) Sustainable management technology for conversion of cropland to forest in loess area. Beijing:Chinese Forestry Press. pp. 160–165.
5. Ong CK, Huxley P (1996) Tree-crop interactions: a physiological approach. Wallingford:CAB International. pp. 386.
6. Friday JB, Fownes JH (2002) Competition for light between hedgerows and maize in an alley cropping system in Hawaii, USA. Agroforestry Systems 55: 125–137.
7. Peng X, Zhang Y, Cai J, Jiang Z, Zhang S (2009) Photosynthesis, growth and yield of soybean and maize in a tree-based agroforestry intercropping system on the Loess Plateau. Agroforestry Systems 76: 569–577.
8. Hall DJM, Sudmeyer RA, McLernon CK, Short RJ (2002) Characterisation of a windbreak system on the south coast of Western Australia. 3. Soil water and hydrology. Australian Journal of Agricultural Research 42: 729–738.
9. Unkovich M, Blott K, Knight A, Mock I, Rab A, et al. (2003) Water use, competition, and crop production in low rainfall, alley farming systems of south-eastern Australia. Australian Journal of Agricultural Research 54: 751–762.
10. Kowalchuk TE, Jong E (1995) Shelterbelts and their effect on crop yield. Canadian Journal of Soil Science 75: 543–550.
11. Jose S, Gillespie AR, Seifert JR, Biehle DJ (2000) Defining competition vectors in a temperate alley cropping system in the midwestern USA: 2. Competition for water. Agroforestry Systems 48: 41–59.
12. Miller AW, Pallardy SG (2001) Resource competition across the crop-tree interface in a maize-silver maple temperate alley cropping stand in Missouri. Agroforestry Systems 53: 247–259.
13. Thevathasan NV, Gordon AM, Simpson JA, Reynolds PE, Price G, et al. (2004) Biophysical and ecological interactions in a temperate tree-based intercropping system. Journal of Crop Improvement 12: 339–363.
14. Newman SM, Bennett K, Wu Y (1997) Performance of maize, beans and ginger as intercrops in Paulownia plantations in China. Agroforestry Systems 39: 23–30.
15. Yun L, Bi H, Gao L, Zhu Q, Ma W, et al. (2012) Soil moisture and soil nutrient content in walnut-crop intercropping systems in the Loess Plateau of China. Arid Land Research and Management 26: 285–296.
16. Meng P, Zhang J, Fan W (2003) Research on agroforestry in china. Beijing:Chinese Forestry Press. pp.235.
17. Meng P, Zhang J (2004) Effects of pear-wheat inter-cropping on water and land utilization efficiency. Forest Research 17: 167–171.
18. Zhang J, Meng P, Yin C (2002) Spatial distribution characteristics of apple tree roots in the apple-wheat intercropping. Scientia Silvae Sinicae 38: 30–33.
19. Zhang J, Meng P (2004) Model on wheat potential evapotranspiration in apple-wheat intercropping. Forest Research 17: 284–290.

20. Yun L, Bi H, Ren Y, Ma W, Tian X (2009) Soil moisture distribution at fruit-crop intercropping boundary in the Loess region of Western Shanxi. Journal of Northeast Forestry University 37: 70–78.
21. Reynolds PE, Simpson JA, Thevathasan NV, Gordon AM (2007) Effects of tree competition on corn and soybean photosynthesis, growth, and yield in a temperate tree-based agroforestry intercropping system in southern Ontario, Canada. Ecological Engineering 29: 362–371.
22. Peng X, Cai J, Jiang Z, Zhang Y, Zhang S (2008) Light competition and productivity of agroforestry system in loess area of Weibei in Shaanxi. Chinese Journal of Applied Ecology 19: 2414–2419.
23. Zhang J, Meng P, Yin C, Cui G (2003) Summary on the water ecological characteristics of agroforestry system. World Forestry Research 16: 10–14.
24. Hirota I, Sakuratani T, Sato T, Higuchi H, Nawata E (2004) A split-root apparatus for examining the effects of hydraulic lift by trees on the water status of neighbouring crops. Agroforestry Systems 60: 181–187.
25. Lott JE, Howard SB, Ong CK, Black CR (2000) Long-term productivity of a Grevillea robusta-based overstorey agroforestry system in semi-arid Kenya: II. Crop growth and system performance. Forest Ecology and Management 139: 187–201.
26. Lehmann J, Peter I, Steglich C, Gebauer G, Huwe B, et al. (1998) Below-ground interactions in dryland agroforestry. Forest Ecology and Management 111:157–169
27. Thomas J, Kumar BM, Wahid PA, Kamalam NV, Fisher RF (1998) Root competition for phosphorus between ginger and Ailanthus triphysa in Kerala, India. Agroforestry Systems 41: 293–305.
28. Cheng D (1994) Resource microbiology. Harbin:Northeast Forestry University Press. pp. 32.
29. Wani SP, Rupela OP, Lee KK (1995) Sustainable agriculture in the semi-arid tropics through biological nitrogen fixation in grain legumes. Plant Soil 174: 29–49.
30. Salvagiotti F, Cassman KG, Specht J E, Walters DT, Weiss A, et al. (2008). Nitrogen uptake, fixation and response to fertilizer N in soybeans: A review. Field Crops Research 108:1–13.
31. Yun L, Bi H, Tian X, Cui Z, Zhou H, et al. (2011) Main interspecific competition and land productivity of fruit-crop intercropping in Loess Region of West Shanxi. Chinese Journal of Applied Ecology 22:1225–1232
32. Zhang J, Meng P, Song Z, Gao J (2004) An overview on micro-climatic effects of agro-forestry systems in plain agricultural areas in China. Agricultural Meteorology 25:52–55
33. Namirembe S (1999) Tree management and resource utilization in agroforestry systems with Senna spectabilis in the drylands of Kenya. Bangor:University of Wales. pp. 206.
34. Singh RP, Saharan N, Ong CK (1989) Above and below ground interactions in alley-cropping in semi-arid India. Agroforestry Systems 9: 259–274.
35. Hou Q, Brandle J, Hubbard K, Schoeneberger M, Nieto C, et al. (2003) Alteration of soil water content consequent to root-pruning at a windbreak/crop interface in Nebraska, USA. Agroforestry Systems 57: 137–147.

20

How to Design a Targeted Agricultural Subsidy System: Efficiency or Equity?

Rong-Gang Cong[1]*, **Mark Brady**[1,2]

1 Centre for Environmental and Climate Research (CEC), Lund University, Lund, Sweden, **2** AgriFood Economics Centre, Department of Economics, Swedish University of Agricultural Sciences, Lund, Sweden

Abstract

In this paper we appraise current agricultural subsidy policy in the EU. Several sources of its inefficiency are identified: it is inefficient for supporting farmers' incomes or guaranteeing food security, and irrational transfer payments decoupled from actual performance that may be negative for environmental protection, social cohesion, etc. Based on a simplified economic model, we prove that there is "reverse redistribution" in the current tax-subsidy system, which cannot be avoided. To find a possible way to distribute subsidies more efficiently and equitably, several alternative subsidy systems (the pure loan, the harvest tax and the income contingent loan) are presented and examined.

Editor: Aldo Rustichini, University of Minnesota, United States of America

Funding: This research is funded by "Biodiversity and Ecosystem services in a Changing Climate – BECC" and SAPES. The funders had no role in study design, data collection and analysis, decision to publish, or preparation of the manuscript. The contents are the responsibility of the authors and do not necessarily reflect the views of Centre for Environmental and Climate Research (CEC), Lund University.

Competing Interests: The authors have declared that no competing interests exist.

* E-mail: cascong@126.com

Introduction

Payments to farmers are a central part of the EU's Common Agricultural Policy (CAP) which dictates agricultural policy in all 27 member states. It accounts for almost half of the EU's budget and almost half of the legislation [1]. Initially the objectives of CAP were to (Rome Treaty in 1955): (1) increase agricultural productivity; (2) ensure a fair standard of living for those engaged in agriculture; (3) stabilize agricultural markets; (4) assure the availability of food; and (5) ensure reasonable prices for consumers.

In recent years, the role of CAP has been further broadened to (Article 4 of Council Regulation (EC) No 1698/2005): (1) provide high-quality food and non-food products; (2) protect the environment; and (3) promote the harmonious development of different regions. The objectives of CAP can be understood from the perspectives of Economy, Environment and Society (EES), as shown in Figure 1.

The most important instrument of CAP is the Single Payment Scheme (SPS) which is not related to the volume of commodity output; so called decoupled payments. To qualify for subsidies (i.e. payments in the language of the European Commision), farmers are required to keep their land in "Good Agricultural and Environmental Condition" (GAEC) and respect relevant statutory management requirements, together referred to as cross-compliance.

The SPS alone accounts for almost 75% of the CAP budget (€54 billion annually) or 32% of the total EU budget [2]. Member States (MS) were given some freedom to choose how to implement the SPS in 2005. They could choose a regionalized payment with farmers receiving an identical payment per hectare within a region (regional model), a farm-specific payment which is based on each

farm's historical production level (historical model) or a combination of both (hybrid model).

The main advantages of SPS are that farmers' output decisions are now guided by consumer demand and not distorted by output subsidies, and its benefits for the environment. However, due to the heterogeneity of agricultural and socio-economic conditions in the EU, SPS may also have some disadvantages as follows:

1. The SPS has limited potential for supporting farmers' incomes which is the original motivation of the support [3]. Current support is highly concentrated to a few large farms, whereas many small farms that are more dependent on support receive only a relatively small share of the total payment.

2. The SPS's contribution to food security is not as large as imagined because the bulk of the payments are paid to the most fertile regions where market prices are sufficient to guarantee food production [1].

3. SPS has the tendency to be distributed to richer regions and farmers, which may be harmful to social cohesion [4].

4. In practice, there is inadequate feedback between levels of public goods provided by agriculture and payments received by individual farms. Farmers are usually remunerated for carrying out particular management tasks rather than being rewarded directly for measured environmental performance, and payment levels are not related to actual costs.

Future EU agricultural policy should aim to enhance the overall competitiveness of agriculture, protect the environment and promote rural development. However, in general SPS has weak rationale in terms of environmental externalities and social cohesion. So the question as to how to distribute payments

Economy:
1. Guarantee long-term food security
2. Provide quality, value and diversity of food sustainably
3. Income support

Society:

1. Create rural employment

2. Eradicate poverty and

improve social cohesion

Environment:
1. Actively manage the natural resources by farming
2. Contribute to mitigate and adapt to climate change

Figure 1. The objectives of CAP from the perspective of Economy, Environment and Society (EES).

reasonably, i.e. to maximize social welfare in terms of the stated goals, is an important issue to study.

The underlying motivation for the current distribution of SPS payments is compensation for historical reductions in agricultural price support (first in 1992 as a result of the MacSharry Reform). One possible justification for SPS could be capital market imperfections which prevent farmers from borrowing for financing investment [5]. Another justification is environmental externalities which cannot be reflected in the market prices [6].

In regard to income redistribution, the current system also has some severe drawbacks. First, farmers with the highest yielding land and hence who are competitive in the market, receive the highest payments per ha. Accordingly, farmers with less fertile land receive lower payments per ha, and hence their farms face marginalization and abandonment, which could have irreversible and detrimental impact on European agricultural production and its cultural landscapes [7]. Secondly, since payments are based on area, the largest farms receive the largest total payments. Consequently, the current CAP and its SPS may imply reverse redistribution, i.e. redistribution from the poor to the rich [8].

In summary, SPS should be better targeted to poorer farmers (small farms) and the environment. In principle, high-income households should receive a low subsidy (if at all) and low-income households a high subsidy [9]. As public goods, environmental products should be subsidized by CAP to cover relevant costs.

However, reality may be more complex. On the one hand, more direct payments to marginal regions and poorer farmers will affect investment and not be conducive to economies of scale [10]. On the other hand, the current subsidy system is hurting small farms and poor farmers through reverse distribution, which will be

harmful for social cohesion and environmental protection [11]. Therefore, there is a tradeoff between efficiency and equity.

While the 2003 reform may be the most radical reform of the CAP to date, the concept of SPS or decoupled agricultural support is not new. Decoupling was first proposed more than 50 years ago. Beard and Swinbank provided a comprehensive review of the early proposals to decouple agricultural support in the US in the 1950s and in Europe in the 1960s [12]. Josling further argued that a direct income payment unrelated to output should be a way of ensuring reasonable standards of living for rural people [13].

However, decoupled payments as a form of government intervention have been roundly criticized from different perspectives since its inception. Criticism has been wide-ranging, and even the European Commission has long been persuaded of the numerous defects of decoupled payments. The main opposing viewpoints include:

(1) Production Distortion

While the payments under SPS may be decoupled from production, they are still a source of income for the farm households and may indirectly affect production decisions through the "wealth effect". Hennessy studied the relationships between decoupled payments, farmers' risk preferences and production decisions [14]. He found that if farmers' risk aversions declined as incomes increased, an increase in wealth as a consequence of the decoupled payment could induce them to take riskier production decisions, and thus increase outputs compared with the situation in which no decoupled payment was made.

Decoupled payments also relax the individuals' capital constraints thus lowering the cost of capital [15]. Revell and

Oglethorpe also suggested the possibility that decoupled payments could affect production through an expectations effect [16]. They claimed that producers might adopt a 'safety first' strategy and make only minimal changes to production plans in case future payments were reassessed and again related to production or agricultural activity. Therefore, one of the decoupling's objectives, not to distort production, is not fully fulfilled in the current policy framework.

(2) Environmental Problems

The cross-compliance effect is the other objective of decoupled payments. However, some empirical results show that this effect is also complicated. Based on two micro-economic models (AgriPolis and MODAM), Uthe et al. found that in the case of grassland, decoupling led to improvement of the environment as a result of the cross-compliance obligations [17]. However, with respect to arable land, decoupling led to negative environmental effects due to changes in the crop mix, with less cereals and a greater area of more intensive winter rape and row crops being grown.

(3) Hurting Small Farms

Although most policy makers in Europe agree that they want to promote "family farms" and small scale production, decoupled payments in fact benefit large farms much more than small farms, because decoupled payments are linked to farm size. So while subsidies allow small farms to persist, large farms tend to receive the greatest share of the subsidies. Within the 2008 Health Check of the CAP [18], a first step was taken to limit decoupled payments to very large landowners.

There are also some other criticisms such as the equity among member states [19], the unfair competition with developing countries [20] and so on. Among them, the tradeoff between efficiency and equity is the source of many controversies.

Motivated by the experiences of EU and the United States, we attempt to construct a simplified economic model and answer the following questions:

(1) How can we define efficiency and equity in regard to agricultural policy? What are the underlying conflicts between efficiency and equity in the current SPS?

(2) Is there an integrated subsidy system that can achieve a balance between efficiency and equity?

The results may shed light on the wisdom of the current CAP and the proposed 2013 reform. Also, some conclusions regarding general rules for designing agricultural subsidy systems are provided. The article is structured as follows: firstly, a theoretical analysis of efficiency and equity in the current SPS is presented; secondly, we compare the efficiency and equity of three novel options for agricultural subsidies; finally, conclusions and policy recommendations according to the current direction of CAP reform are given.

Analysis

1. Models

The model presented in this paper is inspired by the work of García-Peñalosa and Wälde [8]. In this section, a small open economy and relevant assumptions are initially described. Next, the perfect market and its operating mechanism are presented for reference. The main variables and functions are listed in Appendix S1.

1.1 Description of the economy. A small open economy with a population of constant size N is considered here. All individuals live for two periods and are identical in all respects except for their initial wealth which is denoted by n. Its frequency distribution is given by $f(n)$. At the beginning of the first period, people choose whether to work in the agricultural sector or in other sectors. If people choose to work in other sectors, they will receive an income in both periods. Otherwise they enroll in the agricultural sector. In this case, they do some farming in the first period and harvest (realize their income) in the next. If they choose to work in the agricultural sector, they also need to decide whether to produce only agricultural products or to produce both agricultural and environmental products (e.g. landscape). In this paper, farmers are assumed to produce both agricultural and environmental products because the latter is currently profitable.

Farmers are categorized based on the sizes of their farms. It is assumed that there are two sizes of farms based on the area of agricultural land: small farms (s) and large farms (l). The exogenous rental prices for these are r_s and r_l respectively, where $r_s < r_l$. Economies of scale and heterogeneous soil fertility are not considered in this paper.

Consequently the total labor force can be divided into three categories: (1) Individuals who are employed in other sectors (o); (2) Farmers who produce both crops and environmental products on small farms (s); and (3) Farmers who produce both crops and environmental products on large farms (l). The three types of labor (o, s and l) are illustrated in Figure 2. Given that L_i ($i = o$, s or l) is the size of the labor force of each type i, then $L_o + L_s + L_l = N$.

The costs of the different types of labor are given in Table 1. For individuals in other sectors, they don't need to invest any money for their employment. For individuals in the agricultural sector, they need to pay the rent for the farm's land (r_s or r_l) and also the costs of generating environmental products in period one (i.e. prior to receiving any income).

It is assumed that borrowing in order to finance costs for individuals in the agricultural sector is not possible in reality as agriculture is high risk and not satisfactory collateral for private lenders. Hence, in the absence of government intervention, an individual can only enroll in the agricultural sector if his (her) initial wealth (n) is large enough to cover his (her) costs.

The economy produces three types of products: agricultural products (A), environmental products (E) and other products (O). Since we desire to study individuals' career choices, we omit other production factors and only focus on the amount of labor. Define $I()$, $AL()$, $AH()$ and $G()$ as the production functions for other products, agricultural products from small and large farms, and environmental products respectively as shown in equations (1)–(3).

$$O = I(L_o) \qquad (1)$$

$$A = AL(L_s) + AH(L_l) \qquad (2)$$

$$E = G(L_s, L_l) \qquad (3)$$

The purpose of introducing production functions is to determine returns to labor in the different sectors. In a perfect market, returns should be equalized across all sectors. Otherwise, individuals in the low income sectors will move to the high income sectors. In this paper, other endowment differences between individuals, such as abilities, are not considered. Consequently, for a given total population, there is an optimal labor structure for which there is a maximum total income for the entire population.

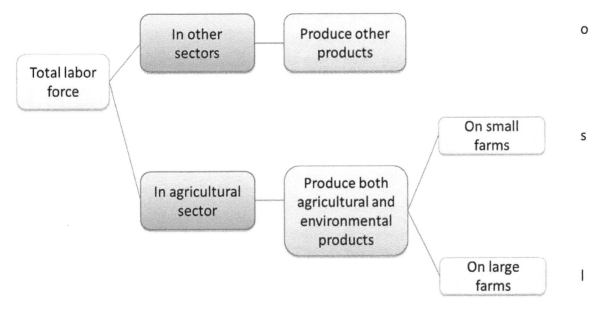

Figure 2. The structure of the labor force.

More specifically, as a production factor, the marginal output of labor should be positive but show a decreasing trend when the amount of labor is increasing. Therefore, first derivatives of production functions are positive while second derivatives are negative. In a small open economy, P_O, P_A and P_E are prices for other products, agricultural products and environmental products respectively which are determined exogenously. The return (Py) of each labor force is the product of its marginal output and corresponding price, as shown in equations (4)–(7):

$$Py_o = I_{L_0}(L_o) \times P_O \tag{4}$$

$$Py_s = AL_{L_s}(L_s) \times P_A + G_{L_s}(L_s,L_l) \times P_E \tag{5}$$

$$Py_l = AH_{L_l}(L_l) \times P_A + G_{L_l}(L_s,L_l) \times P_E \tag{6}$$

$$L_o + L_s + L_l = N \tag{7}$$

where equation (7) is the population constraint.

The role of the government is to buy environmental products, subsidize farmers if necessary and levy taxes to maintain a balanced budget. We will examine the characteristics of several possible subsidy systems below.

1.2 The benchmark case of perfect capital markets. We assume a situation where capital markets are perfect as our benchmark. All people can borrow and there is no need for government intervention. Neither taxes and subsidies, nor uncertainty and risk aversion are considered in this section. The only role of government in the perfect capital market is to set the price of environmental products and buy them. The lifetime income (W) of an individual who is employed in other sectors is the present value of payments from other products for two periods, as shown in equation (8).

$$W_o = (1+R) \times I_{L_o}(L_o) \times P_O \tag{8}$$

where R is the exogenous discount rate.

The lifetime income of an individual in the agricultural sector is equal to the sum of the discounted revenues from agricultural and environmental products minus the land rent (r) and the cost of producing environmental products (EC) which are payable in the first period, as shown in equations (9)–(10).

$$W_s = -r_s - EC + R \times [AL_{L_s}(L_s) \times P_A + G_{L_s}(L_s,L_l) \times P_E] \tag{9}$$

Table 1. Costs of different types of labor.

| Type of labor | Land rental costs | | Costs of producing environmental products |
	Small farm	Large farm	
o			
s	√		√
l		√	√

$$W_l = -r_l - EC + R \times [AH_{L_l}(L_l) \times P_A + G_{L_l}(L_s, L_l) \times P_E] \quad (10)$$

The cost for producing environmental products, EC, is fixed and identical for both farm types.

Since there are not any barriers for employment, individuals with lower incomes can move to the high-income sector, which would reduce the high-income sector's returns to labor and boost those in the low-income sector. Consequently, in an equilibrium economy, all individuals' incomes will be equalized ($W_o^* = W_s^* = W_l^*$). Equations (8)–(10) are simultaneously solved to obtain the optimal structure of labor $L^* = (L_o^*, L_s^*, L_l^*)$, which is shown in equation (11).

$$(1 + R) \times I_{L_o}(L_o^*) \times P_O$$
$$= -r_s - EC + R \times [AL_{L_s}(L_s^*) \times P_A + G_{L_s}(L_s^*, L_l^*) \times P_E] \quad (11)$$
$$= -r_l - EC + R \times [AH_{L_l}(L_l^*) \times P_A + G_{L_l}(L_s^*, L_l^*) \times P_E]$$

The optimal labor structure depends on the factor prices (r_s, r_l and EC) and the product prices (P_O, P_A and P_E). The labor structure L^* is also called the efficient level of labor and will be our point of reference (as shown in Figure 3).

In the next section, the implications of three subsidy systems under imperfect markets are analyzed. They are the current SPS system, subsidies to achieve the efficient level of labor in each sector (i.e. that maximizes output as shown in Figure 3) and subsidies to achieve the equitable level of lifetime income (all individuals have an equal lifetime income).

2. The Working of Traditionally used Tax-subsidy System

In reality, farmers might have difficulty obtaining loans to finance their costs for renting land and producing environmental products as shown in Section 1.2. Therefore, a subsidy is needed to aid poor farmers while a tax is collected from the entire population to keep a balanced budget. The aim of this section is to examine the current SPS and determine the optimal labor structure in the context of the chosen subsidy and taxation policy (Section 2.1). Also two specific examples of the SPS (subsidizing the efficient level of labor and the equitable level of lifetime income) are discussed.

2.1 The current SPS system. We assume that T is a lump-sum tax levied on all individuals in the first period of their lives. In the current SPS system, every farmer who produces certain environmental products is eligible for the subsidy. Small farms receive subsidy $S1$, while large farms receive $S2$. In practice, SPS provides subsidies according to the size of the farms. Therefore, in general there is a relationship that $S1 < S2$.

The government chooses the subsidy rate and then sets the lump-sum tax so as to maintain a balanced budget, as shown in equation (12), which means that the required level of tax is equal to total subsidies.

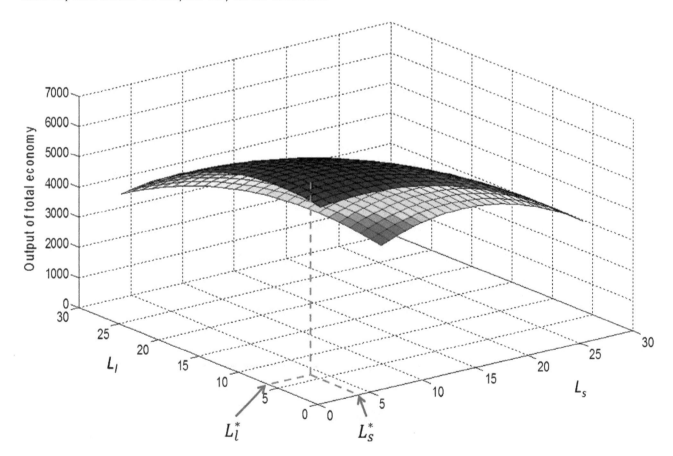

Figure 3. The efficient size of the agricultural work force with small and large farms when borrowing is possible. Note to Figure 3: Although we have three types of labor (s, l and o), the amount of the third type of labor, other sectors, is determined by the amounts of the other two types of labor because the total population is fixed, i.e. $L_o^* = N - (L_s^* + L_l^*)$.

$$T \times N = S1 \times L_s + S2 \times L_l \qquad (12)$$

Individuals in other sectors (o) pay a tax but receive no subsidy. Farmers (s and l) pay the same tax but receive different subsidies. This implies a net transfer to farmers because there are more individuals paying the tax than receiving the subsidy ($N > L_s + L_l$).

Define $\underline{n}_i$ ($i = o, s, l$) as the minimum level of initial wealth in order to be able to cover i's costs. In the absence of borrowing, an individual's initial wealth must be large enough to cover the subsidized costs plus the tax. For example, for farmers with small farms (s), their required minimum initial wealth, $\underline{n}_s$, can be calculated based on equation (12), as shown in equation (13).

$$\underline{n}_s = T + r_s + EC - S1$$
$$= \frac{S1 \times L_s + S2 \times L_l}{N} + r_s + EC - S1 \qquad (13)$$
$$= S1 \times \frac{L_s - N}{N} + S2 \times \frac{L_l}{N} + r_s + EC$$

For farmers with large farms (l), their required minimum initial wealth, $\underline{n}_l$, can be calculated as shown in equation (14).

$$\underline{n}_l = T + r_l + EC - S2$$
$$= \frac{S1 \times L_s + S2 \times L_l}{N} + r_l + EC - S2 \qquad (14)$$
$$= S2 \times \frac{L_l - N}{N} + S1 \times \frac{L_s}{N} + r_l + EC$$

It is assumed that $\underline{n}_l$ is larger than $\underline{n}_s$ because farmers with large farms need more investment (i.e. pay higher land rents). Clearly, the higher the subsidy, the lower the level of initial wealth required for being in the agriculture sector. For clarity, it is assumed that the labor structure under the current SPS is $L^a = (L_o^a, L_s^a, L_l^a)$. Suppose that under the subsidies of $S1$ and $S2$, the capital market constraint is still binding for some individuals, so that more individuals want to be in the agriculture sector than can afford to be ($L_s^a < L_s^*, L_l^a < L_l^*$), which means that the lifetime income of farmers may exceed the lifetime income of people in other sectors. In this case, the number of individuals having small (large) farms is equal to the number of individuals whose initial wealth is between $\underline{n}_s$ and $\underline{n}_l$ (larger than $\underline{n}_l$) as shown in equations (15) and (16).

$$L_s^a = \sum_{\underline{n}_l > n \geq \underline{n}_s} f(n) \qquad (15)$$

$$L_l^a = \sum_{n \geq \underline{n}_l} f(n) \qquad (16)$$

Equation (15) and (16) can be used with equations (13) and (14) to jointly determine the sizes of s and l, and the minimum initial wealth levels required for entering the agricultural sector ($\underline{n}_s$ and $\underline{n}_l$), as a function of the subsidies $S1$ and $S2$, and the distribution of initial wealth $f(n)$.

2.2 Subsidizing the efficient level of labor.

Suppose that the purpose of the government is to maximize the economic output of a given generation; then they should design a subsidy system to achieve the optimal labor structure $L^* = (L_o^*, L_s^*, L_l^*)$. In this case, there should be exactly L_s^* individuals whose initial wealth is between $T^b + r_s + EC - S1^b$ and $T^b + r_l + EC - S2^b$, and L_l^* individuals whose initial wealth is larger than $T^b + r_l + EC - S2^b$ where T^b, $S1^b$ and $S2^b$ are the tax, subsidy for farmers with small farms and subsidy for farmers with large farms to achieve the optimal labor structure respectively.

Figure 4 examines graphically the effects of the subsidies $S1^b$ and $S2^b$ on the lifetime incomes of the three types of labor. The solid lines represent lifetime incomes in the absence of subsidies, where $L^0 = (L_o^0, L_s^0, L_l^0)$ is the labor structure in this case. L_s^0 and L_l^0, which are numbers of s and l whose initial wealth can afford agricultural costs, are smaller than L_s^* and L_l^* respectively. At L^0, lifetime incomes are higher for s and l but lower for o than under perfect markets. Under the tax-subsidy system (subsidizing the efficient level), the lifetime incomes of o are the incomes received in two periods minus the tax, i.e. $W_o^b = -T^b + (1+R) \times I_{L_o}(L_o) \times P_O$. Hence, the introduction of the tax represents a downward shift of the curve W_o. The lifetime incomes of s and l are
$$W_s^b = S1^b - T^b - r_s - EC + R \times$$
$$[AL_{L_s}(L_s) \times P_A + G_{L_s}(L_s, L_l) \times P_E] \qquad \text{and}$$
$$W_l^b = S2^b - T^b - r_l - EC + R \times$$
$$[AH_{L_l}(L_l) \times P_A + G_{L_l}(L_s, L_l) \times P_E]$$ respectively. Because $S1^b$ and $S2^b$ are larger than T^b, the lifetime incomes of s and l increase, which means upward shifts of W_s and W_l. The subsidies can then be set to $S1^b$ and $S2^b$ so that the distribution of labor is exactly $L^* = (L_o^*, L_s^*, L_l^*)$, as with perfect capital markets.

The efficient subsidies ($S1^b$ and $S2^b$) have two distributional consequences:

Firstly, all individuals are paying taxes that are distributed only among those farmers with higher incomes, implying that there is a transfer of resources from the poor to rich individuals. This is what is called "reverse redistribution".

Secondly, the introduction of efficient subsidies leads to a situation where those who have large farms enjoy larger incomes. The efficient subsidy does not remove inequality. And it also fails to provide an equality of chances. The difference in the lifetime incomes of individuals with large (small) farms and individuals in other sectors is $S2^b$ ($S1^b$). Therefore, there is reverse redistribution under the SPS scheme, not only from the non-agricultural sectors to the agricultural sector, but also from all farmers to farmers with large farms.

The efficient subsidies not only fail to equalize life-time incomes, but also fail to provide ex ante equality of chances. Even though some relatively poor individuals can now afford to enter the agricultural sector, the greatest opportunity is still offered to the richest individuals. As a result, poorer individuals are still systematically excluded from agriculture.

2.3 Subsidizing the equitable level of lifetime income.

In this section, subsidies which can guarantee the equality of lifetime incomes are examined. The labor structure in this case is termed the equitable labor structure, $L^c = (L_o^c, L_s^c, L_l^c)$. The equitable subsidies ($S1^c$ and $S2^c$) and their corresponding tax (T^c) are defined by equation (17).

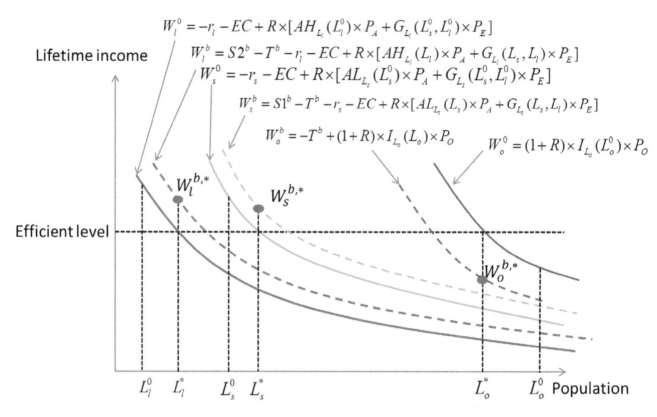

Figure 4. The efficient subsidy rate.

$$-T^c + (1+R) \times I_{L_o}(L_o^c) \times P_O$$

$$= S1^c - T^c - r_s - EC + R$$

$$\times [AL_{L_s}(L_s^c) \times P_A + G_{L_s}(L_s^c, L_l^c) \times P_E] \qquad (17)$$

$$= S2^c - T^c - r_l - EC + R$$

$$\times [AH_{L_l}(L_l^c) \times P_A + G_{L_l}(L_s^c, L_l^c) \times P_E]$$

What then is the equitable labor structure L^c? In the section above, it is known that for the efficient labor structure, $L^* = (L_o^*, L_s^*, L_l^*)$, the lifetime incomes of s and l are larger than that of o. In order to reduce the incomes of the former, the subsidies should be further increased to increase the numbers of s and l, thus reducing farmers' incomes and raising incomes in other sectors. Therefore, $L_o^c \leq L_o^*$, $L_s^c \geq L_s^*$ and $L_l^c \geq L_l^*$ are all satisfied.

The increases in the numbers of s and l imply that their marginal products are lower than their marginal costs. Therefore, the equitable labor structure, L^c, is not efficient since too many people enter the agricultural sector.

Consequently the SPS, as a form of tax-subsidy system, is characterized by a trade-off between efficiency and equity: efficient subsidies imply inequality in lifetime incomes; while equitable subsidies induce an excessively large number of farmers and thus reduction in the total value of economic output. Despite the equality of lifetime income, there is still not equality of opportunity, as those with a very low level of initial wealth will still not be able to afford agricultural costs. The only way around this is to provide a subsidy that covers full costs; but this will further increase the efficiency loss.

3. Some Possible Solutions–the Pure Loan, the Harvest Tax or the Income Contingent Loan?

As the previous discussion has shown, SPS as a traditional tax-subsidy system that implies reverse distribution cannot achieve the targets of efficiency and equity simultaneously. Therefore, it raises the question whether there is a better solution for an agricultural subsidy system. Firstly, the pure loan system is presented; secondly, to solve the problems caused by uncertainty and risk-aversion, the harvest tax and the income contingent loan systems are presented. Finally, some possible policy options for EU in the future are given for reference.

3.1 The pure loan scheme. A straightforward solution, which removes the constraints imposed by imperfect capital markets without generating reverse redistribution, is to abolish all subsidies and introduce a government loan system. Agricultural costs would be fully financed, and the capital market imperfections would be overcome by loans provided by the government. In our highly stylized economy, all individuals would then have identical lifetime incomes, the allocation of resources would be efficient, and any individuals would have the opportunity of entering into agriculture.

However, the pure loan scheme neglects an important aspect which we have so far not taken into account: the risks related to agricultural production. Agricultural production is a risky investment, particularly risk stemming from variations in the weather. Hence, from the farmers' perspectives, there is risk associated with farming. A simple form of uncertainty is presented in what follows assuming the possibility of successful agricultural production is set exogenously.

It is supposed that farmers, irrespective of farm size, will harvest in the second period with probability p ($p \in (0,1)$) and fail to harvest with probability $1-p$. If farmers fail, they will enter other sectors and receive a salary in the second period. Individuals are assumed to be risk averse and have utility functions denoted by $U()$. Assume that the expected returns of farmers with large farms are larger than those of farmers with small farms which are in turn larger than those of individuals in other sectors, as shown in equation (18).

$$-r_l - EC + R \times (p \times (AH_{L_l}(L_l) \times P_A + G_{L_l}(L_s,L_l) \times P_E)$$

$$+ (1-p) \times I_{L_o}(L_o) \times P_O) >$$

$$-r_s - EC + R \times (p \times (AL_{L_s}(L_s) \times P_A + G_{L_s}(L_s,L_l) \times P_E) \quad (18)$$

$$+ (1-p) \times I_{L_o}(L_o) \times P_O) >$$

$$(1+R) \times I_{L_o}(L_o) \times P_O$$

Different from the analysis in sections 1 and 2, is that the expected returns of farmers are larger than individuals in other sectors due to individuals' risk aversions, as in reality. If there is no difference in lifetime incomes between farmers and individuals in other sectors, people would prefer the latter to obtain certain incomes.

Under a pure loan system, individuals who work in other sectors have total wealth of $n + (1+R) \times I_{L_o}(L_o) \times P_O$; farmers with small farms who succeed have total wealth of $n - r_s - EC + R \times (AL_{L_s}(L_s) \times P_A + G_{L_s}(L_s,L_l) \times P_E)$; farmers with small farms who fail have total wealth of $n - r_s - EC + R \times I_{L_o}(L_o) \times P_O$; farmers with large farms who succeed have total wealth of $n - r_l - EC + R \times (AH_{L_l}(L_l) \times P_A + G_{L_l}(L_s,L_l) \times P_E)$; and farmers with large farms who fail have total wealth of $n - r_l - EC + R \times I_{L_o}(L_o) \times P_O$. For simplicity, the marginal outputs of production functions are assumed to be constant in the two periods.

Function $Ge_s(n,0)$ is defined as the difference between the expected utility of farmers with small farms and that of individuals in other sectors in the case of no subsidy, as shown in equation (19) where n is the individual's initial wealth and 0 stands for no subsidy.

$$Ge_s(n,0) = p \times U(n - r_s - EC + R \times (AL_{L_s}(L_s) \times P_A$$

$$+ G_{L_s}(L_s,L_l) \times P_E))$$

$$+ (1-p) \times U(n - r_s - EC + R \times I_{L_o}(L_o) \times P_O) \quad (19)$$

$$- U(n + (1+R) \times I_{L_o}(L_o) \times P_O)$$

Function $Ge_l(n,0)$ is defined as the difference between the expected utility of farmers with large farms and that of individuals in other sectors in the case of no subsidies, as shown in equation (20).

$$Ge_l(n,0) = p \times U(n - r_l - EC + R \times (AH_{L_l}(L_l)$$

$$\times P_A + G_{L_l}(L_s,L_l) \times P_E))$$

$$+ (1-p) \times U(n - r_l - EC + R \times I_{L_o}(L_o) \times P_O) \quad (20)$$

$$- U(n + (1+R) \times I_{L_o}(L_o) \times P_O)$$

When individuals are sufficiently risk averse and their initial wealth, n, is small enough, $Ge_s(n,0)$ and $Ge_l(n,0)$ are both negative which implies that poor individuals will be very sensitive to the risks in the agricultural sector when income represents a large proportion of their wealth. However, when their initial wealth, n, is relatively large, $Ge_s(n,0)$ and $Ge_l(n,0)$ will be positive, implying that rich individuals will be more willing to enter the agricultural sector. Define $\underline{n}_s^{Ge}$ and $\underline{n}_l^{Ge}$ as threshold levels where $Ge_s(\underline{n}_s^{Ge},0) = 0$ and $Ge_l(\underline{n}_l^{Ge},0) = 0$, as shown in Figure 5. Individuals, whose initial

wealth is larger than $\underline{n}_l^{Ge}$, will invest in large farms ($L_l^d = \sum_{n \geq \underline{n}_l^{Ge}} f(n)$); while individuals whose initial wealth is between $\underline{n}_s^{Ge}$ and $\underline{n}_l^{Ge}$ will invest in small farms ($L_s^d = \sum_{\underline{n}_l^{Ge} \geq n \geq \underline{n}_s^{Ge}} f(n)$). Remaining individuals will enter the other sectors.

If there is no uncertainty, i.e. the case of equation (18), the socially optimal labor structure is $L^d = (0,0,N)$. Therefore, when there are uncertainty and risk aversion, the pure loan scheme won't result in an efficient allocation. It is also not equitable due to ex post differences between the lifetime incomes of different types of labor, as shown in equation (18). Finally there is no equality of chance. Although all individuals can get loans to cover the costs of agriculture, only rich individuals will choose to invest in agriculture because they can afford to take the risk.

3.2 The harvest tax system. The harvest tax system, as defined here, has two components. Firstly, there is a public loan scheme, so that any individual can obtain a loan that has to be fully paid back. In addition to making loans available, the government can finance part of the agricultural costs through a subsidy. The total subsidy is then repaid by levying a tax on those who make a profit from agricultural production. Those who don't make a profit from agriculture don't need to pay the harvest tax. Unsuccessful farmers, hence, receive a net subsidy, while successful farmers have to pay back not only their own loans but also the subsidy received by those who fail.

For clarity and simplicity, it is assumed that the government gives the same subsidy S^h to s and l. In the second period, only the farmers who succeed pay the tax T^h. To keep a balanced budget, the following relationship should be satisfied as shown in equation (21).

$$S^h \times (L_s + L_l) = R \times T^h \times (p \times L_s + p \times L_l) \quad (21)$$

Farmers who succeed will have an expense for the harvest tax system, as shown in equation (22).

$$T^h - S^h = \frac{S^h \times (L_s + L_l)}{R \times (p \times L_s + p \times L_l)} - S^h = S^h \times (\frac{1-Rp}{Rp}) \quad (22)$$

Farmers who fail will achieve a net income S^h for the harvest tax system. Compared with the pure loan system, the gap between successful and unsuccessful farmers becomes smaller. The variance of farmers' lifetime incomes however becomes lower, which shows an insurance property of the harvest tax system.

Under the harvest tax system, function $Ge_s(n,S^h)$ becomes:

$$Ge_s(n,S^h) = p \times U(-S^h \times (\frac{1}{Rp} - 1) + n - r_s - EC$$

$$+ R \times (AL_{L_s}(L_s) \times P_A$$

$$+ G_{L_s}(L_s,L_l) \times P_E)) + (1-p) \quad (23)$$

$$\times U(S^h + n - r_s - EC + R \times I_{L_o}(L_o) \times P_O)$$

$$- U(n + (1+R) \times I_{L_o}(L_o) \times P_O)$$

and function $Ge_l(n,S^h)$ becomes:

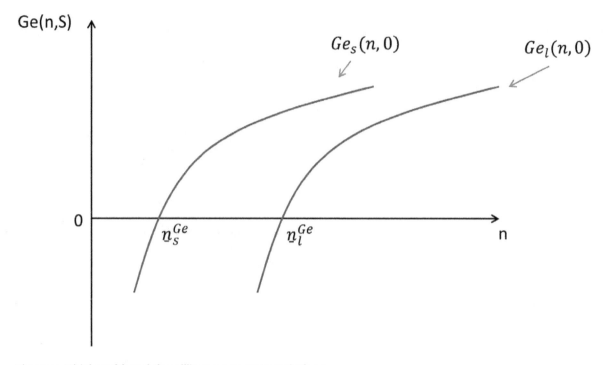

Figure 5. Initial wealth and the willingness to enter agriculture.

$$Ge_l(n,S^h) = p \times U(-S^h \times (\frac{1}{Rp} - 1) + n - r_l - EC + R$$
$$\times (AH_{L_l}(L_l) \times P_A + G_{L_l}(L_s, L_l) \times P_E)) \qquad (24)$$
$$+ (1-p) \times U(S^h + n - r_l - EC + R \times I_{L_o}(L_o) \times P_O)$$
$$- U(n + (1+R) \times I_{L_o}(L_o) \times P_O)$$

$Ge_s(n,S^h)$ and $Ge_l(n,S^h)$ are both functions of subsidy S^h. When $S^h = 0$, it returns to the pure loan system. For any levels of initial wealth, a higher subsidy S^h implies a smaller gap (risk) between successful and unsuccessful farmers. An increase in S^h thus shifts $Ge_s(n, 0)$ and $G_l(n, 0)$ upwards, as shown in Figure 6. For a given subsidy S^h, an individual whose initial wealth is larger than $\underline{n}_l^{Ge,S^h}$ will choose to enter the agricultural sector with a large farm; an individual whose initial wealth is between $\underline{n}_s^{Ge,S^h}$ and $\underline{n}_l^{Ge,S^h}$ will choose to enter agriculture with a small farm; remaining individuals will choose to work in other sectors. However, the socially optimal labor structure L^d ($L^d = (0,0,N)$) cannot still be achieved as long as $\underline{n}_l^{Ge,S^h} > \underline{n}_s^{Ge,S^h} > 0$.

The harvest tax has two desirable equity implications: it does not imply reverse redistribution, and it reduces the differences between the ex post lifetime incomes of successful and unsuccessful (i.e. unlucky) farmers. However, there remain differences between individuals as far as their willingness to undertake risk is concerned. Individuals with a large initial wealth are more likely to enter agriculture than poorer individuals. The harvest tax, by providing some degree of insurance, weakens this effect but does not eliminate it. Equality of opportunity is still not achieved.

3.3 The income contingent loans. Another possible policy option is a system of income contingent loans. An income contingent loan is a loan such that: (1) repayment only takes place in the event that an individual's income exceeds a pre-specified level; (2) annual repayment doesn't constitute more than a certain proportion of an individual's income; (3) repayment ceases once the loan plus interest has been repaid [8].

For clarity and simplicity, farmers will borrow S^I in the first period. The successful farmers will repay their own loans in the second period. To keep a balanced budget, a lump-sum tax T^I is levied on all individuals to cover the costs of unsuccessful farmers, as shown in equation (25):

$$(1-p) \times L_s \times S^I + (1-p) \times L_l \times S^I = R \times N \times T^I \qquad (25)$$

The difference between the expected utility of farmers with small farms and that of individuals in other sectors, $Ge_s(n,S^I)$, is shown in equation (26):

$$Ge_s(n,S^I) = p \times U(n - RT^I - r_s - EC + R$$
$$\times (AL_{L_s}(L_s) \times P_A + G_{L_s}(L_s, L_l) \times P_E))$$
$$+ (1-p) \times U(n - RT^I + S^I - r_s - EC + R \times I_{L_o}(L_o) \times P_O) \qquad (26)$$
$$- U(n - RT^I + (1+R) \times I_{L_o}(L_o) \times P_O)$$

The difference between the expected utility of farmers with large farms and that of individuals in other sectors, $Ge_l(n,S^I)$, is shown in equation (27):

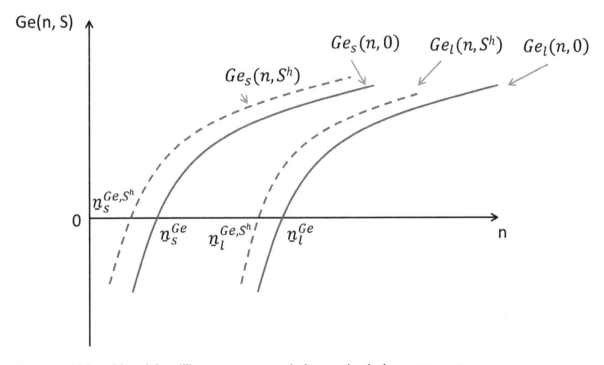

Figure 6. Initial wealth and the willingness to enter agriculture under the harvest tax system.

$$Ge_l(n,S^I) = p \times U(n - RT^I - r_l - EC + R$$
$$\times (AH_{L_l}(L_l) \times P_A + G_{L_l}(L_s, L_l) \times P_E))$$
$$+ (1-p) \times U(n - RT^I + S^I - r_l - EC \qquad (27)$$
$$+ R \times I_{L_o}(L_o) \times P_O) - U(n - RT^I + (1+R)$$
$$\times I_{L_o}(L_o) \times P_O)$$

The difference between expected lifetime income of s (l) under the income contingent loans and under the harvest tax system is:

$$-RT^I + (1-p) \times S^I = -\frac{(1-p) \times L_s + (1-p) \times L_l}{N} \times S^I$$
$$+ (1-p) \times S^I \qquad (28)$$
$$= S^I \frac{(1-p)L_o}{N} > 0$$

This reflects the fact that farmers are subsidized by individuals in other sectors. In the income contingent loan system, it is less attractive to enter other sectors compared with the harvest system, which will produce a more efficient labor structure.

The higher expected lifetime incomes of farmers in the income contingent loan system are at the expense of individuals in other sectors who become worse off because they have to pay the tax, RT^I. Successful farmers earn as much as under the pure loan system and more than under the harvest tax system. Unsuccessful farmers will earn more than under the pure loan system (as they don't have to repay their loans but only RT) and either more or less than under the harvest tax system. They earn less if

$RT^I - S^I > - S^h$ which holds if the subsidy S^h under the harvest tax system is large enough.

Therefore, on the one hand, it is possible with the harvest tax system to outperform a contingent loan system, provided that the subsidy is large enough; on the other hand, the harvest tax system is more equitable as it avoids the reverse distribution to some extent.

In summary, the income contingent loan system is similar to the harvest tax system and therefore is characterized by most of its advantages over the traditional system. In particular, it also provides insurance and results in a more efficient number of farmers. As the maximum repayment under the income contingent loan system is limited to the loans of successful farmers, some general taxes are still needed to subsidize unsuccessful farmers. In contrast to a harvest tax system, it again implies that reverse redistribution occurs. However, it may be more practical to implement because successful farmers don't have to pay more than their loans.

3.4 Some policy options for EU in the future. According to the European Commission, there are three main directions for the CAP to take in the future: (1) keep the current direct payment system unchanged; (2) introduce more equity in the distribution of direct payments. Decoupled payments would be composed of a basic rate serving as income support and a compulsory additional payment for specific "greening" public goods; or (3) phase-out decoupled payments in the current form and provide instead limited payments for environmental products. According to the results in this paper, three alternative policy options are given in Table 2.

Results and Discussion

The motivation of this paper is to prove that the tax-subsidy system currently used to finance the agricultural sector in the EU is characterized by "reverse redistribution"–rich farmers are subsidized by poor farmers and the rest of society. Assuming that

Table 2. Policy options.

Option	No.	Name	Description
1	Reference	Current policy (2010)	Decoupled payments
2	I	Payment based on farmer's income and a subsidy for environmental products	The subsidy should primarily be distributed to poor farmers.
	II	Dynamic hybrid model (Ref and I)	The combination of reference and I which implies slower reform.
3	III	A pure loan plus a subsidy for environmental products	Farmers can take up a loan and repay it after they harvest. Any individual has the opportunity to enter agriculture.
	IV	An income contingent loan system plus a subsidy for environmental products	A system of income contingent loans makes repayments conditional on whether the income of the farmer exceeds a pre-specified level and computes repayments as a percentage of their earnings. The maximum amount to be paid is the loan plus interests.
	VI	A harvest tax plus a subsidy for environmental products	A harvest tax system makes repayment contingent on income. Repayments from successful farmers exceed the cost of their loans. The difference between repayment and cost is used to subsidize unsuccessful farmers.

government intervention is needed due to the difficulty of obtaining private loans to finance the agricultural sector, some possible policy options that can avoid reverse redistribution are analyzed: the pure loan system, the harvest tax system and the income contingent loan system.

The three systems are identical when agricultural output is certain. When there is agricultural risk, the systems differ in how the subsidy should be financed. Whereas under a pure loan system the farmers repay in full their loans plus interest, a harvest tax system makes the repayments of loan costs contingent on whether the farmers make a profit from agricultural production. Farmers making losses (or not achieving the minimum income threshold) don't need to repay their loans. Successful farmers, on the other hand, are required to repay their loans plus an extra amount to cover the costs of unsuccessful farmers. In the case of the income contingent loan, it still requires some general taxation to subsidize the less successful farmers and hence has some reverse redistribution effects.

These loan systems may seem unrealistic given the current CAP and the general acceptance of providing farmers with subsidies. The major issues driving CAP reform (section 3.4) seem however to be uncontentious A) there should be more equity in the distribution of agricultural support; and B) farmers should be paid for environmental provisioning. The relevant economic question is therefore the means necessary to achieve these ends. One needs to look no further than the education sector to find examples of loan systems being used in practice to deal with concerns of efficiency and equity [21–23]. As our analysis shows, reverse redistribution is unavoidable given the current tax-subsidy basis of the CAP; so the answer cannot be found in the current thinking. However, if payments are offered for environmental products according to demand then a loan system will ensure that sufficient farmers enter or stay in the sector–despite an imperfect capital market–to deliver efficient quantities of these services, e.g. small or low-income farms can finance themselves and repay loans after outcomes.

Since differences in abilities between individuals are not considered in this paper (only differences in initial wealth), an equal distribution of income is also the equitable distribution of income because all individuals will have the same productive capacity. What the solution might be when considering heterogeneity in individuals' abilities may be another interesting issue. Further the harvest tax system and income contingent loan system presented in this paper are based on the actual incomes obtained by the farmers, hence moral hazard–which implies that farmers may conceal their real incomes to obtain financial advantage–is a relevant issue and a potential subject for future work. Another extension of this paper is to analyze the effect of environmental costs on the desirable distribution of subsidies.

Acknowledgments

The authors gratefully acknowledge the good research environments provided by the Centre for Environmental and Climate Research (CEC), Lund University and AgriFood Economics Centre. We thank Alfons Balmann, Katarina Hedlund and Henrik Smith for fruitful discussions. We also thank Aldo Rustichini and four anonymous referees for their helpful suggestions and corrections on the earlier draft of our paper which improved the contents.

Author Contributions

Conceived and designed the experiments: RC MB. Performed the experiments: RC. Analyzed the data: RC MB. Contributed reagents/materials/analysis tools: RC MB. Wrote the paper: RC.

References

1. Brady M, Höjgård S, Kaspersson E, Rabinowicz E (2009) The CAP and Future Challenges. Europ Pol Anal 11: 1–13.

2. EU (2009) General Budget: Agriculture and Rural Development. Available: http://eur-lex.europa.eu/budget/data/D2009_VOL4/EN/nmc-titleN123A5/index.html (Accessed 2009 May 11.

3. Happe K, Kellermann K, Balmann A (2006) Agent-Based Analysis of Agricultural Policies: An Illustration of the Agricultural Policy Simulator AgriPoliS, Its Adaptation and Behavior. Ecolog Society 11: 49.

4. Baldwin RE (2005) Who Finances the Queen's CAP Payments? CEPS Policy Briefs, No. 88, Brussels, Centre for European Policy Studies.

5. Shah A (2006) Exploring sustainable production systems for agriculture: Implications for employment and investment under north–south trade scenario. Ecolog Econ 59(2): 237–241.

6. Cong RG, Wei YM (2010) Potential impact of (CET) carbon emissions trading on China's power sector: A perspective from different allowance allocation options. Energy 35: 3921–3931.

7. Brady M, Kellermann K, Sahrbacher C, Jelinek L (2009) Impacts of Decoupled Agricultural Support on Farm Structure, Biodiversity and Landscape Mosaic: Some EU Results. J Agri Econ 60(3): 563–585.

8. García-Peñalosa C, Wälde K (2000) Efficiency and equity effects of subsidies to higher education. Oxford Econ Pap 52(4): 702–722.

9. Ahmed AU, Bouis HE, Gutner T, Löfgren H (2001) The Egyptian food subsidy system: Structure, performance, and options for reform. Research report 119. International Food Policy Research Institute. Washington, D.C.

10. Dunne W, O'Connell JJ (2002) A Multi-commodity EU Policy Framework Incorporating Public Good Criteria into the Direct Payment System in Agriculture. 2002 International Congress, August 28–31, 2002, Zaragoza, Spain.

11. Magnani E (2000) The Environmental Kuznets Curve, environmental protection policy and income distribution. Ecolog Econ 32: 431–443.

12. Beard N, Swinbank A (2001) Decoupled payments to facilitate CAP reform. Food Pol 26: 121–145.

13. Josling TE (1974) Agricultural policies in developed countries: a review. J Agr Econ 25: 229–264.

14. Hennessy DA (1998) The production effects of agricultural income support policies under uncertainty. Amer J Agr Econ 80: 346–357.

15. Andersson FCA (2004) Decoupling: the concept and past experiences. SLI Working Paper 2004, Swedish Institute for Food and Agricultural Economics: 1.

16. Revell B, Oglethorpe D (2003) Decoupling and UK Agriculture: A Whole Farm Approach. Report commissioned by DEFRA.

17. Uthes S, Sattler C, Reinhardt FJ, Piorr A, Zander P, et al. (2008) Ecological effects of payment decoupling in a case study region in Germany. J Farm Manage 13 (3): 219–230.

18. EC (2008) The Health Check of the CAP reform: Impact Assessment of alternative Policy Option, Maggio.

19. Zahrnt V (2009) Public Money for Public Goods: Winners and Losers from CAP Reform. ECIPE working paper, No.08.

20. Akalpler E (2006) The Impact of the Support System of the CAP on Free Trade in the Light of the Turkey's EU Membership. Doctoral thesis, WU Vienna University of Economics and Business.

21. Chapman B (1997) Conceptual issues and the Australian experience with income continent charges for higher education. Econ J 107(442): 738–751.

22. Chapman B (2006) Income Contingent Loans for Higher Education: International Reforms. Handbook of the Economics of Education. Amsterdam, North-Holland.

23. Vandenberghe V, Debande O (2008) Refinancing Europe's higher education through deferred and income-contingent fees: An empirical assessment using Belgian, German and UK data. Eur J Polit Econ 24(2): 364–386.

21

Environmental Fate of Soil Applied Neonicotinoid Insecticides in an Irrigated Potato Agroecosystem

Anders S. Huseth[1], Russell L. Groves[2]*

1 Department of Entomology, Cornell University, New York State Agricultural Experiment Station, Geneva, New York, United States of America, **2** Department of Entomology, University of Wisconsin-Madison, Madison, Wisconsin, United States of America

Abstract

Since 1995, neonicotinoid insecticides have been a critical component of arthropod management in potato, *Solanum tuberosum* L. Recent detections of neonicotinoids in groundwater have generated questions about the sources of these contaminants and the relative contribution from commodities in U.S. agriculture. Delivery of neonicotinoids to crops typically occurs as a seed or in-furrow treatment to manage early season insect herbivores. Applied in this way, these insecticides become systemically mobile in the plant and provide control of key pest species. An outcome of this project links these soil insecticide application strategies in crop plants with neonicotinoid contamination of water leaching from the application zone. In 2011 and 2012, our objectives were to document the temporal patterns of neonicotinoid leachate below the planting furrow following common insecticide delivery methods in potato. Leaching loss of thiamethoxam from potato was measured using pan lysimeters from three at-plant treatments and one foliar application treatment. Insecticide concentration in leachate was assessed for six consecutive months using liquid chromatography-tandem mass spectrometry. Findings from this study suggest leaching of neonicotinoids from potato may be greater following crop harvest in comparison to other times during the growing season. Furthermore, this study documented recycling of neonicotinoid insecticides from contaminated groundwater back onto the crop via high capacity irrigation wells. These results document interactions between cultivated potato, different neonicotinoid delivery methods, and the potential for subsurface water contamination via leaching.

Editor: Christopher J. Salice, Texas Tech University, United States of America

Funding: This research was supported by the Wisconsin Potato Industry Board and the National Potato Council's State Cooperative Research Program FY11-13. The funders had no role in study design, data collection and analysis, decision to publish, or preparation of the manuscript.

Competing Interests: RLG has received research funding, not related to this project, from Bayer CropScience, DuPont, Syngenta, and Valent U.S.A.

* E-mail: groves@entomology.wisc.edu

Introduction

The neonicotinoid group of insecticides is among the most broadly adopted, conventional management tools for insect pests of annual and perennial cropping systems [1]. Benefits of the neonicotinoid group of compounds include flexibility of application, diversity of active ingredients, and broad spectrum activity [2]. Moreover, growers have readily adopted neonicotinoids for two specific reasons: first, these compounds are fully systemic in plants after soil application and second, several new generic formulations have recently become available which have incentivized their continued use in many crops [1–3]. Since 2001, the United States Environmental Protection Agency (EPA) has classified several neonicotinoids as either conventional, reduced-risk pesticides, or as organophosphate alternatives [4],[5]. EPA certification often requires replacement of older, broad-spectrum pesticides with newer, more specific products for management of key economic pests. Critical attributes of replacement insecticides include documented reductions in human and environmental risk when compared to older, broad-spectrum pesticides [5]. Despite acceptance of neonicotinoid insecticides as reduced-risk by growers and regulatory agencies, nearly two decades of widespread, repetitive use has resulted in several insecticide resistance issues, impacts on native and domestic pollinators, and unanticipated environmental impacts [6–9].

The environmental fate of several neonicotinoid active ingredients have been previously assessed. Previous studies focused on degradation and movement processes in soil, leachate, and runoff [10–15]. The leaching potential of the neonicotinoids into groundwater, as well as persistence in the plant canopy, is related to properties of the chemicals and delivery method of the compound to the crop (Fig. S1)[12],[15],[16]. Soil application (e.g., seed treatment or in-furrow) has been adopted as the principal form of insecticide delivery in potato production as it provides the longest interval of pest control, while also reducing non-target impacts, and limits exposure to workers when compared to foliar application methods. Since 1995, soil-applied neonicotinoids (i.e., clothianidin, imidacloprid, thiamethoxam) have been the most common pest management strategy used to control infestation of Colorado potato beetle, *Leptinotarsa decemlineata* Say; potato leafhopper, *Empoasca fabae* Harris; green peach aphid, *Myzus persicae* Sulzer; and potato aphid, *Macrosiphium euphorbiae* Thomas. The now widespread and extensive use of these systemic neonicotinoid insecticides, coupled with the recent detection of thiamethoxam in groundwater [17],[18], supports the hypothesis that potato pest management may contribute a portion of the documented neonicotinoid contaminants reported in

Wisconsin, USA. Furthermore, we hypothesized that neonicotinoid insecticides applied to potato are most vulnerable to leaching in the spring season when the root system of the plant has yet to fully exploit all of the active ingredient applied directly in the seed furrow. Large rain events at this time could drive insecticide leaching from potato and subsequent groundwater contamination at large scales. In this study, we examined how neonicotinoid concentrations in leachate were altered in response to different insecticide delivery methods using potatoes grown under commercial production practices. We also report the patterns of historic neonicotinoid insecticide detections in groundwater using water quality surveys collected by the Wisconsin Department of Agriculture, Trade and Consumer Protection-Environmental Quality Section (WI DATCP-EQ). Second, using potato as a model system, we analyzed leachate captured below different seed treatments, soil-applications, and foliar delivery treatments for thiamethoxam using liquid chromatography-tandem mass spectrometry (LC/MS/MS) over two consecutive field seasons. In this experiment, thiamethoxam was chosen as one representative insecticide in a broader group of water-soluble neonicotinoids. Moreover, this active ingredient represented the majority of positive neonicotinoid detections in groundwater monitoring surveys conducted by the WI DATCP-EQ [17], [18]. Third, using identical quantitative methods, we measured thiamethoxam concentration in irrigation water collected from operating, high-capacity irrigation wells at two time points in each sampling year. And finally, we characterize irrigation use and production trends of crops that may contribute to neonicotinoid detection in groundwater. Results of this study increase our understanding about the influence of insecticide delivery method on the neonicotinoid insecticides leaching from potato into the surrounding environment.

Materials and Methods

Ethics Statement

No specific permits were required for the field study described here. Access to field sites was granted by the private landholder to conduct leaching experiments. No specific permissions were needed to present publically available records provided by Wisconsin Department of Agriculture, Trade and Consumer Protection or Wisconsin Department of Natural Resources. Field studies did not involve any endangered or protected species.

Groundwater Contamination

Permanent groundwater monitoring wells, maintained by the WI DATCP-EQ, were used to measure neonicotinoid contamination of subsurface water resources as one component of an ongoing study documenting agrochemical (e.g., insecticides, herbicides, nutrients) impact on groundwater quality. Beginning in 2006, analytical water quality assessments for neonicotinoid contamination were conducted by the Wisconsin Department of Agriculture Trade and Consumer Protection-Bureau of Laboratory Services. Concentrations of acetamiprid, clothianidin, dinotefuran, imidacloprid, and thiamethoxam were monitored in 20–30 different monitoring well locations from 2006–2012. Presented are positive detections of those insecticides in different monitoring wells from 2006–2012 [17],[18]. Data provided by WI DATCP-EQ characterize the temporal and spatial profile of thiamethoxam and other neonicotinoid detections that occurred between 2008–2012. These data are presented in summary as a foundation for following objectives (Table 1).

Experimental Site and Design

In 2011 and 2012, leaching experiments were conducted 6 km west of Coloma, Wisconsin. Experiments were planted in two different fields approximately 0.5 km apart on 20 May 2011 and 11 May 2012. The soil at both sites consisted of Richford loamy sand (sandy, mixed, mesic, Typic Udipsamments) [19]. Soil composition was 7% clay, 82% sand, and 11% silt. Organic matter was 0.53 percent by weight. Study sites soils had a high infiltration rate (Hydrological Soil Group A), a high saturated hydraulic conductivity (K_{sat}) at 28 micrometers per second, and an available water capacity rating of 0.1 cm per cm [19]. No restrictive layer that would impede water movement through the soil has been documented [19]. Study site soil was formed in the bed of glacial Lake Wisconsin from parent material of glacial till overlain by glacial outwash [20]. Upper soil horizons (A and B) are sand with minimal structure. Subsurface soil (C horizon) had no structure. Irrigation pivots in sample fields withdrew water at a depth of 37 m and the water table depth (static water level) was approximately 6 m for both sites [21].

A randomized complete block design with four insecticide delivery treatments and an untreated control was established using the potato cultivar, 'Russet Burbank'. Plots were 0.067 ha in size and planted at a rate of one seed piece per 0.3 m with 0.76 m spacing between rows. Each year, experiments were nested within a different ~32 ha commercial potato field, and maintained under commercial management practices by the producer (e.g., nutrient application timing, chemical usage, tillage practices, etc.), with the exception of insecticide inputs. The decision to locate these experiments in commercial fields was, in part, based upon access to a center pivot irrigation system to best duplicate water inputs used to produce commercial potato in Wisconsin. All other inputs and production strategies (e.g. tillage, fumigation, fertility, and disease management) were conducted by the producer with equipment and products in a manner consistent with the best management practices for potato production in Wisconsin. Prior to planting in each season, a tension plate lysimeter (25.4×25.4×25.4 cm) was buried at a depth of 75 cm below the soil surface. Lysimeters were constructed of stainless steel with a porous stainless steel plate affixed to the top to allow water to flow into the collection basin over each sampling interval. Experimental blocks were connected with 9.5 mm copper tubing to a primary manifold and equipped with a vacuum gauge. A predefined, fixed suction was maintained under regulated vacuum at 107 ± 17 kPa (15.5 ± 2.5 lb per in^2) with a twin diaphragm vacuum pump (model UN035.3 TTP, KnF, Trenton, NJ) connected to a 76 L portable air tank. Each treatment block was equipped with a datalogging rain gauge (Spectrum Technologies, Inc. model # 3554WD1) recording daily water inputs at a five minute interval. Data was offloaded with Specware 9 Basic software (Spectrum Technologies, Inc., Plainfield, IL, USA) and aggregated into daily irrigation or rain event totals using the *aggregate* and *dcast* function in R (package: reshape2, [22]). Irrigation event records were obtained from the grower to identify days and estimated inputs of water application throughout the growing season.

Insecticides and Application

Thiamethoxam treatments (Platinum 75SG, 75% thiamethoxam per formulated unit, Syngenta, Greensboro, NC) were selected to represent a common, soil-applied insecticide in potato. A second formulation of thiamethoxam was selected to represent a common pre-plant insecticide seed treatment in potato (Cruiser 5FS, 47.6% thiamethoxam per formulated unit, Syngenta, Greensboro, NC). Each insecticide formulation is used to manage early season infestations of Colorado potato beetle, potato

Table 1. Positive (means±SD) neonicotinoid detections in groundwater from 2008–2012, State of Wisconsin Department of Agriculture Trade and Consumer Protection.

Year	County	Area potato (ha)[a]	Row crops (ha)[b]	Percent potato[c]	Well ID	N positive samples	Insecticide concentration (μg/L)[d]		
							clothianidin	imidacloprid	thiamethoxam
2008	Adams	2,617	21,385	10.9	6	2	-	-	4.34 (4.97)
	Grant	0	47,827	0.0	10	1	-	-	1.25
	Iowa	18	25,795	0.1	11,12,13	9	-	-	1.50 (0.67)
	Richland	29	9,582	0.3	16	1	-	-	0.69
	Sauk	30	31,931	0.1	17	2	-	-	2.41 (1.32)
	Waushara	2,630	29,447	8.2	20	2	-	-	0.67 (0.05)
2009	Adams	3,989	24,894	13.8	6	2	-	-	5.31 (5.12)
	Dane	22	101,527	0.0	9	1	-	-	1.61
	Iowa	343	33,375	1.0	11,12	3	-	-	1.31 (0.68)
	Richland	87	14,402	0.6	16	1	-	-	1.26
	Sauk	328	40,571	0.8	17	2	-	-	3.00 (0.94)
2010	Adams	4,188	24,871	14.4	6	4	3.43	-	2.97 (2.04)
	Brown	1	39,322	0.0	7	1	-	-	0.52
	Dane	34	110,979	0.0	8,9	4	0.54 (0.24)	0.54	1.08
	Grant	49	74,566	0.1	10	1	0.73	-	-
	Iowa	356	38,840	0.9	11,12,13	7	-	-	1.25 (1.02)
	Sauk	188	45,309	0.4	17	5	0.41	-	1.81 (0.88)
	Waushara	4,184	33,576	11.1	19,20	2	-	2.77 (0.81)	-
2011	Adams	4,066	27,693	12.8	2,5,6	9	0.63 (0.36)	0.33	0.63 (0.26)
	Brown	7	38,309	0.0	7	1	-	-	0.21
	Dane	33	107,214	0.0	8	2	0.62 (0.19)	-	-
	Grant	13	75,436	0.0	10	1	0.30	-	-
	Iowa	47	40,138	0.1	12	4	-	0.34 (0.09)	0.88 (0.23)
	Portage	7,364	45,324	14.0	15	1	-	-	0.32
	Sauk	213	46,686	0.5	17,18	5	0.54 (0.10)	-	1.92 (0.43)
	Waushara	4,536	36,676	11.0	19,20,21,23	23	0.25 (0.03)	0.78 (0.69)	1.40 (0.56)
2012	Adams	4,263	27,037	13.6	1,3,4,6	6	0.52 (0.30)	0.51 (0.26)	0.27
	Dane	11	115,501	0.0	8	1	0.67	-	-
	Grant	4	72,920	0.0	10	1	0.26	-	-
	Iowa	369	40,764	0.9	12	2	0.24	0.28	0.44
	Juneau	907	28,542	3.1	14	2	0.42 (0.18)	-	0.20
	Portage	7,622	46,337	14.1	15	2	-	0.47	0.47
	Waushara	5,904	38,999	13.1	21,22,23	13	-	0.68 (0.88)	1.51 (0.72)
	summary			N=23		67	25	30	68

Table 1. Cont.

Year	County	Area potato (ha)[a]	Row crops (ha)[b]	Percent potato[c]	Well ID	N positive samples	Insecticide concentration (μg/L)[d]		
							clothianidin	imidacloprid	thiamethoxam
						Average	0.62 (0.63)	0.79 (0.83)	1.59 (1.51)
						Range	0.21–3.34	0.26–3.34	0.20–8.93

[a]Acreage estimates generated from USDA National Agricultural Statistics Service – Cropland Data Layer, 2008–2012 [26].

[b]Row crops class is the sum of the following crop areas (ha): maize, soy, small grains, wheat, peas, sweet corn, and miscellaneous vegetables and fruits.

[c]Percent potato calculated as the potato area grown annually divided by total arable row crop acreage (other row crops + potato).

[d]Positive neonicotinoid detections extracted from long-term, groundwater wells maintained by the WI-DATCP-EQ Program.

leafhopper, and colonizing aphid in Wisconsin potato crops. Commercially formulated insecticides were applied at maximum labeled rates for in-furrow (140 g thiamethoxam ha^{-1}) and seed treatment (112 g thiamethoxam ha^{-1} at planting density of 1,793 kg seed ha^{-1}) for potato [23]. A calibrated CO_2 pressurized, backpack sprayer with a single nozzle boom was used to deliver an application volume of 94 liters per hectare at 207 kPa through a single, extended range, flat-fan nozzle (TeeJet XR80015VS, Spraying Systems, Wheaton, IL) for in-furrow applications. Spray applications were directed onto seed pieces in the furrow at a speed of one meter per second and furrows were immediately closed following application. Seed treatments were applied using a calibrated CO_2 pressurized backpack sprayer with a single nozzle boom delivering an application volume of 102.2 L per hectare at 207 kPa through a single, extended range, flat-fan nozzle (TeeJet XR80015VS, Spraying Systems, Wheaton, IL) was used for delivery of thiamethoxam in water (130 mL) directly to suberized, cut seed pieces (23 kg) 24 hours prior to planting. Seed treatments were allowed to dry in the absence of light at 20°C during that pre-plant period. A novel soil application method, impregnated copolymer granules, was included as another treatment in an attempt to stabilize applied insecticide in the soil. Polyacrylamide horticultural copolymer granules (JCD-024SM, JRM Chemical, Cleveland, OH) were impregnated at an application rate of 16 kg per hectare. The polyacrylamide treatment was included as a novel delivery method to stabilize insecticide in the rooting zone and possibly reduce leaching in the early season. Thiamethoxam (0.834 g, Platinum 75SG) was initially diluted in 250 mL of deionized water and 100 μL of blue food coloring was incorporated into solution to ensure uniform mixing (brilliant blue FCF). Insecticide solutions were mixed with 75 g polyacrylamide then stirred until the liquid was absorbed and a uniform color was observed. Impregnated granules were vacuum dried in the absence of light for 24 hours at 20°C. Treated granules were divided into even quantities per row and evenly distributed into the four treatment rows for each polyacrylamide plot. A single untreated flanking row was planted between plots. All soil-applied insecticides were applied on 20 May 2011 and 11 May 2012 at the time of planting.

Two foliar applications of thiamethoxam (Actara 25WG, 25% thiamethoxam per formulated unit, Syngenta, Greensboro, NC) sprayed on the same plot were included as a fourth delivery treatment. Two successive neonicotinoid applications are recommended for foliar control of pests in potato [23]. Foliar thiamethoxam was applied using a calibrated CO_2 pressurized backpack sprayer delivering an application volume of 187.1 liters per hectare at 207 kPa through four, extended range flat-fan nozzles (TeeJet XR80015VS, Spraying Systems, Wheaton, IL) spaced at 45.2 cm. The first foliar application was followed approximately seven days later with a second equivalent rate of thiamethoxam to total the season-long maximum labeled rate (105 g thiamethoxam ha^{-1}) [23] and were timed to coincide with the appearance of 1st and 2nd instar larvae of native populations of *L. decemlineata*. Foliar applications of thiamethoxam were applied on 28 June and 5 July in 2011 and 15 and 22 June in 2012. Although total amounts of active ingredient differ by formulation, these rates are identical to registered label recommendations [23] and reflect the maximum amount of active ingredient used on an average hectare of cultivated potato. Specific chemical properties of formulated thiamethoxam that affect solubility and leaching potential in soil can be found in Gupta et al. [15] and the references therein (Fig. S1).

Chemical Extraction and Quantification

Lysimeter leachate was sampled twice monthly beginning on June 1 of each year and concluding in October of 2011 and November of 2012. Total leachate volume was recorded for each plot. A 500 mL subsample was taken from each plot into a 0.5 L glass vessel and immediately placed on ice and refrigerated at 4–6°C in the laboratory prior to analysis. Samples were homogenized into a 400 mL monthly (i.e., two samples per month) sample as percent volume per volume dependent on total catch measured in the field. Neonicotinoid residues from monthly water samples were extracted using automated solid phase extraction (AutoTrace SPE workstation, Zymark, Hopkinton, MA) with LiChrolut EN SPE columns (Merk KGaA, Darmstadt, Germany). If visual inspection of sample found excessive sediment contamination, samples were filtered through a 0.45 μm filter prior to extraction. Columns were conditioned prior to extraction with 3 mL of methanol (MeOH) and 3 mL of water. 210 mL of sample were loaded onto columns and rinsed with 10 mL of water then dried under flowing nitrogen for 15 minutes (N-evap, Organomation, Berlin, MA). Samples were eluted using a 50% ethyl acetate (EtOAc) and 50% methanol solution to collect a 2 mL sample fraction. Sample extract fractions were analyzed using a Waters 2690 HPLC/Micromass Quattro LC/MS/MS (Waters Corporation, Milford, MA). All thiamethoxam residues were identified, quantified, and confirmed using LC/MS/MS by the Wisconsin Department of Agriculture Trade and Consumer Protection-Bureau of Laboratory Services. The method detection limit (MDL) of the extraction procedure was $0.2 \ \mu g \ L^{-1}$. Specific conditions for all quantitative procedures follow WI-DATCP Standard Operating Procedure #1009 developed from Seccia et al. [24] and references therein.

Irrigation Use and Crop Area

To determine the extent of irrigated agriculture present within the watershed, we utilized current high capacity well pumping data and irrigated agriculture estimates derived from digital imagery. Publically available operator reporting data for high capacity agricultural pivots were obtained from the Wisconsin Department of Natural Resources Bureau of Drinking Water and Groundwater. Records included location information and pumping volume for the year 2012. High capacity wells service several irrigated fields and often these fields are further divided into individual crop management units each with unique irrigation requirements. We digitized the area watered by all identifiable center pivot, linear move, and traveling gun irrigation systems using digital aerial photography to measure the total number of management units present within the greater Central Wisconsin Water Management Unit watershed [25] (ArcGIS version 10.1, Redlands, CA). Fields were subdivided into management units using the consistent divisions in crop types with a sequence of National Agricultural Statistics Service Cropland Data Layer (NASS-CDL) [26] thematic data and aerial photography images [25] from 2010–2012.

To determine agronomic trends in the Central Sands vegetable production region of Wisconsin, we used a combination of publically available land use data and current neonicotinoid registration information. A geospatial watershed management boundary layer delineated by the Wisconsin Department of Natural Resources [27] was used to generally define the spatial extent where agriculture could be contributing to the detection of neonicotinoid insecticides in subsurface water. The Central Wisconsin Water Management Unit extent was used to estimate annual crop composition using the NASS-CDL [26] from 2006–2012 using ArcGIS. From these data, we selected major crops that

frequently receive either seed or in-furrow soil-applied neonicotinoid insecticide treatments. Application rates were identical for several similar crops (e.g. soybean and green bean), and so, we chose to aggregate crops based on insecticide rate and crop type into three primary groups: maize, beans, and potato [23],[28–30]. These crop groups comprise the majority of production area in the Central Wisconsin Water Management Unit extent. To our knowledge, limited information exists documenting the proportion of different soil-applied neonicotinoid active ingredients that are used on a per crop basis in the Central Wisconsin Water Management Unit. Based on this level of uncertainty, we chose not to extend tabulated crop areas to a direct calculation or estimate of neonicotinoid active ingredients applied.

Data Analysis

To determine the impact of different insecticide delivery treatments on thiamethoxam leachate detected over time, we reported the mean concentration over a period of several months. All lysimeter analyses included samples where neonicotinoid insecticides were not detected (i.e., zero detections). All data manipulation and statistical analyses of leachate concentrations were performed in R, version 2.15.2 [31] using the base distribution package. Functions used in the analysis are available in the base package of R unless otherwise noted. Observed concentration for time points in each year were subjected to a repeated-measures analysis of variance (ANOVA) using a linear mixed-effects model to determine significant delivery (i.e. treatment), date, and delivery×date effects ($P<0.05$). Because the agronomic conditions differed between years and given that our comparison of interest was at the insecticide delivery treatment level, insecticide concentrations were analyzed separately for each year. Mixed-effects models (i.e., repeated-measures analysis of variance) were fit using the *lme* function (package nlme, [32]). Empirical autocorrelation plots from unstructured correlation model residuals were examined using the *ACF* function (package nlme, [32]). Correlation among within-group error terms were structured and examined in three ways: first, unstructured correlation, second, with compound symmetry using the function *corCompSymm* and third, with autoregressive order one covariance using the function *corAR1* (package nlme, [32])[33]. Since models were not nested, fits of unstructured, compound symmetry, and autoregression order one covariance were compared using Akaike's information criterion statistic with the function *anova* (test = "F"). Data were transformed with natural logarithms before analysis to satisfy assumptions of normality, however untransformed means are graphically presented. In 2012, a single lysimeter in the polyacrylamide treatment of the leachate study malfunctioned and these observations were dropped from subsequent analyses leading to an unbalanced replicate number for that treatment (N = 3) in 2012. Water input data collected from tipping bucket samplers were averaged across block by day and aggregated as cumulative water inputs using the *cumsum* function. All summary statistics and model estimates were extracted using *aggregate*, *summary*, and *anova* functions.

Results and Discussion

Groundwater Detections

Neonicotinoid insecticides were detected at 23 different well monitoring well locations by WI-DATCP-EQ surveys between the years 2008 and 2012 (Table 1). These annual surveys, administered by WI-DATCP-EQ, occur at sensitive geologic or hydrogeologic locations that are at high risk of non-point source agrochemical leaching. Specifically, two agriculturally intensive

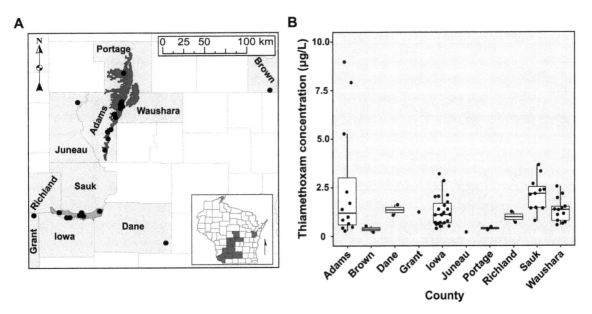

Figure 1. Positive thiamethoxam residue detections in groundwater 2008–2012. Points in the map (A) correspond to positive detection locations. Dark grey shaded region indicates the Central Sands potato production region. Light grey delimits the Lower Wisconsin River potato production region. Positive detections were obtained from established agrochemical monitoring wells collected by the Wisconsin Department of Agriculture, Trade and Consumer Protection (DATCP)-Environmental Quality division in collaboration with the Wisconsin DATCP Bureau of Laboratory Services. Boxplots (B) indicate average concentration detected from 2008–2012. Points show individual measured concentrations.

production regions of the state, the Central Sands and Lower Wisconsin River valley, are classified as high-risk areas for groundwater contamination and are frequently monitored for the presence of common agrochemicals (Fig. 1A). These regions have well-drained, sandy soils and easily accessible groundwater for irrigation that has driven agricultural intensification focused on vegetable production. Commercial potato is a key component in the agricultural production sequence, but is also rotated with many other specialty crops such as: carrots, onions, peas, pepper, processing cucumber, sweet corn, and snap beans. Unfortunately, the unique soil and water characteristics supporting a profitable

specialty crop production system are also particularly vulnerable to groundwater contamination with water-soluble agricultural products [34–36]. Regulatory exceedences of nitrates and herbicide products (e.g. triazines, triazinones, and chloroacetamide) have been commonplace for several years [34–37], but recent detections of neonicotinoid contaminants have created new groundwater quality concerns. Beginning in the spring of 2008, two wells had detections of 1.25 and 1.47 μg L^{-1} thiamethoxam in Grant and Sauk Counties, WI (Fig. 1B, Table 1). Subsequent sampling later that season identified six additional locations for a total of 17 independent positive thiamethoxam detections that year. Since

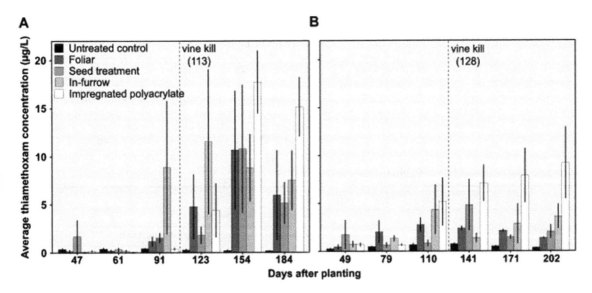

Figure 2. Thiamethoxam concentration in leachate from potato. Average thiamethoxam (±SD) recovered from in-furrow and foliar treatments in (A) 2011 an (B) 2012. Dotted lines indicate the date that the producer applied vine desiccant prior to harvest. Lysimeter studies continued in undisturbed soil following vine kill.

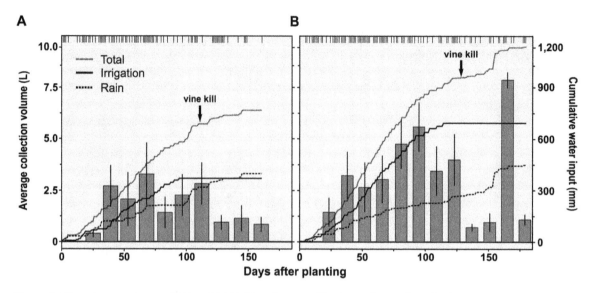

Figure 3. Water input volumes, 2011 and 2012. Water inputs and leachate volume collected in lysimeter studies in (A) 2011 and (B) 2012. Lines indicate cumulative water measured in tipping bucket rain gauges installed in plots each season. Bar plots indicate average leachate volume ($\pm$SD) collected in lysimeters on a bi-monthly sampling frequency. Hash marks at the top of each figure indicate days that overhead irrigation or rainfall occurred in each season.

these early detections, the WI-DATCP-EQ [17],[18] has repeatedly detected thiamethoxam, imidacloprid, and clothianadin residues at 23 different monitoring well locations over a five-year period (Table 1). Although the sampling effort was not uniformly distributed within the state, neonicotinoid detections often correspond to areas where intensive irrigated agricultural production occurs (Fig. 1A). As an indication of specialty crop production intensity, we used county-level potato abundance to better describe trends in historical neonicotinoid detections. Observed frequency and magnitude of neonicotinoid detections did not consistently correspond to potato abundance (Table 1). Although the contribution of potato production to the observed detections was not clear, regulatory agencies have continued to pursue this interaction by sampling where potato occurs at a high density, specifically the Central Sands and Wisconsin River Valley. Groundwater sampling strategies have provided a useful timeline of non-point source agrochemical pollution events in subsurface water resources. Identifying the origin of pollutants in the state is complicated by the diversity of neonicotinoid registrations, application methods and formulations; currently Wisconsin has 164 different registrations for field, forage, tree fruit, vegetable, turf, and ornamentals crops (6 acetamiprid, 18 clothianadin, 4 dinotefuran, 108 imidacloprid, 1 thiacloprid, 26 thiamethoxam) [38].

Neonicotinoid Losses and Concentrations in Leachate

The neonicotinoid insecticide thiamethoxam was included in field experiments to investigate the potential for leaching losses associated with different types of pesticide delivery. Specifically, formulations of thiamethoxam were applied as foliar and as at plant systemic treatments in commercial potato over two years and at two different irrigated fields. We hypothesized that thiamethoxam would be most vulnerable to leaching early in the season when plants were small and episodic heavy rains can be common. Interestingly, we observed the greatest insecticide losses following vine-killing operations which occurred more than 100 days after planting (Fig. 2). Detections of thiamethoxam in lysimeters varied between insecticide delivery treatments through time in 2011 (delivery$\times$date interaction, $F=2.1$; d.f. $=20,88$; $P=0.0131$) and again in 2012 (delivery$\times$date interaction, $F=1.8$; d.f. $=20,87$; $P=0.0384$). Moreover, the impregnated polyacrylamide delivery produced the greatest amount of thiamethoxam leachate late in each growing season (Fig. 2) when compared with other types of insecticide delivery.

Early season rainfall was not exceptionally heavy in either year of this experiment (Fig. 3). The accumulation of leachate detections in lysimeters likely is reflected by the steady application of irrigation water and rainfall. One clear exception to this pattern occurred in 2012 at 155–156 days after planting when 89 mm of

Table 2. Neonicotinoid concentration from irrigation water, 2011 and 2012.

Date	Days after planting	Insecticide concentration (μg/L)[a]	
		clothianidin	thiamethoxam
28 June 2011	39	-	0.310
1 September 2011	114	-	0.327
10 July 2012	60	-	0.533
15 August 2012	96	0.225	0.580

[a]Samples obtained from irrigation pivots while under operation in potato fields containing lysimeter experiments.

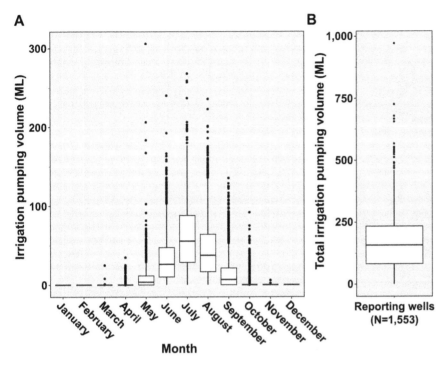

Figure 4. Reported irrigation inputs in the Central Wisconsin River Water Management Unit. Average reported agricultural pumping (megaliters, ML) in the Central Wisconsin River Water Management Unit for 2012. Monthly pumping records were reported by growers to the Wisconsin Department of Natural Resources Bureau of Drinking Water and Groundwater. Upper and lower whiskers extend to the values that are within 1.5*Inter-quartile range beyond the first (25%) and third (75%) percentiles. Data beyond the end of whiskers indicate outlier values and have been plotted as points.

rain fell within a 24-hour period. Peak detections of thiamethoxam in 2012 began to trend upward following this rain event, however the timing of similar detections across treatments in 2011 occurred at about the same time. One additional explanation may be that increased levels of pesticide losses are associated with plant death or senescence. In each year of this study, the largest proportion of pesticide detections in leachate occurred after vine killing with herbicide in the potato crop. Vine killing in commercial potato production is a common practice designed to aid the tubers in developing a periderm. Perhaps the rapid loss in root function following plant death permits excess pesticide to be solubilized and washed through the soil profile more quickly in root channels. In both seasons of this study, however, large episodic rain events did not occur early in the growing season. These results do appear, however, to document low to moderate levels of leaching losses that occur throughout the season even when the crop is managed at nominal evapo-transpirative need.

Untreated control plots also yielded low-level detections of thiamethoxam throughout both seasons. To better understand these insecticide detections in control plots, we sampled water directly from the center pivot irrigation system providing irrigation directly to the potato crop. Samples were taken while the systems were operational from lateral spigots mounted on the well casings. In both years, samples revealed low concentrations of thiamethoxam present in the groundwater at two time points in each sample season (Table 2) from which irrigation water was being drawn. Clothianidin was also present at a single time point in 2012 (Table 2). These positive detections of low-dose thiamethoxam were obviously being unintentionally applied directly to the crop through irrigation and this information is new to the producers in the Central Sands of Wisconsin. Although systemic neonicotinoids have recently been detected from surface water runoff and catch

basins associated with irrigated orchards [10], [39], to our knowledge no other study has documented the occurrence of neonicotinoids in subsurface groundwater being recycled through operating irrigation wells. Currently, the known exposure pathways for insecticide residues are most often associated with direct application or systemic movement of insecticides in floral structure and guttation water [8],[9],[40].

The implications for non-target effects resulting from these groundwater contaminants is currently unknown, but could be important considering the scale of irrigation ongoing in the Central Sands potato agroecosystem in Wisconsin (Fig. S2). Using a combination of aerial photography and NASS Cropland Data Layers, we identified 2,530 different irrigated field units distributed within the Central Wisconsin River Water Management Unit (Fig. S2). In all, 71,864 hectares of irrigated cropland were identified within the extent of the water management unit. Average irrigated field unit size was 28.4 ± 17.7 hectares (min. 1, max 138). Irrigation use patterns demonstrated clear increases in the summer months of the 2012 growing season (Fig. 4). Average annual pumping volume reported to the Wisconsin Department of Natural Resources in 2012 was 170.6 ± 115.6 megaliters (ML) of irrigation water (min. 0.00001, max 972.1) distributed over 1,553 reporting wells. Peak pumping volumes occurred in the month of July, averaging 61 ± 43.3 ML (min. 0, max 286.4). The timing of peak pumping correspond with crop demands for and reproductive phases of common open and closed pollination crops grown in the region.

While considerable attention has been focused on the positive attributes of the neonicotinoids [1–3], an increasing body of research suggests substantial negative impacts not only in terms of pest resistance development (e.g., Colorado potato beetle), but also impacts on non-target organisms and surrounding ecosystems

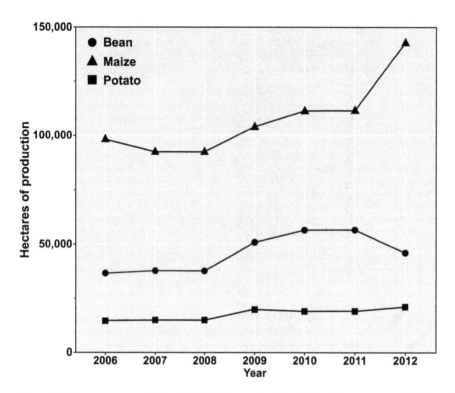

Figure 5. Crop area grown in the Central Wisconsin River Water Management Unit. Cropping trends in the Central Wisconsin River Water Management Unit from 2006–2012. Crop groups are often planted with a soil-applied neonicotinoid insecticide for insect pest management. Crop totals within the water management unit were tabulated from annual USDA-NASS Cropland Data Layers [26].

[8],[10],[41–44]. Recent studies have documented the negative influence of neonicotinoids on pollinator population health (both native and managed) which, in turn, created substantial concern about the long-term sustainability of these pesticides in agriculture [7],[11],[43],[45–49]. Exposures to pollinators reportedly occur through chronic, sub-lethal contact with low concentrations of neonicotinoid residues in pollen, nectar, waxes, and guttation drops of common crop plants [50–53]. Gill et al. [43] and Whitehorn et al. [54] found that low concentrations ($\leq 10~\mu g~L^{-1}$) of imidacloprid significantly reduced colony-level health in bumblebee (*Bombus terrestris* L.). Imidacloprid residues measured by those authors are consistent with insecticide concentrations found in nectar and pollen of flowering crops, further supporting the direct crop-pollinator toxicological pathway hypothesis [47],[52],[54],[55]. Though they have received much less attention, many closed pollination crops also provide resources for pollinators (e.g., pollen, water)[56],[57]. These crops also rely on neonicotinoids and may have currently undescribed risks for non-target organisms through indirect contaminant pathways in the agroecosystem [51],[58].

Possible exposure related to a high frequency of irrigation could drive the exposure of non-target arthropods to low concentrations of neonicotinoid insecticides in irrigation water. Although such impacts have yet to be documented directly, new comprehensive reviews of neonicotinoid environmental impacts have demonstrated numerous unanticipated impacts occurring at the ecosystem scale [9],[58]. In the Wisconsin agroecosystem, neonicotinoids are used on a large proportion of crops grown with irrigation [28],[29]. Trends in production show increased maize production over the past six years in the Central Wisconsin River Water Management Unit (Fig. 5). As a result of common neonicotinoid seed treatment on maize, accelerating production may partially explain the increased frequency of neonicotinoid detection in

groundwater. Unfortunately, little crop-specific pesticide information exists for individual neonicotinoids at the watershed scale [26]. Although measurement of specific contributions of crops to measured insecticide contamination is currently not available, this study demonstrates a research approach to better understand leaching from different application methods. Improved understanding of crops and insecticide delivery that results in greater risk of insecticide leaching will inform targets to reduce aquifer contamination and recirculation of soil-applied insecticides. Area-wide application of neonicotinoid insecticides through irrigation water applications may have considerable unanticipated or undocumented environmental impacts for non-target organisms through chronic low-dose exposure to insecticides.

Conclusions

To gain a better understanding of the seasonal cycle of neonicotinoids moving from the potato system, this study used an experimental approach to document the leaching potential of common neonicotinoid application methods. Results presented here benefit both potato producers and regulators by identifying trends in leachate losses for these commonly used, water-soluble insecticides. Lysimeter experiments documented loss of thiamethoxam following the application of vine desiccants at the conclusion of the potato production season. Leachate losses did vary among the different delivery methods over time indicating some variability in the patterns of pesticide leachate throughout the season. Quantification of crops commonly using neonicotinoid soil applications in the Central Wisconsin Water Management Unit highlights the need to research leaching potential from soil-applied neonicotinoids in other commodities. Documentation of several neonicotinoids in irrigation water suggests a new candidate pathway for non-target environmental impacts of insecticides.

Supporting Information

Figure S1   Chemical structures and properties of common neonicotinoid insecticides. Chemical structures were drawn using ChemDraw (version 13, Perkin Elmer Inc., Waltham, MA). Properties of each active ingredient were accessed from the National Center for Biotechnology Information PubChem online interface. Available: https://pubchem.ncbi.nlm.nih.gov/. Accessed 2014 Mar 20.

Figure S2   Irrigated field locations in the Central Wisconsin River Water Management Unit. Distribution of fields irrigated with high capacity wells (n = 2530) in the Central Wisconsin River Water Management Unit [27]. Points indicate locations of individual irrigation units identified from aerial photography using ArcGIS.

References

1. Jeschke P, Nauen R, Schindler M, Elbert A (2010) Overview of the status and global strategy for neonicotinoids. J Agric Food Chem 59: 2897–2908.
2. Elbert A, Haas M, Springer B, Thielert W, Nauen R (2008) Applied aspects of neonicotinoid uses in crop protection. Pest Manag Sci 64: 1099–1105.
3. Jeschke P, Nauen R (2008) Neonicotinoids–from zero to hero in insecticide chemistry. Pest Manag Sci 64: 1084–1098.
4. United States Environmental Protection Agency (2003) Imidacloprid; pesticide tolerances. Fed Regist 68: 35303–35315.
5. United States Environmental Protection Agency (2012) What is the conventional reduced risk pesticide program? Available: http://www.epa.gov/opprd001/workplan/reducedrisk.html. Accessed 2012 Oct 17.
6. Szendrei Z, Grafius E, Byrne A, Ziegler A (2012) Resistance to neonicotinoid insecticides in field populations of the Colorado potato beetle (Coleoptera: Chrysomelidae). Pest Manag Sci 68: 941–946.
7. Cresswell JE, Desneux N, vanEngelsdorp D (2012) Dietary traces of neonicotinoid pesticides as a cause of population declines in honey bees: An evaluation by Hill's epidemiological criteria. Pest Manag Sci 68: 819–827.
8. Blacquiere T, Smagghe G, Van Gestel CA, Mommaerts V (2012) Neonicotinoids in bees: A review on concentrations, side-effects and risk assessment. Ecotoxicology 21: 973–992.
9. Goulson D (2013) An overview of the environmental risks posed by neonicotinoid insecticides. J Appl Ecol 50: 977–987.
10. Starner K, Goh KS (2012) Detections of the neonicotinoid insecticide imidacloprid in surface waters of three agricultural regions of California, USA, 2010–2011. Bull Environ Contam Toxicol 88: 316–321.
11. Miranda GR, Raetano CG, Silva E, Daam MA, Cerejeira MJ (2011) Environmental fate of neonicotinoids and classification of their potential risks to hypogean, epygean, and surface water ecosystems in Brazil. Human and Ecological Risk Assessment: An International Journal 17: 981–995.
12. Gupta S, Gajbhiye V, Agnihotri N (2002) Leaching behavior of imidacloprid formulations in soil. Bull Environ Contam Toxicol 68: 502–508.
13. Papiernik SK, Koskinen WC, Cox L, Rice PJ, Clay SA, et al. (2006) Sorption-desorption of imidacloprid and its metabolites in soil and vadose zone materials. J Agric Food Chem 54: 8163–8170.
14. Chiovarou ED, Siewicki TC (2008) Comparison of storm intensity and application timing on modeled transport and fate of six contaminants. Sci Total Environ 389: 87–100.
15. Gupta S, Gajbhiye V, Gupta R (2008) Soil dissipation and leaching behavior of a neonicotinoid insecticide thiamethoxam. Bull Environ Contam Toxicol 80: 431–437.
16. Juraske R, Castells F, Vijay A, Muñoz P, Antón A (2009) Uptake and persistence of pesticides in plants: measurements and model estimates for imidacloprid after foliar and soil application. J Hazard Mater 165: 683–689.
17. Wisconsin Department of Agriculture, Trade and Consumer Protection (2010) Fifteen years of the DATCP exceedence well survey. WI-DATCP, Madison, WI.
18. Wisconsin Department of Agriculture, Trade and Consumer Protection (2011) Agrichemical Management Bureau annual report – 2011. Available: http://datcp.wi.gov/Environment/Water_Quality/ACM_Annual_Report/. Accessed 2012 Jul 10.
19. United States Department of Agriculture - Natural Resources Conservation Soil Service (2013) Web Soil Survey. USDA-NRCS, Washington, DC. Available: http://websoilsurvey.sc.egov.usda.gov. Accessed 2014 Jan 8.
20. Cooley ET, Lowery B, Kelling KA, Speth PE, Madison FW, et al. (2009) Surfactant use to improve soil water distribution and reduce nitrate leaching in potatoes. Soil Sci 174: 321–329.
21. Wisconsin Department of Natural Resources (2013) DNR drinking water system: high capacity wells. WI DNR, Madison, WI. Available: http://dnr.wi.gov/topic/wells/highcapacity.html. Accessed 2013 Aug 22.
22. Wickham H (2007) Reshaping data with the reshape package. Journal of Statistical Software 21: 1–20. Available: http://www.jstatsoft.org/v21/i12/. Accessed 2011 Jan 15.
23. Bussan A, Colquhoun J, Cullen E, Davis V, Gevens A, et al. (2012) Commercial vegetable production in Wisconsin. Publication A3422. University of Wisconsin-Extension, Madison WI.
24. Seccia S, Fidente P, Barbini DA, Morrica P (2005) Multiresidue determination of nicotinoid insecticide residues in drinking water by liquid chromatography with electrospray ionization mass spectrometry. Anal Chim Acta 553: 21–26.
25. United States Department of Agriculture - National Agricultural Imagery Program (2010) Wisconsin NAIP. USDA-NAIP, Washington, DC. Available: http://datagateway.nrcs.usda.gov/. Accessed 2011 Jan 15.
26. United States Department of Agriculture - National Agricultural Statistics Service Cropland Data Layer (2012) Wisconsin Cropland data layer. USDA-NASS, Washington, DC. Available: http://nassgeodata.gmu.edu/CropScape/. Accessed 2013 May 10.
27. Wisconsin Department of Natural Resources (2002) Wisconsin DNR 2003 watersheds. Wisconsin DNR, Madison, WI. Available: http://dnr.wi.gov/maps/gis/documents/dnr_watersheds.pdf. Accessed 2013 Jun 12.
28. Thelin GP, Stone WW (2013) Estimation of annual agricultural pesticide use for counties of the conterminous United States, 1992–2009. US Department of the Interior, US Geological Survey.
29. Stone WW (2013) Estimated annual agricultural pesticide use for counties of conterminous United States, 1992–2009. U.S. Geological Survey Data Series 752, 1-p. pamphlet, 14 tables.
30. Cullen EM, Davis VM, Jensen B, Nice GRW, Renz M (2013) Pest management in Wisconsin field crops. Publication A3646.University of Wisconsin-Extension, Madison WI. Available: http://learningstore.uwex.edu/pdf/A3646.PDF as of 08/18/2013. Accessed 2013 Aug 23.
31. Team R Core (2011) R: A language and environment for statistical computing (Version 2.15.2). Vienna, Austria: R foundation for statistical computing; 2012. Available: http://cran.r-project.org. Accessed 2012 Jun 15.
32. Pinheiro J, Bates D, DebRoy S, Sarkar D (2007) Linear and nonlinear mixed effects models. R package version 3.1–108.
33. Pinheiro J, Bates D (2000) Mixed-effects models in S and S-PLUS. New York: Springer.
34. Mossbarger Jr W, Yost R (1989) Effects of irrigated agriculture on groundwater quality in corn belt and lake states. J Irrig Drain Eng 115: 773–790.
35. Kraft GJ, Stites W, Mechenich D (1999) Impacts of irrigated vegetable agriculture on a humid North-Central US sand plain aquifer. Ground Water 37: 572–580.
36. Saad DA (2008) Agriculture-related trends in groundwater quality of the glacial deposits aquifer, central Wisconsin. J Environ Qual 37: 209–225.
37. Postle JK, Rheineck BD, Allen PE, Baldock JO, Cook CJ, et al. (2004) Chloroacetanilide herbicide metabolites in Wisconsin groundwater: 2001 survey results. Environ Sci Technol 38: 5339–5343.
38. Agrian Inc. (2013) Advanced product search. Available: http://www.agrian.com/labelcenter/results.cfm. Accessed 2013 Mar 21.
39. Hladik ML, Calhoun DL (2012) Analysis of the herbicide diuron, three diuron degradates, and six neonicotinoid insecticides in water–Method details and application to two Georgia streams: U.S. Geological Survey Scientific Investigations Report 2012–5206.
40. Hopwood J, Vaughan M, Shepherd M, Biddinger D, Mader E, et al. (2012) Are neonicotinoids killing bees? A review of research into the effects of neonicotinoid insecticides on bees, with recommendations for action. Xerces Society for Invertebrate Conservation, USA.
41. Casida JE (2012) The greening of pesticide–environment interactions: Some personal observations. Environ Health Perspect 120: 487–493.

Acknowledgments

We thank the cooperating growers for generously allowing us to conduct lysimeter studies on their farm. We thank Amy DeBaker, Rick Graham, Jeff Postle, Wendy Sax, Stan Senger, and Steve Sobek of Wisconsin DATCP for their support of this project. We thank Dave Johnson and Robert Smail of Wisconsin DNR-Water Bureau for providing 2012 irrigation use data. We thank Birl Lowery and Mack Naber for input on lysimeter design and installation. We thank Scott Chapman, Ken Frost, and David Lowenstein for their help installing lysimeters. We thank Claudio Gratton, George Kennedy, Jessica Petersen, and Wesley Stone for their insightful comments on earlier versions of this manuscript. We thank the Wisconsin Potato and Vegetable Growers Association for continued support of our research efforts.

Author Contributions

Conceived and designed the experiments: ASH RLG. Performed the experiments: ASH. Analyzed the data: ASH. Wrote the paper: ASH RLG

42. Krupke CH, Hunt GJ, Eitzer BD, Andino G, Given K (2012) Multiple routes of pesticide exposure for honey bees living near agricultural fields. PLoS ONE 7: e29268.
43. Gill RJ, Ramos-Rodriguez O, Raine NE (2012) Combined pesticide exposure severely affects individual-and colony-level traits in bees. Nature 491: 105–108.
44. Seagraves MP, Lundgren JG (2012) Effects of neonicotinoid seed treatments on soybean aphid and its natural enemies. J Pest Sci 85: 125–132.
45. Cresswell JE, Page CJ, Uygun MB, Holmbergh M, Li Y, et al. (2012) Differential sensitivity of honey bees and bumble bees to a dietary insecticide (imidacloprid). Zoology 115: 365–371.
46. Henry M, Beguin M, Requier F, Rollin O, Odoux J, et al. (2012) A common pesticide decreases foraging success and survival in honey bees. Science 336: 348–350.
47. Stoner KA, Eitzer BD (2012) Movement of soil-applied imidacloprid and thiamethoxam into nectar and pollen of squash (*Cucurbita pepo*). PloS ONE 7: e39114.
48. Tapparo A, Marton D, Giorio C, Zanella A, Soldà L, et al. (2012) Assessment of the environmental exposure of honeybees to particulate matter containing neonicotinoid insecticides coming from corn coated seeds. Environ Sci Technol 46: 2592–2599.
49. Tomé HVV, Martins GF, Lima MAP, Campos LAO, Guedes RNC (2012) Imidacloprid-induced impairment of mushroom bodies and behavior of the native stingless bee *Melipona quadrifasciata anthidioides*. PloS ONE 7: e38406.
50. Chauzat M, Faucon J, Martel A, Lachaize J, Cougoule N, et al. (2006) A survey of pesticide residues in pollen loads collected by honey bees in France. J Econ Entomol 99: 253–262.
51. Girolami V, Mazzon L, Squartini A, Mori N, Marzaro M, et al. (2009) Translocation of neonicotinoid insecticides from coated seeds to seedling guttation drops: A novel way of intoxication for bees. J Econ Entomol 102: 1808–1815.
52. Laurent FM, Rathahao E (2003) Distribution of (14C) imidacloprid in sunflowers (*Helianthus annuus* L.) following seed treatment. J Agric Food Chem 51: 8005–8010.
53. Mullin CA, Frazier M, Frazier JL, Ashcraft S, Simonds R, et al. (2010) High levels of miticides and agrochemicals in North American apiaries: Implications for honey bee health. PLoS ONE 5: e9754.
54. Whitehorn PR, O'Conner S, Wackers FL, Goulson D (2012) Neonicotinoid pesticide reduces bumble bee colony growth and queen production. Science 336: 351–352.
55. Dively GP, Kamel A (2012) Insecticide residues in pollen and nectar of a cucurbit crop and their potential exposure to pollinators. J Agric Food Chem 60: 4449–4456.
56. Free JB (1993) Insect pollination of crops. London: Academic Press.
57. Klein A, Vaissière BE, Cane JH, Steffan-Dewenter I, Cunningham SA, et al. (2007) Importance of pollinators in changing landscapes for world crops. Proc R Soc B 274: 303–313.
58. Sánchez-Bayo F, Tennekes HA, Goka K (2013) Impact of systemic insecticides on organisms and ecosystems. *In* Stanislav T, editor, Insecticides - Development of Safer and More Effective Technologies. Rijeka: InTech. 367–416.

22

Effects of Transgenic Cry1Ac + CpTI Cotton on Non-Target Mealybug Pest *Ferrisia virgata* and Its Predator *Cryptolaemus montrouzieri*

Hongsheng Wu[1,2], Yuhong Zhang[1], Ping Liu[1], Jiaqin Xie[1], Yunyu He[1], Congshuang Deng[1], Patrick De Clercq[2]*, Hong Pang[1]*

1 State Key Laboratory of Biocontrol, School of Life Sciences, Sun Yat-sen University, Guangzhou, China, 2 Department of Crop Protection, Faculty of Bioscience Engineering, Ghent University, Ghent, Belgium

Abstract

Recently, several invasive mealybugs (Hemiptera: Pseudococcidae) have rapidly spread to Asia and have become a serious threat to the production of cotton including transgenic cotton. Thus far, studies have mainly focused on the effects of mealybugs on non-transgenic cotton, without fully considering their effects on transgenic cotton and trophic interactions. Therefore, investigating the potential effects of mealybugs on transgenic cotton and their key natural enemies is vitally important. A first study on the effects of transgenic cotton on a non-target mealybug, *Ferrisia virgata* (Cockerell) (Hemiptera: Pseudococcidae) was performed by comparing its development, survival and body weight on transgenic cotton leaves expressing Cry1Ac (Bt toxin) + CpTI (Cowpea Trypsin Inhibitor) with those on its near-isogenic non-transgenic line. Furthermore, the development, survival, body weight, fecundity, adult longevity and feeding preference of the mealybug predator *Cryptolaemus montrouzieri* Mulsant (Coleoptera: Coccinellidae) was assessed when fed *F. virgata* maintained on transgenic cotton. In order to investigate potential transfer of Cry1Ac and CpTI proteins via the food chain, protein levels in cotton leaves, mealybugs and ladybirds were quantified. Experimental results showed that *F. virgata* could infest this bivalent transgenic cotton. No significant differences were observed in the physiological parameters of the predator *C. montrouzieri* offered *F. virgata* reared on transgenic cotton or its near-isogenic line. Cry1Ac and CpTI proteins were detected in transgenic cotton leaves, but no detectable levels of both proteins were present in the mealybug or its predator when reared on transgenic cotton leaves. Our bioassays indicated that transgenic cotton poses a negligible risk to the predatory coccinellid *C. montrouzieri* via its prey, the mealybug *F. virgata*.

Editor: Nicolas Desneux, French National Institute for Agricultural Research (INRA), France

Funding: This research was supported by grants from National Basic Research Program of China (973) (2013CB127605), National Natural Science Foundation of China (Grant No. 31171899) and the Youth Scientific Research Foundation of Guangdong Academy of Sciences (No. qnjj201206). The funders had no role in study design, data collection and analysis, decision to publish, or preparation of the manuscript.

Competing Interests: The authors have declared that no competing interests exist.

* E-mail: lsshpang@mail.sysu.edu.cn (HP); patrick.declercq@ugent.be (PDC)

Introduction

Genetically modified (GM) crops hold great promise for pest control [1–4]. Most popular GM crops express one or more toxin genes from bacteria such as *Bacillus thuringiensis* (Bt), trypsin inhibitors such as cowpea trypsin inhibitor (CpTI), plant lectins, ribosome-inactivating proteins, secondary plant metabolites, vegetative insecticidal proteins and small RNA viruses [5–7]. So far Bt-cotton has been commercialized in the United States (1996), Mexico (1996), Australia (1996), China (1997), Argentina (1998), South Africa (1998), Colombia (2002), India (2002), Brazil (2005), and Burkina Faso (2008) and occupies 49% of the total global cotton area [8,9]. To delay the development of pesticide resistance in the major cotton pests [7], the bivalent transgenic cotton cultivar (CCRI41) expressing Cry1Ac and CpTI, has been commercially available since 2002 in China [10]. Currently, the cotton cultivar CCRI41 is planted at a large scale in the Yellow river cotton area in China [11]. However, with the rapid expansion in the commercial use of GM plants, there is an increasing need to understand their possible impact on non-target

organisms [12–14]. Non-target effects of several cultivars (Cry1Ac + CpTI cotton) on beneficial arthropods including pollinator insects have been recently studied [11,15–21].

Most studies on the potential ecological impacts of transgenic plants on phloem-feeding insects have focused on aphids or whiteflies [4,22–27]. Studies on the interactions between mealybugs and GM crops have not been previously reported. Like aphids and whiteflies, mealybugs are obligate phloem feeders. Several species of mealybugs have caused considerable economic damage to agricultural and horticultural plants in the tropics in the last few decades [28]. They also have the potential to become major cotton pests which is evident from the severe damage reported in different parts of Asia [29–31]. Particularly, *Phenacoccus solenopsis* Tinsley (Hemiptera: Pseudococcidae) has attracted much attention worldwide because of its harmful effects on cotton [30,32–35]. Indeed, this pest can successfully thrive on both Bt-cotton and non-Bt cultivars of cotton [36]. However, *P. solenopsis* is not the only mealybug species that infests cotton in Asia. Also *Maconellicoccus hirsutus* (Green) has increasingly been reported infesting cotton in India and Pakistan [37,38]. Mealybugs are attacked by a range of specialist predators and parasitoids. These non-target species can thus be exposed to GM toxins by feeding on or parasitizing their prey or host [39–41] and there may be side effects on the behavior of these natural enemies [12,42]. Therefore, there is a need to evaluate the potential effects of transgenic cotton on mealybugs and their key natural enemies.

The striped mealybug, *Ferrisia virgata* (Cockerell) (Hemiptera: Pseudococcidae), is also a cosmopolitan and polyphagous species that attacks a wide variety of crops including cotton [34,43,44]. The adult female is wingless, and has an elongated body covered by a powdery white wax, with a pair of dark longitudinal stripes on the dorsum and white wax threads extending from the posterior end resembling tails [34]. In cotton, *F. virgata* occurs in patches and feeds on all parts of a plant, particularly on growing tips or on leaves [33]. The species has been found infesting colored fiber cotton and has emerged as a serious pest in the Northeast of Brazil [34]. Given that mealybugs like *P. solenopsis*, *M. hirsutus* and *F. virgata* are aggressive invasive pests that seriously threaten cotton production, significant concern over their potential effects on transgenic cotton should be raised. At present, only the cotton mealybug *P. solenopsis* has been reported to damage Bt cotton. However, whether other mealybug species can infest transgenic cotton is yet to be determined.

The mealybug destroyer, *Cryptolaemus montrouzieri* Mulsant (Coleoptera: Coccinellidae), is a ladybird native to Australia and has been used in many biological control programs as one of the most efficient natural enemies to suppress mealybug outbreaks around the world [45–47]. Both the adults and larvae of the ladybird prey on a variety of mealybugs [47]. *C. montrouzieri* has also been used as a biological control agent in areas where outbreaks of *F. virgata* and *P. solenopsis* occur [38,48–50]. These predators can encounter transgene products expressed by plants (Bt toxins) when feeding on plant material such as pollen, nectar, or leaf exudates and when preying on organisms that have consumed transgenic plant tissue or toxin-loaded prey [51–53].In the present study, bioassays were performed to assess the development, reproduction and feeding choices of *C. montrouzieri* presented with mealybugs reared on the cotton cultivar CCRI41 versus its near-isogenic non-transgenic line. To study whether Cry1Ac and CpTI proteins can pass through the trophic chain up to a natural enemy, quantification of Cry1Ac and CpTI proteins in leaves, mealybugs and ladybirds was also done.

This study is the first report on tritrophic relationships involving a non-target pest mealybug (*F. virgata*), its predator (*C. montrouzieri*) and a transgenic cotton cultivar expressing Cry1Ac (Bt toxin) and CpTI (Cowpea Trypsin Inhibitor).

Materials and Methods

Plants

Bivalent transgenic cotton cultivar CCRI 41 (Bt+CpTI cotton) and non-transgenic cotton cultivar CCRI 23 (control) were used as the host plants in all experiments. CCRI 41 was bred by introducing the synthetic Cry1Ac gene and modified CpTI (cowpea trypsin inhibitor) gene into the elite cotton cultivar CCRI 23 by way of the pollen tube pathway technique [54]. Seeds of transgenic Cry1Ac and CpTI cotton cultivar CCRI 41 and its near-isogenic CCRI 23 were obtained from the Institute of Cotton Research, Chinese Academy of Agricultural Sciences. Both cultivars were planted singly in plastic pots (16×13 cm) with the same soil. All plants were individually grown from seeds in climate chambers ($25 \pm 1°C$, $75 \pm 5\%$ RH, 16: 8 h (L: D)) and they were

five weeks old (about five to eight true leaves) at the start of experiments.

Insects

Stock cultures of *C. montrouzieri* and *F. virgata* were originally obtained from the State Key Laboratory of Biocontrol, Sun Yat-sen University, Guangzhou, China. Cultures of *C. montrouzieri* were reared on *Planococcus citri* Risso (Hemiptera: Pseudococcidae) and *F. virgata*, which were both produced on pumpkin fruits (*Cucurbita moschata* (Duch.ex Lam.) Duch. ex Poiretand) in metal frame cages ($45 \times 36 \times 33$ cm) covered with fine-mesh nylon gauze. The colony of *F. virgata* was maintained on plastic trays (40×30 cm) containing pumpkins as food. Environmental conditions at the insectarium were $26 \pm 2°C$, $50 \pm 10\%$ RH and a photoperiod of 16: 8 h (L: D). Both *C. montrouzieri* and *F. virgata* cultures used in these experiments had been maintained at our facilities for at least six years.

Bioassay with F. Virgata

Effects of transgenic Cry1Ac and CpTI cotton on development and survival of F. virgate. Development and survival of *F. virgata* on the leaves of transgenic and non-transgenic cotton plants was studied in climate chambers ($25 \pm 1°C$, $75 \pm 5\%$ RH, 16: 8 h (L: D)). The experiment was subdivided into two stages: crawlers (first instars) of the mealybug were reared for the first 5 days in 6-cm diameter plastic containers to preclude escape, whereas in a second stage larger plastic bags were used to accommodate the later instars. In the first stage of the experiment 20 newly emerged first-instar nymphs (<24 h) springing from the same female were placed in a plastic container (6.0×1.5 cm) covered with a fine-mesh nylon gauze using a soft paintbrush. Each plastic container had a small hole in it allowing a leaf to be inserted. A piece of cotton wool was wrapped around the petiole to prevent *F. virgata* from escaping through the hole in the container. To encourage crawlers to settle, the environmental chamber was maintained in complete darkness for 24 h [55,56]. All plastic containers were fixed on live cotton plants by small brackets. Mealybugs on each cotton plant represented a cohort or a replicate. A total of 15 cohorts (replicates) were prepared for both the treatments with transgenic and control cotton plants.

In the second stage of the experiment the mealybugs were kept in transparent plastic bags (15×10 cm) with several small holes for ventilation. The transparent plastic bag together with cotton wool wrapped around the petiole could also prevent mealybugs from escaping or dropping off. The mealybug cohorts on each leaf (still attached to the plant) were examined every 12 h, and the development and survival of each nymphal instar were recorded. Successful development from one instar to the next was determined by the presence of exuviae. Survival rate of each stage was calculated as the percentage of individuals that successfully developed to the next stage in a cohort [56]. The sex of individual mealybugs could not be determined at the crawler stage. Therefore, sex was determined during the latter part of the second instar when males change their color from yellow to dark. At this point, the developmental times of males and females were recorded separately [55].

Effects of transgenic Cry1Ac and CpTI cotton on body weight of F. virgate. To assess the body weights of *F. virgata*, 200 second-instar nymphs were collected at the same time from stock cultures reared on pumpkin. Ten mealybugs per cotton plant were placed as a cohort on the leaves of 10 non-transgenic or transgenic cotton plants using a soft paintbrush. Thus, a total of 10 cohorts (replicates) were prepared for both the treatments with transgenic and control cotton plants. To prevent mealybugs from escaping or dropping off, each leaf infested with *F. virgata* was placed in a transparent plastic bag (15×10 cm) with several small holes for ventilation. To encourage the nymphs to settle, the environmental chamber was maintained in complete darkness for 24 h. Thereafter, the plants were kept in an environmental chamber as described above. Surviving mealybugs from the initial 10 individuals on each plant were weighed individually after 10 and 20 days using an electronic balance (Sartorius BSA124S, Germany) with a precision of 0.1 mg.

Tritrophic Bioassay with C. Montrouzieri

Effects of transgenic Cry1Ac and CpTI cotton on the development and survival of immature C.montrouzieri. Two plastic boxes (12.0×5.0×4.0 cm, covered with fine-mesh nylon gauze for ventilation) each containing 50 C. montrouzieri eggs (<12 h old) collected from the stock colony were placed in a climate chamber (25±1°C, 75±5% RH, 14:10 h (L:D) photoperiod). The eggs were observed carefully every 12 h and numbers of larvae that hatched were recorded. Newly hatched first-instar larvae from 50 C. montrouzieri eggs (<12 h old) were individually transferred to the leaves of non-transgenic (45 larvae) or transgenic cotton (46 larvae), which were previously infested with F. virgata (~60–100 mealybugs per leaf). Each cotton plant received two or three C. montrouzieri larvae which were distributed on different leaves. Pieces of cotton wool were wrapped around the stem or petiole to prevent the larvae from leaving the cotton leaves. Predator larvae were randomly moved to newly infested plants when mealybug prey was depleted. In total, about 60 non-transgenic or transgenic cotton plants were used for the experiment. Larvae of C. montrouzieri were checked every 12 h for molting, which was determined by the presence of exuviae. The developmental time and survival of each immature stage of C. montrouzieri were also recorded up to adulthood.

Effects of transgenic Cry1Ac and CpTI cotton on reproduction and adult longevity. After adult emergence, C. montrouzieri females and males were single paired and each pair was transferred to a transparent plastic bag (15×10 cm) with several small holes for ventilation. A total of 12 and 16 pairs (replicates) were set up for non-transgenic and transgenic cotton plants, respectively. A piece of cotton was placed in the bag for oviposition. A leaf of non-transgenic or transgenic cotton infected with F. virgata (~60–100 mealybugs per leaf) was also placed in this bag. The bag containing C. montrouzieri adults was transferred to a new freshly infested leaf on the same plant every 3 days. The pre-oviposition period, number of eggs and survival of the mating pairs of C. montrouzieri were checked every day until the death of all adults.

Effects of transgenic Cry1Ac and CpTI cotton on body weight of C. montrouzieri. In order to determine fresh body weight during each developmental stage, 50 newly hatched first instar C. montrouzieri (<12 h old) were individually transferred to the leaves of non-transgenic or transgenic cotton using a soft hairbrush and placed in close vicinity to the prey. The leaf with mealybugs (~60–100 mealybugs per leaf) was replaced every 3 days and C. montrouzieri larvae were checked every 12 h for molting and development. Newly emerged 1st, 2nd, 3rd, and 4th instar larvae, pupae and adults of C. montrouzieri were weighed individually after 24 h using an electronic balance (Sartorius BSA124S, Germany) with a precision of 0.1 mg to record their body mass.

Feeding performance of C. montrouzieri on mealybugs reared on non-transgenic versus transgenic cotton leaves. Metal frame cages (45×36×33 cm) covered with fine-mesh nylon gauze were used in these experiments with five cages or replicates each. In each cage, 20 C. montrouzieri adults (10 males

and 10 females, <1 month old) were taken from the laboratory stock and starved for 24 h. Three pots each of non-transgenic and transgenic cotton (with one cotton plant per pot) were placed in a cage. Each non-transgenic or transgenic cotton plant was previously infested with 20 similar-sized female adult mealybugs. Every day, the plants infected with 20 mealybugs were replaced with newly infested plants. The experiment continued for 9 days and the numbers of consumed mealybugs were recorded every day.

Quantification of Toxins in Leaves, F. Virgata and C. Montrouzieri

To confirm Cry1Ac and CpTI expression of the transgenic cotton plants (8-leaf stage) used in both bioassays, five leaf samples were collected from five different cotton plants. Each sample was obtained from a middle-upper leaf of a transgenic or control plant [57]. Approximately 100 mg fresh weight (f.w.) of the transgenic or control cotton leaves was collected.

To quantify the level of Cry1Ac and CpTI in F. virgata, a group of approximately sixty gravid females from the laboratory culture were allowed to settle on cotton leaves and reproduce. After 24 h, about 100 newborn nymphs were brushed carefully onto each transgenic or control cotton leaf and the leaf was covered with a transparent plastic bag (15×10 cm) with several small holes for ventilation. A piece of cotton wool was wrapped around the petiole to prevent F. virgata from escaping from the leaf. To encourage crawlers to settle, the environmental chamber was maintained in complete darkness for 24 h. Three weeks later, five samples of F. virgata larvae (with a total fresh weight of 60–100 mg) were collected from plants of either variety.

To assess the potential transfer of Cry1Ac and CpTI proteins via the food chain, a transgenic or control cotton leaf (still attached to the plant) which was previously infested with F. virgata as described above and a newly molted 2nd instar larva or an adult (< 1 month old) of C. montrouzieri were kept in a ventilated plastic bag (15×10 cm). Ten transgenic or control cotton plants were used. After 3 days, five samples of individual C. montrouzieri larvae or adults were collected for analysis.

All experiments described above were conducted in a growth chamber at 25±1°C, 75±5% RH and a photoperiod of 16:8 h (L:D). All samples were weighed and transferred to 1.5-ml centrifuge tubes. Samples were kept at −20°C until quantification of Cry1Ac and CpTI proteins.

The amount of Cry1Ac protein in the leaf and insect material was measured using an enzyme linked immuno-sorbent assay (ELISA). Envirologix Qualiplate Kits (EnviroLogix Quantiplate Kit, Portland, ME, USA) were used to estimate Cry1Ac quantities. The quantitative detection limit of the Cry1Ac kit was 0.1 ng ml^{-1}. The ELISA polyclonal kits used to detect CpTI protein were obtained from the Center for Crop Chemical Control, China Agricultural University (Beijing, China). The method has been validated [58] and the limit of detection and working range of the assay were 0.21 and 1–100 ng ml^{-1}, respectively [59]. Prior to analysis, all insects were washed in phosphate buffered saline with Tween-20 (PBST) buffer to remove any Cry1Ac and CpTI toxin from their outer surface. After adding PBST to the samples at a ratio of about 1:10 (mg sample: μl buffer) in 1.5 ml centrifuge tubes, the samples were fully ground by hand using a plastic pestle. To detect Cry1Ac protein, samples were centrifuged for 5 min at 13,000×g and leaf samples were diluted to 1:10 with PBST (insect samples were not diluted). For analysis of CpTI protein, the tubes were centrifuged at 10,000×g for 15 min. The supernatants were used to detect targeted proteins. ELISA was performed based on the manufacturer's instructions.

ODs were calibrated by a range of concentrations of Cry1Ac or CpTI made from purified toxin solution.

Data Analysis

For the studied parameters in the bioassay with *F. virgata*, the average values of each cohort were used as replicates for the data analyses. The duration of the immature stages, survival and weight on transgenic and non-transgenic cotton were compared using independent t-tests. For the tritrophic bioassay with *C. montrouzieri*, a Mann–Whitney U test was performed for the duration of the immature stages and preoviposition period. Weights, fecundity, oviposition period, and adult longevity were analyzed using independent t-tests. The percentages of total survival and egg hatch were compared by logistic regression, which is a generalized linear model using a probit (log odds) link and a binomial error function [60]. Each test consists of a regression coefficient that is calculated and tested for being significantly different from zero, for which P-values are presented [61]. Consumption rates in the feeding performance test were compared using a general linear model for repeated measures analysis of variance (ANOVA) followed by a LSD test. All datasets were first tested for normality and homogeneity of variances using a Kolmogorov-Smirnov test and Levene test, respectively, and transformed if necessary. SPSS software (IBM SPSS Statistics, Ver. 20) was used for all statistical analyses. For all tests, the significance level was set at $P \leq 0.05$.

Results

Bioassay with *F. Virgata*

Effects of transgenic Cry1Ac and CpTI cotton on the developmental duration of *F. virgate*. *F. virgata* nymphs completed their development when reared on non-transgenic cotton CCRI 23 and its near-isogenic transgenic cotton CCRI 41 (Table 1). However, there was no significant difference in the developmental duration of female or male *F. virgata* larvae reared on transgenic or non-transgenic cotton except during the first and fourth instars. The duration of first instar development was longer on transgenic cotton. In contrast, fourth instar males reared on transgenic cotton had shorter development compared to those reared on non-transgenic cotton. No significant differences were observed in the developmental durations of the second instar, third instar and in cumulative developmental time.

Effects of transgenic Cry1Ac and CpTI cotton on nymphal survival of *F. virgate*. No significant difference was observed in the survival rate of female or male *F. virgata* nymphs reared on transgenic or non-transgenic cotton except in the first instar (Table 2). The survival rate of the first instars was lower when reared on transgenic cotton. No significant differences in the survival rates of the second instar, third instar, fourth instar of male and in cumulative survival rate were observed.

Effects of transgenic Cry1Ac and CpTI cotton on body weight of *F. virgate*. The weight of all *F. virgata* nymphs increased when reared on transgenic or non-transgenic cotton leaves for 10 or 20 days. However, nymphal weights were not significantly influenced by cotton variety (P>0.05, independent t-tests). Mean weights ($\pm$ SE) of adult *F. virgata* reared on non-transgenic cotton (77 and 52, respectively) and transgenic cotton (87 and 48, repectively) leaves for 10 days were 1.29 ± 0.15 mg and 1.30 ± 0.10 mg ($t = -0.091$; df = 18; P = 0.928), and for 20 days were 2.30 ± 0.22 mg and 2.06 ± 0.21 mg ($t = 0.799$; df = 18; P = 0.434), respectively.

Table 1. Mean number of days ($\pm$SE) for each developmental stage of *F. virgata* reared non-transgenic or transgenic cotton leaves.

Developmental time per stage (days)								Cumulative (days)	
	First[†]	Second		Third		Fourth*			
Cotton cultivar		Female	Male	Female	Male	Male		Female	Male
Non-transgenic cotton	8.24±0.16a	5.31±0.19a	6.34±0.28a	6.43±0.07a	2.48±0.09a	5.68±0.08a		19.98±0.23a	22.55±0.24a
Transgenic cotton	8.72±0.14b	5.10±0.16a	6.62±0.23a	6.59±0.14a	2.39±0.07a	5.26±0.14b		20.41±0.36a	22.87±0.22a
t	−2.218	0.846	−0.776	−0.982	−0.063	2.654		−1.007	−1.008
df	28	28	28	28	28	28		28	28
P	0.035	0.405	0.444	0.334	0.405	0.013		0.323	0.322

Means ± SE within a column followed by the same letter are not significantly different (P>0.05; independent t-test). The experiment was started with 15 cohorts (replicates) per treatment.
[†]Sex could not be determined before the second instar.
*Female mealybugs have only three nymphal instars while males have four nymphal instars.

Table 2. Mean (±SE) survival rate (%) of each developmental stage of *F. virgata* reared on non-transgenic or transgenic cotton leaves.

Cotton cultivar	First[†]	Second	Third		Fourth*	Total survival
			Female	Male	Male	
Non-transgenic cotton	75.67±3.68a	83.56±3.71a	94.93±3.06a	87.98±4.53a	97.38±1.86a	57.00±4.05a
Transgenic cotton	61.33±4.15b	86.96±4.20a	89.79±2.93a	97.22±1.94a	97.00±2.06a	49.00±4.37a
t	2.583	−0.592	1.213	1.878	0.006	1.343
df	28	28	28	28	28	28
P	0.015	0.501	0.235	0.071	0.892	0.190

Means ± SE within a column followed by the same letter are not significantly different (P>0.05; independent t-test). The experiment was started with 15 cohorts (replicates) per treatment.
[†]Sex could not be determined before the second instar.
*Female mealybugs have only three nymphal instars while males have four nymphal instars.

Tritrophic Bioassay with *C. Montrouzieri*

Effects of transgenic Cry1Ac and CpTI cotton on development and survival of immature *C. montrouzieri*. The developmental time of all immature stages and total survival did not differ when reared on transgenic or its near-isogenic non-transgenic cotton (Table 3). There was no significant difference in immature stages and survival.

Effects of transgenic Cry1Ac and CpTI cotton on body weight of *C. montrouzieri*. When reared on transgenic cotton, first instar (t = −1.579; df = 8; P = 0.153), second instar (t = 1.941; df = 98; P = 0.055), third instar (t = −0.343; df = 97; P = 0.733) and fourth instar larvae (t = 0.782; df = 95; P = 0.436), pupae (t = 0.659; df = 90; P = 0.512), and male (t = −1.795; df = 39; P = 0.080) and female (t = −0.421; df = 34; P = 0.677) adults showed no significant difference in their body weight upon emergence compared with their counterparts reared on non-transgenic cotton (Figure 1).

Reproduction and longevity of *C. montrouzieri* reared on non-transgenic or transgenic cotton leaves. Preoviposition period (U = 68; df = 1; P = 0.906), fecundity (t = 0.390; df = 21; P = 0.700), number of eggs laid per female per day (t = 1.581; df = 21; P = 0.129), egg hatch (χ^2 = 1.753; df = 1; P = 0.185), male longevity (t = 0.148; df = 26; P = 0.883) and female longevity (t = −1.183; df = 26; P = 0.247) were not significantly affected by treatment (Table 4).

Feeding performance of *C. montrouzieri* on mealybugs reared on non-transgenic versus transgenic cotton leaves. Daily consumption of mealybugs by *C. montrouzieri* adults on non-transgenic cotton was not different from that on transgenic cotton during the entire 9-day test period (F = 0.111; df = 1; P = 0. 748) (Figure 2). The interaction between the factors cotton type and time was also not significant, meaning that differential consumption of mealybugs between transgenic cotton and non-transgenic cotton was not a function of time (F = 0.692; df = 8; P = 0.697). However, *C. montrouzieri* consumed a decreasing number of mealybugs on both cotton varieties over the course of the experiment (F = 5.098; df = 8; P<0.001).

Quantification of Toxins in Leaves, *F. Virgata* and *C. Montrouzieri*

Expressed levels of the Cry1Ac and CpTI proteins in CCRI41 cotton leaves averaged 5.76±0.33 µg Cry1Ac/g f.w. and 14.28±1.70 ng CpTI/g f.w. (means ± SE), respectively. ELISA revealed that *F. virgata* maintained on transgenic cotton did not contain detectable amounts of the Cry1Ac and CpTI proteins. Similarly, no Cry1Ac or CpTI protein was detected in *C.*

montrouzieri larvae and adults. None of the non-transgenic cotton leaves, or of the mealybug and ladybird samples reared on control plants were found to contain any Cry1Ac or CpTI protein.

Discussion

F. virgata is a widely spread mealybug and is reported in more than 100 countries around the world, including the USA, Argentina, Canada, India, China, Brazil, and Pakistan [44], where transgenic cotton is being cultivated. Our results demonstrate that *F. virgata* nymphs completed their development when reared on leaves of both non-transgenic and transgenic cotton. Overall, no significant differences were detected in the total survival, cumulative developmental duration and body weight of the immature stages of *F. virgata* reared on transgenic and non-transgenic cotton. Higher mortality was observed during the first instar on transgenic cotton but the difference was small and total mortality from first instar to adult did not differ between treatments. These results indicate that the transgenic Bt+CpTI cotton had negligible adverse effects on the development of *F. virgata*, which is consistent with previous reports by Dutt [36] and Zhao et al. [18] stating that the mealybug *P. solenopsis* was able to infest Bt and Bt+CpTI transgenic cotton without negative effects on its fitness.

Further, ELISA analyses revealed that none of the mealybug samples from the Bt+CpTI cotton contained detectable Bt protein despite high expression levels in leaves. Like aphids and whiteflies, mealybugs are obligate phloem sap feeding insects. We postulate that *F. virgata* was not exposed to the Bt endotoxins expressed in the cotton plants given its phloem feeding habit. In previous studies on transgenic maize, Bt toxins were not detected or only in negligible amounts in the phloem sap, or in aphids that had fed on the maize [62,63]. In transgenic Bt cotton fields the density of sap-feeding insects, such as whiteflies, aphids and leafhoppers, has been reported to be higher than in non-transgenic cotton fields [64,65]. Lawo et al. [26] noted that Indian Bt cotton varieties had no effect on aphids, leading them to conclude that Bt cotton poses a negligible risk for aphid antagonists and that the aphids should remain under natural control in Bt cotton fields.

On the other hand, it was expected that any impact of transgenic Bt+CpTI cotton on mealybugs may be largely attributed to the CpTI gene encoding the cowpea trypsin inhibitor, which acts on insect gut digestive enzymes and inhibits protease activity [66]. The cysteine protease inhibitor, oryzacystatin I (OC-I), was detected in both leaves and phloem sap of

Table 3. Developmental time (days) and total survival rate (%) of the immature stages of *C. montrouzieri* reared non-transgenic or transgenic cotton leaves.

Cotton cultivar	Developmental time per stage (days)*						Total survival (%)†
	1st instar	2nd instar	3rd instar	4th instar	Pupa	Total immature	
Non-transgenic cotton	3.12±0.04	2.74±0.05	3.22±0.06	5.50±0.07	8.71±0.07	23.40±0.13	80.00±0.06
Transgenic cotton	3.11±0.03	2.81±0.05	3.27±0.04	5.49±0.07	8.66±0.09	23.31±0.09	86.96±0.05
U/χ^2	942.5	776.0	820.0	878.0	640.5	672.5	0.798
df	1	1	1	1	1	1	1
P	0.793	0.293	0.538	0.973	0.610	0.610	0.372

No significant difference was observed between the control and treated groups within the same column (means ± SE) (P>0.05; *Mann-Whitney U test or †Wald χ^2 test); 45and 46 larvae were initially tested for non-transgenic and transgenic cotton plants, respectively.

transgenic oilseed rape, which significantly inhibited growth of *Aphis gossypii* Glover, *Acyrthosiphon pisum* (Harris), and *Myzus persicae* (Sulzer) in vitro, despite low levels of proteolysis in the guts of these homopterans [67]. Although in the present study no CpTI protein could be detected by ELISA in *F. virgata* samples, the effects of the CpTI protein on the mealybug cannot be fully excluded. Low amounts of the cowpea trypsin inhibitors (CpTI) ingested by *F. virgata*, could act as an anti-feedant to the mealybugs, which may explain lower survival rates in the first instar. In fact, Han et al. [11] demonstrated an antifeedant effect of CCRI41 cotton pollen (Bt+CpTI) on the honey bee *Apis mellifera* L. Feeding behaviour of the bees was disturbed and they consumed significantly less CCRI41 cotton pollen than in the control group given conventional cotton pollen. The antifeedant affect may have led to insufficient food uptake and malnutrition for the larvae and newly emerged bees [11,68,69]. Further, according to an EPG (Electric Penetration Graph) signal, Liu et al.[23] found that the frequencies of moving and searching for feeding sites, and probing activity of the aphid *A. gossypii* reared on CCRI 41cotton were significantly higher than those on control cotton. Given their high mobility 1st instar mealybugs are responsible for plant colonization in the field [70]. When 1st instars of *F. virgata* select their feeding site on transgenic cotton a succession of walks and stops is observed. Consequently, the 1st instar mealybugs in our study may have spent more energy in finding and probing for food on transgenic cotton leaves than on non-transgenic leaves, which might have negatively affected the survival rates in the first instar. However, if present, this antifeedant effect to *F. virgata* appears limited because no significant difference was found in total survival and developmental duration. Besides, the *F. virgata* clones used in the present study were not resistant to the transgenic plants, as they had been maintained exclusively on pumpkin for at least 6 years without any contact with cotton. Due to inadvertent adaptations to laboratory conditions, host finding and acceptance behaviors of mass produced insects may be changed over the generations [71–73]. Colonization effects may therefore have influenced the responses of the mealybug to cotton as a host plant and it may be warranted to investigate the interactions between transgenic cotton and wild or recently colonized mealybugs.

The mealybug destroyer, *C. montrouzieri*, might ingest toxins expressed by transgenic plants that accumulate in the mealybugs feeding on these plants. In this context, we conducted tritrophic bioassays to investigate the potential effects of CCRI 41 cotton on *C. montrouzieri* by using *F. virgata* as prey. These experiments did not reveal any adverse effects on the fitness of *C. montrouzieri* after ingestion of *F. virgata* that fed on Bt+CpTI cotton leaves compared with those that fed on the corresponding non-transgenic cotton leaves. Besides a longer oviposition period on transgenic cotton than on non-transgenic cotton, there were no differences in reproductive parameters. This finding is consistent with other studies which reported no or little adverse effects on various predators or parasitoids after feeding on different Bt + CpTI cottons, including a ladybird [74] and two hymenopteran parasitoids [15,75].

Several possible mechanisms can explain the observed results. Firstly, *C. montrouzieri* may not be sensitive to Cry1Ac proteins. Porcar et al. [76] reported no statistical differences in mortality of *C. montrouzieri* adults and *Adalia bipunctata* L. larvae fed on artificial diets with or without Cry1Ab and Cry3Aa toxins. Duan et al. [77] and Lundgren and Wiedenmann [78] found no significant adverse effects when Bt maize pollen were fed to larvae of the ladybird *Coleomegilla maculata* DeGeer. The same ladybird species was also found to be unaffected by Bt cotton or higher amounts of Cry2Ab and Cry1Ac proteins indicating that Bt cotton poses a negligible

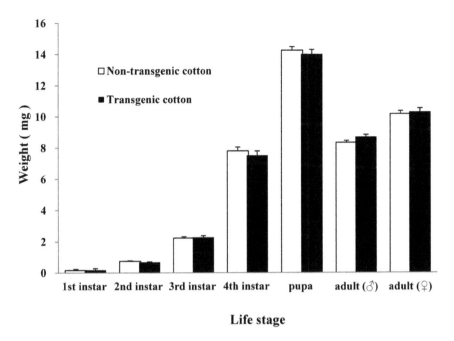

Figure 1. Weight upon molting (means ± SE) of different life stages of *C. montrouzieri* **reared on non-transgenic or transgenic cotton leaves.** No significant difference was observed between the control and treated groups in each life stage (P>0.05; independent t-test). The experiment was started with 50 larvae per treatment.

risk to *C. maculata* [57]. In addition, no negative effects of Bt-transgenic plants were observed on the development, survival, and reproduction of the ladybirds *Hippodamia convergens* (Guérin-Méneville) and *Propylea japonica* (Thunberg) through their aphid prey that fed on the Bt plants [25,79].

In the field, no significant differences were observed in the abundance of coccinellid beetles on Bt-transgenic and non-transgenic cottons [80]. Pollen from Cry1Ac+CpTI transgenic cotton (CCRI41) did not affect the pollinating beetle *Haptoncus luteolus* (Erichson) in the field and in the laboratory [19]. Xu et al. [81] found that CCRI41 cotton did not affect the population dynamics of non-target pests and predators including ladybirds and spiders in Xinjiang, China. Zhang et al. [82] observed negative effects on the ladybird *P. japonica* when offered young *Spodoptera litura* (F.) larvae reared on Bt-transgenic cotton expressing Cry1Ac toxin; however, adverse effects on the ladybird were attributed to poor prey quality. Lumbierres et al. [83] investigated the effects of Bt maize on aphid parasitism and the aphid–parasitoid complex in field conditions on three transgenic varieties and found that Bt maize did not alter the aphid–parasitoid associations and had no effect on aphid parasitism and hyperparasitism rates.

Ramirez-Romero et al. [84] concluded that Bt-maize did not affect the development of the non-target aphid *Sitobion avenae* (F.) and Cry1Ab toxin quantities detected in these aphids were nil, indicating that none or negligible amounts of Cry1Ac are passed on from the aphids to higher trophic levels. Probably, the amount of Cry1Ac/CpTI proteins ingested by the mealybugs in our study was too low to be effective. Indeed, ELISA measurements indicated that Bt+CpTI cotton-reared *F. virgata* and its predator did not contain detectable amounts of the Cry1Ac and CpTI protein. Because the commercial ELISA kit for determining CpTI expression was not available [19] or the amount of CpTI proteins was lower than the lowest limit of quantification [11,21] there are few earlier reports on tritrophic interactions involving CpTI protein. On the contrary, many studies related to the transfer of Bt toxic proteins to higher trophic levels have been carried out. For example, ELISA analyses revealed no or only trace amounts of Bt protein in sap-sucking insects of the order Hemiptera after feeding on different Bt plants, including maize [63,84–86] and cotton [26,87]. Trace amounts of Bt toxins were detected in *A. gossypii* feeding on Bt cotton cultivars and ladybirds preying on Bt-fed aphids [41]. Another possible reason for the weak effect of Cry1Ac/CpTI proteins is that ladybirds may digest or excrete the

Table 4. Reproduction and longevity of *C. montrouzieri* females reared on non-transgenic or transgenic cotton leaves.

Cotton cultivar	Preoviposition period (days)[‡]	Fecundity (eggs/♀)*	Oviposition rate (eggs/♀/day)*	Egg hatch (%)[†]	Longevity (days)*	
					♂	♀
Non-transgenic cotton	7.00±0.93	823.80±84.25	7.21±0.83	90.60±0.01	160.96±17.36	131.42±9.82
Transgenic cotton	7.18±0.58	766.46±110.99	5.44±0.74	90.80±0.11	157.88±11.62	152.91±13.85

No significant difference was observed between the control and treated groups within the same column (Means ± SE) (P>0.05; *independent t-test, [‡]Mann-Whitney U test or [†]Wald χ^2 test); 12 and 16 pairs of *C. montrouzieri* were used for non-transgenic and transgenic cotton plants, respectively.

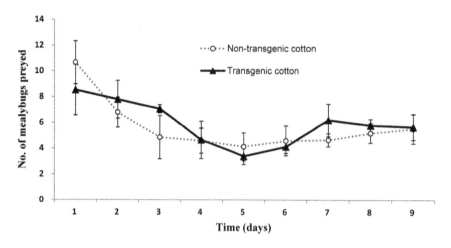

Figure 2. Feeding performance of *C. montrouzieri* **on** *F. virgata* **mealybugs reared on non-transgenic or transgenic cotton.** The data represent numbers (±SE) of mealybugs consumed per individual predator over a 9 day period (P>0.05; repeated measures analysis of variance (ANOVA) followed by a LSD test).

toxins taken up via their prey. For example, Li and Romeis [88] fed the ladybird *Stethorus punctillum* (Weise) with spider mites, *Tetranychus urticae* (Koch), reared on Cry3Bb1-expressing Bt maize. Subsequent bioassays revealed that the Cry protein concentrations in the ladybird beetle larvae and adults were 6- and 20-fold lower, respectively, than the levels in the spider mite prey. Cry1 proteins were also detected in *C. maculata* when offered *Trichoplusia ni* (Hübner) larvae reared on Bt-cotton, but the Bt protein levels were 21-fold lower for Cry2Ab and 6-fold lower for Cry1Ac compared to the concentrations in the prey [57].

In summary, our study indicates that *F. virgata* can successfully develop on bivalent transgenic cotton CCRI41expressing Cry1Ac+CpTI and thus can pose a risk for this crop. The finding that not only *P. solenopsis* but also other mealybugs like *F. virgata* can easily infest transgenic cotton plants has important implications for pest management in this cropping system. Further, our study demonstrates that transgenic cotton poses a negligible risk to the predatory coccinellid *C. montrouzieri* via its mealybug prey.

However, further field studies assessing the impact of transgenic cotton on the mealybug pest and its key natural enemies are needed.

Acknowledgments

We wish to thank our colleagues of Sun Yat-sen University for support and input during all stages of the work. Dr. Fenglong Jia, Dandan Zhang and Binglan Zhang provided useful suggestions on an earlier design of the experiment. Ruixin Jiang helped with the statistical analysis and Lijun Ma is thanked for assistance with insect rearing and cotton planting.

Author Contributions

Conceived and designed the experiments: HW HP PDC. Performed the experiments: HW YZ PL JX YH CD. Analyzed the data: HW HP PDC. Contributed reagents/materials/analysis tools: HW YZ PL JX YH. Wrote the paper: HW HP PDC.

References

1. Kos M, van Loon JJ, Dicke M, Vet LE (2009) Transgenic plants as vital components of integrated pest management. Trends in biotechnology 27: 621–627.

2. Ferry N, Edwards M, Gatehouse J, Capell T, Christou P, et al. (2006) Transgenic plants for insect pest control: a forward looking scientific perspective. Transgenic Research 15: 13–19.

3. Poppy GM, Sutherland JP (2004) Can biological control benefit from genetically-modified crops? Tritrophic interactions on insect-resistant transgenic plants. Physiological Entomology 29: 257–268.

4. Lu Y, Wu K, Jiang Y, Guo Y, Desneux N (2012) Widespread adoption of Bt cotton and insecticide decrease promotes biocontrol services. Nature 487: 362–365.

5. Hilder VA, Boulter D (1999) Genetic engineering of crop plants for insect resistance - a critical review. Crop Protection 18: 177–191.

6. Sharma H, Ortiz R (2000) Transgenics, pest management, and the environment. Current Science 79: 421–437.

7. Lundgren JG, Gassmann AJ, Bernal J, Duan JJ, Ruberson J (2009) Ecological compatibility of GM crops and biological control. Crop Protection 28: 1017–1030.

8. James C (2009) Global status of commercialized biotech/GM crops: 2009. ISAAA Brief No, 39 (International Service for the Acquisition of Agri-Biotech Applications, Ithaca, NY, USA).

9. Dhillon M, Sharma H (2013) Comparative studies on the effects of Bt-transgenic and non-transgenic cotton on arthropod diversity, seedcotton yield and bollworms control. Journal of Environmental Biology 34: 67–73.

10. Cui JJ (2003) Effects and mechanisms of the transgenic Cry1Ac plus CpTI (cowpea trypsin inhibitor) cotton on insect communities. Dissertation, Chinese Academy of Agricultural Sciences.

11. Han P, Niu CY, Lei CL, Cui JJ, Desneux N (2010) Quantification of toxins in a Cry1Ac + CpTI cotton cultivar and its potential effects on the honey bee *Apis mellifera* L. Ecotoxicology 19: 1452–1459.

12. Faria CA, Wackers FL, Pritchard J, Barrett DA, Turlings TC (2007) High susceptibility of Bt maize to aphids enhances the performance of parasitoids of lepidopteran pests. PLoS One 2: e600.

13. Romeis J, Bartsch D, Bigler F, Candolfi MP, Gielkens MM, et al. (2008) Assessment of risk of insect-resistant transgenic crops to nontarget arthropods. Nature biotechnology 26: 203–208.

14. Desneux N, Bernal JS (2010) Genetically modified crops deserve greater ecotoxicological scrutiny. Ecotoxicology 19: 1642–1644.

15. Liu XX, Sun CG, Zhang QW (2005) Effects of transgenic Cry1A+ CpTI cotton and Cry1Ac toxin on the parasitoid, *Campoketis chlorideae* (Hymenoptera: Ichneumonidae). Insect Science 12: 101–107.

16. Liu B, Shu C, Xue K, Zhou K, Li X, et al. (2009) The oral toxicity of the transgenic Bt+ CpTI cotton pollen to honeybees (*Apis mellifera*). Ecotoxicology and Environmental Safety 72: 1163–1169.

17. Xu Y, Wu KM, Li HB, Liu J, Ding RF, et al. (2012) Effects of Transgenic Bt + CpTI Cotton on Field Abundance of Non-Target Pests and Predators in Xinjiang, China. Journal of Integrative Agriculture 11: 1493–1499.

18. Zhao XN, Cui XH, Chen L, Hung WF, Zheng D, et al. (2012) Ontogenesis and Adaptability of Mealybug *Phenacoccus solenopsis* Tinsley (Hemiptera:Pseudococcidae) on Different Varieties of Cotton. Cotton Science 24: 496–502.

19. Chen L, Cui J, Ma W, Niu C, Lei C (2011) Pollen from Cry1Ac/CpTI-transgenic cotton does not affect the pollinating beetle Haptoncus luteolus. Journal of Pest Science 84: 9–14.

20. Han P, Niu CY, Lei CL, Cui JJ, Desneux N (2010) Use of an innovative T-tube maze assay and the proboscis extension response assay to assess sublethal effects

of GM products and pesticides on learning capacity of the honey bee *Apis mellifera* L. Ecotoxicology 19: 1612–1619.

21. Han P, Niu CY, Biondi A, Desneux N (2012) Does transgenic Cry1Ac + CpTI cotton pollen affect hypopharyngeal gland development and midgut proteolytic enzyme activity in the honey bee *Apis mellifera* L.(Hymenoptera, Apidae)? Ecotoxicology 21: 2214–2221.

22. Ashouri A, Michaud D, Cloutier C (2001) Unexpected effects of different potato resistance factors to the Colorado potato beetle (Coleoptera: Chrysomelidae) on the potato aphid (Homoptera: Aphididae). Environmental entomology 30: 524–532.

23. Liu XD, Zhai BP, Zhang XX, Zong JM (2005) Impact of transgenic cotton plants on a non-target pest, *Aphis gossypii* Glover. Ecological Entomology 30: 307–315.

24. Zhou FC, Du YZ, Ren SX (2005) Effects of transgenic cotton on population of the piercing-sucking mouthparts insects. Entomological Journal of East China 14: 132–135.

25. Zhu S, Su J, Liu X, Du L, Yardim EN, et al. (2006) Development and reproduction of *Propylaea japonica* (Coleoptera: Coccinellidae) raised on *Aphis gossypii* (Homoptera: Aphididae) fed transgenic cotton. Zoological Studies 45: 98–103.

26. Lawo NC, Wäckers FL, Romeis J (2009) Indian Bt cotton varieties do not affect the performance of cotton aphids. PLoS One 4: e4804.

27. Porcar M, Grenier A-M, Federici B, Rahbé Y (2009) Effects of *Bacillus thuringiensis* δ-Endotoxins on the Pea Aphid (*Acyrthosiphon pisum*). Applied and environmental microbiology 75: 4897–4900.

28. Beltrà A, Soto A, Germain J, Matile-Ferrero D, Mazzeo G, et al. (2010) The Bougainvillea mealybug *Phenacoccus peruvianus*, a rapid invader from South America to Europe. Entomol Hell 19: 137–143.

29. Solangi GS, Mahar GM, Oad FC (2008) Presence and abundance of different insect predators against sucking insect pest of cotton. Journal of Entomology 5: 31–37.

30. Wang YP, Watson GW, Zhang RZ (2010) The potential distribution of an invasive mealybug *Phenacoccus solenopsis* and its threat to cotton in Asia. Agricultural and Forest Entomology 12: 403–416.

31. Khuhro S, Lohar M, Abro G, Talpur M, Khuhro R (2012) Feeding potential of lady bird beetle, *Brumus suturalis* Fabricius (Coleopteran: Coccinellidae) on cotton mealy bug *Phenococcus solenopsis* (Tinsley) in laboratory and field. Sarhad J Agric 28: 259–265.

32. Hodgson C, Abbas G, Arif MJ, Saeed S, Karar H (2008) *Phenacoccus solenopsis* Tinsley (Sternorrhyncha: Coccoidea: Pseudococcidae), an invasive mealybug damaging cotton in Pakistan and India, with a discussion on seasonal morphological variation. Zootaxa 1: 1913.

33. Nagrare V, Kranthi S, Kumar R, Dhara Jothi B, Amutha M, et al. (2011) Compendium of cotton mealybugs. Shankar Nagar, Nagpur, India: CICR.

34. Silva-Torres C, Oliveira M, Torres J (2013) Host selection and establishment of striped mealybug, *Ferrisia virgata*, on cotton cultivars. Phytoparasitica 41: 31–40.

35. Hanchinal S, Patil B, Basavanagoud K, Nagangoud A, Biradar D, et al. (2011) Incidence of invasive mealybug (*Phenacoccus solenopsis* Tinsley) on cotton. Karnataka Journal of Agricultural Sciences 24.

36. Dutt U (2007) Mealy Bug Infestation in Punjab: Bt. Cotton Falls Flat. Countercurrents org Available: http://www.countercurrents.org/dutt210807.htm. Accessed 2013 Jun 20.

37. Hanchinal S, Patil B, Bheemanna M, Hosamani A (2010) Population dynamics of mealybug, *Phenacoccus solenopsis* Tinsley and it's natural enemies on Bt cotton. Karnataka Journal of Agricultural Sciences 23: 137–139.

38. Khan HAA, Sayyed AH, Akram W, Raza S, Ali M (2012) Predatory potential of *Chrysoperla carnea* and *Cryptolaemus montrouzieri* larvae on different stages of the mealybug, *Phenacoccus solenopsis*: A threat to cotton in South Asia. Journal of Insect Science 12: 1–12.

39. Azzouz H, Cherqui A, Campan EDM, Rahbé Y, Duport G, et al. (2005) Effects of plant protease inhibitors, oryzacystatin I and soybean Bowman-Birk inhibitor, on the aphid *Macrosiphum euphorbiae* (Homoptera, Aphididae) and its parasitoid *Aphelinus abdominalis* (Hymenoptera, Aphelinidae). Journal of insect physiology 51: 75–86.

40. Ramirez-Romero R, Bernal J, Chaufaux J, Kaiser L (2007) Impact assessment of Bt-maize on a moth parasitoid, *Cotesia marginiventris* (Hymenoptera: Braconidae), via host exposure to purified Cry1Ab protein or Bt-plants. Crop Protection 26: 953–962.

41. Zhang GF, Wan FH, Lövei GL, Liu WX, Guo JY (2006) Transmission of Bt toxin to the predator *Propylaea japonica* (Coleoptera: Coccinellidae) through its aphid prey feeding on transgenic Bt cotton. Environmental entomology 35: 143–150.

42. Desneux N, Ramírez-Romero R, Bokonon-Ganta AH, Bernal JS (2010) Attraction of the parasitoid *Cotesia marginiventris* to host (*Spodoptera frugiperda*) frass is affected by transgenic maize. Ecotoxicology 19: 1183–1192.

43. Schreiner I (2000) Striped mealybug [*Ferrisia virgata* (Cockrell)]. Available: http://wwwadaphawaiiedu/adap/Publications/ADAP_pubs/2000-18pdf. Accessed 2013 Jun 20.

44. Ben-Dov Y, Miller DR, Gibson GAP (2005) ScaleNet: A Searchable Information System on Scale Insects. Available: http://wwwselbarcusdagov/scalenet/scalenethtm. Accessed 2013 Jun 20.

45. Bartlett BR (1974) Introduction into California of cold-tolerant biotypes of the mealybug predator *Cryptolaemus montrouzieri*, and laboratory procedures for testing natural enemies for cold-hardiness. Environmental entomology 3: 553–556.

46. Li LY (1993) The research and application prospects of *Cryptolaemus montrouzieri* in China. Nature Enemies Insects 15: 142–152.

47. Jiang RX, Li S, Guo ZP, Pang H (2009) Research status of *Cryptolaemus montrouzieri* Mulsant and establishing its description criteria. Journal of Environmental Entomology 31: 238–247.

48. Mani M, Krishnamoorthy A, Singh S (1990) The impact of the predator, *Cryptolaemus montrouzieri* Mulsant, on pesticide-resistant populations of the striped mealybug, *Ferrisia virgata*(Ckll.) on guava in India. Insect Science and its Application 11: 167–170.

49. Mani M, Krishnamoorthy A (2008) Biological suppression of the mealybugs *Planococcus citri* (Risso), *Ferrisia virgata* (Cockerell) and *Nipaecoccus viridis* (Newstead) on pummelo with *Cryptolaemus montrouzieri* Mulsant in India. Journal of Biological Control 22: 169–172.

50. Kaur H, Virk J (2012) Feeding potential of *Cryptolaemus montrouzieri* against the mealybug *Phenacoccus solenopsis*. Phytoparasitica 40: 131–136.

51. Harwood JD, Wallin WG, Obrycki JJ (2005) Uptake of Bt endotoxins by nontarget herbivores and higher order arthropod predators: molecular evidence from a transgenic corn agroecosystem. Molecular Ecology 14: 2815–2823.

52. Zwahlen C, Andow DA (2005) Field evidence for the exposure of ground beetles to Cry1Ab from transgenic corn. Environmental Biosafety Research 4: 113–117.

53. Schmidt JE, Braun CU, Whitehouse LP, Hilbeck A (2009) Effects of activated Bt transgene products (Cry1Ab, Cry3Bb) on immature stages of the ladybird *Adalia bipunctata* in laboratory ecotoxicity testing. Archives of environmental contamination and toxicology 56: 221–228.

54. Li FG, Cui JJ, Liu CL, Wu ZX, Li FL, et al. (2000) The studies on Bt+CpTI cotton and its resistance. Scientia Agricultura Sinica 33: 46–52.

55. Amarasekare KG, Mannion CM, Osborne LS, Epsky ND (2008) Life history of *Paracoccus marginatus* (Hemiptera: Pseudococcidae) on four host plant species under laboratory conditions. Environmental entomology 37: 630–635.

56. Chong JH, Roda AL, Mannion CM (2008) Life history of the mealybug, *Maconellicoccus hirsutus* (Hemiptera: Pseudococcidae), at Constant temperatures. Environmental entomology 37: 323–332.

57. Li Y, Romeis J, Wang P, Peng Y, Shelton AM (2011) A comprehensive assessment of the effects of Bt cotton on *Coleomegilla maculata* demonstrates no detrimental effects by Cry1Ac and Cry2Ab. PLoS One 6: e22185.

58. Rui YK, Wang BM, Li ZH, Duan LS, Tian XL, et al. (2004) Development of an enzyme immunoassay for the determination of the cowpea trypsin inhibitor (CpTI) in transgenic crop. Scientia Agricultura Sinica 37: 1575–1579.

59. Tan GY, Nan TG, Gao W, Li QX, Cui JJ, et al. (2013) Development of Monoclonal Antibody-Based Sensitive Sandwich ELISA for the Detection of Antinutritional Factor Cowpea Trypsin Inhibitor. Food Analytical Methods 6: 614–620.

60. Quinn GP, Michael JK (2002) Experimental design and data analysis for biologists. Cambridge, UK: Cambridge University Press.

61. McCullagh P, Nelder JA (1989) Generalized linear models. London, UK: Chapman & Hall.

62. Raps A, Kehr J, Gugerli P, Moar W, Bigler F, et al. (2001) Immunological analysis of phloem sap of *Bacillus thuringiensis* corn and of the nontarget herbivore *Rhopalosiphum padi* (Homoptera: Aphididae) for the presence of Cry1Ab. Molecular Ecology 10: 525–533.

63. Dutton A, Klein H, Romeis J, Bigler F (2002) Uptake of Bt-toxin by herbivores feeding on transgenic maize and consequences for the predator *Chrysoperla carnea*. Ecological Entomology 27: 441–447.

64. Cui JJ, Xia JY (2000) Effects of Bt (*Bacillus thuringiensis*) transgenic cotton on the dynamics of pest population and their enemies. Acta Phytophylacica Sinica 27: 141–145.

65. Lumbierres B, Albajes R, Pons X (2004) Transgenic Bt maize and *Rhopalosiphum padi* (Hom., Aphididae) performance. Ecological Entomology 29: 309–317.

66. Lawrence PK, Koundal KR (2002) Plant protease inhibitors in control of phytophagous insects. Electronic Journal of Biotechnology 5: 5–6.

67. Rahbe Y, Deraison C, Bonade-Bottino M, Girard C, Nardon C, et al. (2003) Effects of the cysteine protease inhibitor oryzacystatin (OC-I) on different aphids and reduced performance of Myzus persicae on OC-I expressing transgenic oilseed rape. Plant science 164: 441–450.

68. Desneux N, Decourtye A, Delpuech J-M (2007) The sublethal effects of pesticides on beneficial arthropods. Annu Rev Entomol 52: 81–106.

69. Decourtye A, Mader E, Desneux N (2010) Landscape enhancement of floral resources for honey bees in agro-ecosystems. Apidologie 41: 264–277.

70. Renard S, Calatayud PA, Pierre JS, Rü BL (1998) Recognition Behavior of the Cassava Mealybug *Phenacoccus manihoti* Matile-Ferrero (Homoptera: Pseudococcidae) at the Leaf Surface of Different Host Plants. Journal of Insect Behavior 11: 429–450.

71. Kölliker-Ott UM, Bigler F, Hoffmann AA (2003) Does mass rearing of field collected *Trichogramma brassicae* wasps influence acceptance of European corn borer eggs? Entomologia experimentalis et applicata 109: 197–203.

72. Geden C, Smith L, Long S, Rutz D (1992) Rapid deterioration of searching behavior, host destruction, and fecundity of the parasitoid Muscidifurax raptor (Hymenoptera: Pteromalidae) in culture. Annals of the Entomological Society of America 85: 179–187.

73. Joyce AL, Aluja M, Sivinski J, Vinson SB, Ramirez-Romero R, et al. (2010) Effect of continuous rearing on courtship acoustics of five braconid parasitoids, candidates for augmentative biological control of *Anastrepha* species. BioControl 55: 573–582.

74. Lu Y, Xue L, Zhou ZT, Dong JJ, Gao XW, et al. (2011) Effects of Transgenic Bt Plus CpTI cotton on Predating Function Response of *Coccinella septempunctata* to *Aphis gossypii* Glover. Acta Agriculturae Boreali-Sinica 26: 163–167.

75. Geng JH, Shen ZR, Song K, Zheng L (2006) Effect of pollen of regular cotton and transgenic Bt+ CpTI cotton on the survival and reproduction of the parasitoid wasp *Trichogramma chilonis* (Hymenoptera: Trichogrammatidae) in the laboratory. Environmental entomology 35: 1661–1668.

76. Porcar M, García-Robles I, Domínguez-Escribà L, Latorre A (2010) Effects of *Bacillus thuringiensis* Cry1Ab and Cry3Aa endotoxins on predatory Coleoptera tested through artificial diet-incorporation bioassays. Bulletin of entomological research 100: 297.

77. Duan JJ, Head G, McKee MJ, Nickson TE, Martin JW, et al. (2002) Evaluation of dietary effects of transgenic corn pollen expressing Cry3Bb1 protein on a non-target ladybird beetle, *Coleomegilla maculata*. Entomologia experimentalis et applicata 104: 271–280.

78. Lundgren JG, Wiedenmann RN (2002) Coleopteran-specific Cry3Bb Toxin from Transgenic Corn Pollen Does Not Affect The Fitness of a Nontarget Species, *Coleomegilla maculata* DeGeer (Coleoptera: Coccinellidae). Environmental entomology 31: 1213–1218.

79. Dogan E, Berry R, Reed G, Rossignol P (1996) Biological parameters of convergent lady beetle (Coleoptera: Coccinellidae) feeding on aphids (Homoptera: Aphididae) on transgenic potato. Journal of Economic Entomology 89: 1105–1108.

80. Sharma HC, Arora R, Pampapathy G (2007) Influence of transgenic cottons with *Bacillus thuringiensis* cry1Ac gene on the natural enemies of *Helicoverpa armigera*. BioControl 52: 469–489.

81. Xu Y, Wu KM, Li HB, Liu J, Ding RF, et al. (2012) Effects of Transgenic Bt+ CpTI Cotton on Field Abundance of Non-Target Pests and Predators in Xinjiang, China. Journal of Integrative Agriculture 11: 1493–1499.

82. Zhang GF, Wan FH, Wan XL, Guo JY (2006) Early Instar Response to Plant Derived Bt-Toxin in a Herbivore (*Spodoptera litura*) and a Predator (*Propylaea japonica*). Crop Protection 25: 527–533.

83. Lumbierres B, Starý P, Pons X (2011) Effect of Bt maize on the plant-aphid–parasitoid tritrophic relationships. BioControl 56: 133–143.

84. Ramirez-Romero R, Desneux N, Chaufaux J, Kaiser L (2008) Bt-maize effects on biological parameters of the non-target aphid *Sitobion avenae* (Homoptera: Aphididae) and Cry1Ab toxin detection. Pesticide Biochemistry and Physiology 91: 110–115.

85. Head G, Brown CR, Groth ME, Duan JJ (2001) Cry1Ab protein levels in phytophagous insects feeding on transgenic corn: implications for secondary exposure risk assessment. Entomologia experimentalis et applicata 99: 37–45.

86. Obrist L, Dutton A, Albajes R, Bigler F (2006) Exposure of arthropod predators to Cry1Ab toxin in Bt maize fields. Ecological Entomology 31: 143–154.

87. Torres JB, Ruberson JR, Adang MJ (2006) Expression of *Bacillus thuringiensis* Cry1Ac protein in cotton plants, acquisition by pests and predators: a tritrophic analysis. Agricultural and Forest Entomology 8: 191–202.

88. Li Y, Romeis J (2010) Bt maize expressing Cry3Bb1 does not harm the spider mite, *Tetranychus urticae*, or its ladybird beetle predator, *Stethorus punctillum*. Biological Control 53: 337–344.

Permissions

The contributors of this book come from diverse backgrounds, making this book a truly international effort. This book will bring forth new frontiers with its revolutionizing research information and detailed analysis of the nascent developments around the world.

We would like to thank all the contributing authors for lending their expertise to make the book truly unique. They have played a crucial role in the development of this book. Without their invaluable contributions this book wouldn't have been possible. They have made vital efforts to compile up to date information on the varied aspects of this subject to make this book a valuable addition to the collection of many professionals and students.

This book was conceptualized with the vision of imparting up-to-date information and advanced data in this field. To ensure the same, a matchless editorial board was set up. Every individual on the board went through rigorous rounds of assessment to prove their worth. After which they invested a large part of their time researching and compiling the most relevant data for our readers.

The editorial board has been involved in producing this book since its inception. They have spent rigorous hours researching and exploring the diverse topics which have resulted in the successful publishing of this book. They have passed on their knowledge of decades through this book. To expedite this challenging task, the publisher supported the team at every step. A small team of assistant editors was also appointed to further simplify the editing procedure and attain best results for the readers.

Apart from the editorial board, the designing team has also invested a significant amount of their time in understanding the subject and creating the most relevant covers. They scrutinized every image to scout for the most suitable representation of the subject and create an appropriate cover for the book.

The publishing team has been an ardent support to the editorial, designing and production team. Their endless efforts to recruit the best for this project, has resulted in the accomplishment of this book. They are a veteran in the field of academics and their pool of knowledge is as vast as their experience in printing. Their expertise and guidance has proved useful at every step. Their uncompromising quality standards have made this book an exceptional effort. Their encouragement from time to time has been an inspiration for everyone.

The publisher and the editorial board hope that this book will prove to be a valuable piece of knowledge for researchers, students, practitioners and scholars across the globe.

List of Contributors

Corinne Vacher
INRA, UMR1202 Biodiversité Génes et Communautés, Cestas, France
Université Montpellier II, UMR5554 Institut des Sciences de l'Evolution, Montpellier, France

Tanya M. Kossler
Department of Ecology and Evolutionary Biology, University of California Irvine, Irvine, California, United States of America
Department of Biology, Duke University, Durham, North Carolina, United States of America

Michael E. Hochberg
Université Montpellier II, UMR5554 Institut des Sciences de l'Evolution, Montpellier, France

Arthur E. Weis
Department of Ecology and Evolutionary Biology, University of California Irvine, Irvine, California, United States of America
Department of Ecology and Evolutionary Biology, University of Toronto, Toronto, Canada

Zhi Liu, Jie Zhao, Yunhe Li, Wenwei Zhang, Guiliang Jian, Yufa Peng and Fangjun Qi
State Key Laboratory for Biology of Plant Diseases and Insect Pests, Institute of Plant Protection, Chinese Academy of Agricultural Sciences, Beijing, People's Republic of China

Zaijian Yuan
School of Economics & Management, Hebei University of Science and Technology, Shijiazhuang, China,
Center for Agricultural Resources Research, Institute of Genetics and Developmental Biology, Chinese Academy of Sciences, Shijiazhuang, China

Yanjun Shen
Center for Agricultural Resources Research, Institute of Genetics and Developmental Biology, Chinese Academy of Sciences, Shijiazhuang, China

Wentao Xu, Kunlun Huang and Yunbo Luo
Laboratory of Food Safety, College of Food Science and Nutritional Engineering, China Agricultural University, Beijing, China

The Supervision, Inspection and Testing Center of Genetically Modified Food Safety, Ministry of Agriculture, Beijing, China

Zhifang Zhai and Yanfang Yuan
Laboratory of Food Safety, College of Food Science and Nutritional Engineering, China Agricultural University, Beijing, China

Nan Zhang and Ying Shang
The Supervision, Inspection and Testing Center of Genetically Modified Food Safety, Ministry of Agriculture, Beijing, China

Assaf Anyamba
National Aeronautics and Space Administration, Goddard Space Flight Center, Biospheric Sciences Laboratory, Greenbelt, Maryland, United States of America
Universities Space Research Association, Columbia, Maryland, United States of America

Jennifer L. Small and Edwin W. Pak
National Aeronautics and Space Administration, Goddard Space Flight Center, Biospheric Sciences Laboratory, Greenbelt, Maryland, United States of America
Science Systems and Applications Incorporated, Lanham, Maryland, United States of America

Seth C. Britch and Kenneth J. Linthicum
United States Department of Agriculture, Agricultural Research Service, Center for Medical, Agricultural, & Veterinary Entomology, Gainesville, Florida, United States of America

Compton J. Tucker and Curt A. Reynolds
National Aeronautics and Space Administration, Goddard Space Flight Center, Biospheric Sciences Laboratory, Greenbelt, Maryland, United States of America

James Crutchfield
United States Department of Agriculture, Foreign Agricultural Service, International Production & Assessment Division, Washington, District of Columbia, United States of America

Lin Niu and Lizhen Chen
Hubei Insect Resources Utilization and Sustainable
Pest Management Key Laboratory, Huazhong
Agricultural University, Wuhan, Hubei, China
College of Plant Science and Technology, Huazhong
Agricultural University, Wuhan, Hubei, China

Yan Ma
Institute of Cotton Research, Chinese Academy of
Agricultural Sciences, Anyang, Henan, China

Amani Mannakkara
Hubei Insect Resources Utilization and Sustainable
Pest Management Key Laboratory, Huazhong
Agricultural University, Wuhan, Hubei, China
Department of Agricultural Biology, Faculty of
Agriculture, University of Ruhuna, Kamburupitiya,
Sri Lanka

Yao Zhao, Weihua Ma and Chaoliang Lei
Hubei Insect Resources Utilization and Sustainable
Pest Management Key Laboratory, Huazhong
Agricultural University, Wuhan, Hubei, China

Matin Qaim
Department of Agricultural Economics and
Rural Development, Georg-August-University of
Goettingen, Goettingen, Germany

Shahzad Kouser
Department of Agricultural Economics and
Rural Development, Georg-August-University of
Goettingen, Goettingen, Germany
Institute of Agricultural and Resource Economics,
University of Agriculture, Faisalabad, Pakistan

Jonathan W. Leff
Cooperative Institute for Research in Environmental
Sciences, University of Colorado, Boulder, Colorado,
United States of America

Noah Fierer
Cooperative Institute for Research in Environmental
Sciences, University of Colorado, Boulder, Colorado,
United States of America
Department of Ecology and Evolutionary Biology,
University of Colorado, Boulder, Colorado, United
States of America

**Xiang Gao, Man Wu, Ruineng Xu, Xiurong Wang,
Ruqian Pan and Hong Liao**
State Key Laboratory for Conservation and
Utilization of Subtropical Agro-bioresources,
Root Biology Center, South China Agricultural
University, Guangzhou, China

Hye-Ji Kim
Department of Tropical Plants and Soil Sciences,
College of Tropical Agriculture and Human
Resources, University of Hawaii at Manoa,
Honolulu, Hawaii, United States of America

Felicity A. Edwards and Keith C. Hamer
School of Biology, University of Leeds, Leeds, West
Yorkshire, United Kingdom

David P. Edwards
Centre for Tropical Environmental and Sustainability
Science (TESS) and School of Marine and Tropical
Biology, James Cook University, Cairns, Queensland,
Australia
Department of Animal and Plant Sciences,
University of Sheffield, Sheffield, South Yorkshire,
United Kingdom

Sean Sloan
Centre for Tropical Environmental and Sustainability
Science (TESS) and School of Marine and Tropical
Biology, James Cook University, Cairns, Queensland,
Australia

**Francisco M. Padilla, Hannie de Caluwe, Annemiek
E. Smit-Tiekstra, Hans de Kroon**
Experimental Plant Ecology, Institute for Water and
Wetland Research, Radboud University Nijmegen,
Nijmegen, The Netherlands

Liesje Mommer
Experimental Plant Ecology, Institute for Water and
Wetland Research, Radboud University Nijmegen,
Nijmegen, The Netherlands

Cornelis A. M. Wagemaker and N. Joop Ouborg
Nature Conservation and Plant Ecology,
Wageningen University, Wageningen, The
Netherlands
Molecular Ecology, Institute for Water and Wetland
Research, Radboud University Nijmegen, Nijmegen,
The Netherlands

Wilhelm Klümper and Matin Qaim
Department of Agricultural Economics and
Rural Development, Georg-August-University of
Goettingen, Goettingen, Germany

**Baoru Sun, Yi Peng, Hongyu Yang, Zhijian Li,
Yingzhi Gao, Chao Wang, Yuli Yan and Yanmei Liu**
Key Laboratory of Vegetation Ecology, Northeast
Normal University, Changchun, China

Qichao Zhao, Minghong Liu, Miaomiao Tan and Zhicheng Shen
State Key Laboratory of Rice Biology, Institute of Insect Sciences, Zhejiang University, Hangzhou, China

Jianhua Gao
College of Life Science, Shanxi Agricultural University, Taigu, China

Mengyi Wang, Cuinan Wu, Zhihui Cheng, Huanwen Meng, Mengru Zhang and Hongjing Zhang
College of Horticulture, Northwest A&F University, Yangling, Shaanxi, China

Juan Luis Jurat-Fuentes and Siva Rama Krishna Jakka
Department of Entomology and Plant Pathology, University of Tennessee, Knoxville, Tennessee, United States of America

Lohitash Karumbaiah
Department of Entomology, University of Georgia, Athens, Georgia, United States of America

Changming Ning, Chenxi Liu and Kongming Wu
State Key Laboratory of Plant Disease and Insect Pests, Institute of Plant Protection, Chinese Academy of Agricultural Science, Beijing, People's Republic of China

Jerreme Jackson
Genome Science and Technology Program, University of Tennessee, Knoxville, Tennessee, United States of America

Fred Gould
Department of Entomology, North Carolina State University, Raleigh, North Carolina, United States of America

Carlos Blanco
Animal and Plant Health Inspection Service, Biotechnology Regulatory Services, United States Department of Agriculture, Riverdale, Maryland, United States of America

Maribel Portilla and Omaththage Perera
Southern Insect Management Research Unit, Agricultural Research Service, United States Department of Agriculture, Stoneville, Mississippi, United States of America

Michael Adang
Department of Entomology, North Carolina State University, Raleigh, North Carolina, United States of America
Department of Biochemistry and Molecular Biology, University of Georgia, Athens, Georgia, United States of America

Kimberly A. Stoner
Department of Entomology, The Connecticut Agricultural Experiment Station, New Haven, Connecticut, United States of America

Brian D. Eitzer
Department of Analytical Chemistry, The Connecticut Agricultural Experiment Station, New Haven, Connecticut, United States of America

Andy Hector
Institute of Evolutionary Biology and Environmental Studies, University of Zurich, Zurich, Switzerland
Microsoft Research, Cambridge, United Kingdom

Stefanie von Felten and Maja Weilenmann
Institute of Evolutionary Biology and Environmental Studies, University of Zurich, Zurich, Switzerland

Yann Hautier
Institute of Evolutionary Biology and Environmental Studies, University of Zurich, Zurich, Switzerland
Department of Ecology, Evolution and Behavior, University of Minnesota, Saint Paul, Minnesota, United States of America

Helge Bruelheide
Institute of Biology/Geobotany and Botanical Garden, Martin Luther University Halle-Wittenberg, Halle (Saale), Germany

Lubo Gao, Huasen Xu, Biao Bao, Xiaoyan Wang, Chao Bi and Yifang Chang
College of Water and Soil Conservation, Beijing Forestry University, Beijing, P.R. China

Huaxing Bi
College of Water and Soil Conservation, Beijing Forestry University, Beijing, P.R. China
Key Laboratory of Soil and Water Conservation, Ministry of Education, Beijing, P.R. China

Weimin Xi
Department of Biological and Health Sciences, Texas A&M University-Kingsville, Kingsville, Texas, United States of America

Rong-Gang Cong
Centre for Environmental and Climate Research (CEC), Lund University, Lund, Sweden

Mark Brady
Centre for Environmental and Climate Research (CEC), Lund University, Lund, Sweden
AgriFood Economics Centre, Department of Economics, Swedish University of Agricultural Sciences, Lund, Sweden

Anders S. Huseth
Department of Entomology, Cornell University, New York State Agricultural Experiment Station, Geneva, New York, United States of America

Russell L. Groves
Department of Entomology, University of Wisconsin-Madison, Madison, Wisconsin, United States of America

Hongsheng Wu
State Key Laboratory of Biocontrol, School of Life Sciences, Sun Yat-sen University, Guangzhou, China
Department of Crop Protection, Faculty of Bioscience Engineering, Ghent University, Ghent, Belgium

Yuhong Zhang, Ping Liu, Jiaqin Xie, Yunyu He, Congshuang Deng and Hong Pang
State Key Laboratory of Biocontrol, School of Life Sciences, Sun Yat-sen University, Guangzhou, China

Patrick De Clercq
Department of Crop Protection, Faculty of Bioscience Engineering, Ghent University, Ghent, Belgium

Index

A

Agricultural Production, 35-43, 54, 78, 95, 158, 160, 163, 168-169, 172, 179

Agricultural Subsidy System, 162-163, 165, 167, 169, 171, 173

Agricultural Water Consumption, 16-17, 19-21, 23

Agroforestry Systems, 154-155, 157-159, 161

Alkaline Phosphatase, 120-122, 125, 127, 130, 132-133, 135-137, 139-140

Apis Mellifera, 44, 52, 141, 145, 190, 192-193

B

Bacillus Thuringiensis, 2, 5, 7, 15, 44, 51-53, 59, 97, 101, 114, 119, 133, 139-140, 185, 193-194

Bacterial Communities, 60-68

Bt-cotton, 44-45, 47, 49, 51, 185, 192

C

Calorie Consumption, 53-58

Canopy Pruning, 154

Continuous Cropping, 102, 112, 120-121, 123-125, 127-131

Crop Monocultures, 78-79, 81, 83, 85

Crop Yields, 35, 43, 83, 95, 98, 113, 154, 159-160

Cry Toxins, 52, 133-134, 137-140

Cryptolaemus Montrouzieri, 185-186, 193

Crystal Protein, 51-52, 114

Cucurbita Pepo, 141-143, 145, 184

Culture-independent Approach, 60

Cylindrocladium Parasiticum, 69-70, 77

D

Depleting Resource Levels, 86

Developmental Stages, 8-9, 11-14

Differentially Expressed Genes (deg), 8

Diverse Populations, 60

Dominance, 89-91, 146-153

Drip Irrigation, 141-144

E

Eco-environmental Security, 102, 111

Ecological Settings, 35

Economic Incomes, 102-103, 109, 111

Economy, Environment And Society (ees), 162-163

Ecosystem Services, 78, 83-85, 162

Efficiency, 9, 23, 26, 33, 37, 105, 111, 116-117, 130, 132, 154, 158, 160-165, 167-169, 171-173

Eggplant, 120-131

Ellenberg's Experimental Water Table, 146-147, 149, 151, 153

Enterobacteriaceae Taxa, 60, 63

Environmental Fate, 174-175, 177, 179, 181, 183

Environmental Gradients, 146

Evapotranspiration, 17, 23, 160-161

Evolution, 1, 3-7, 17, 20, 52, 94, 119, 133, 140, 146

F

Fertilization, 15, 111, 126, 131-132, 154, 160

Flowering Time, 1-5, 7

Food Security, 16, 53-55, 57-59, 84, 101, 103, 112, 162

G

Garlic Relay Intercropping Systems, 120-121, 123, 125, 127, 129-131

Genetic Modification, 8, 11

Genetically Modified Crops, 7, 15, 33, 52-53, 55, 57, 59, 95, 97, 99, 101, 192

Groundwater Resources, 16, 24

H

Hemiptera, 185-186, 191-193

Heterogeneity, 37, 57-58, 69, 78, 84, 93-96, 98, 162

Higher Food Availability, 53

Honey Bee, 44-45, 47, 49, 51-52, 141, 144-145, 184, 190, 192-193

Hunger Problem, 53

I

Insect Resistance, 25-26, 45, 97-98, 100, 114, 117, 119, 133, 139-140

Intercropping, 69-77, 102-113, 120-121, 123-132, 153-161

Interspecific Hybridization, 1, 3-7

Irrigated Grain Production, 16

Irrigated Potato Agroecosystem, 174-175, 177, 179, 181, 183

L
Lepidopteran Strains, 133, 135, 137, 139

M
Medicago Sativa, 62, 102-105, 107, 109, 111, 113
Meta-analysis, 52, 95-99, 101, 145, 151, 153
Multiple Crop Species, 69

N
Neonicotinoid Insecticides, 141-142, 144-145, 174-175, 177-179, 181-184
Non-target Mealybug Pest, 187, 189, 191, 193
Non-uniform Distribution Pattern, 8-9, 11, 13, 15
Normalized Difference Vegetation Index (ndvi), 35
Nutrient-poor Soil Layer, 86

O
Oil Palm Yield, 78-85
Organic Analogs, 60, 62
Over-exploitation, 16, 21

P
Pan Lysimeters, 174
Pastoral Areas, 102
Pesticide Use, 53, 66, 95, 141, 183
Photosynthetically Active Radiation, 154-156, 158
Plantago Lanceolata, 86, 88, 90-92, 94
Polycistronic Transgene, 114-117, 119
Polymerase Chain Reaction (pcr), 25
Productivity, 7, 16-17, 19-20, 23, 35, 38, 53, 70, 77, 94, 96, 101-102, 113, 121, 131-132, 146-149, 151-153, 156, 160-162
Pseudococcidae, 185-186, 193

R
Regression Coefficient, 99, 188
Relative Yield Total, 88, 90

Resource-based Niche Differentiation, 146
Retaining Forest, 78-79, 81, 83, 85
Root Interactions, 69-77, 86, 94
Root Overproduction, 86-87, 89, 91, 93

S
Self-cleavage Peptide, 114-115, 117, 119
Sequence-specific Primers, 25
Sequencing Gel Electrophoresis, 25-27, 29, 31, 33
Silkworm, 44-52, 140
Soil Chemical Property, 120-121, 123, 125, 127, 129, 131
Soil-applied Imidacloprid, 141, 143, 145, 184
Solanum Melongena, 120
Solanum Tuberosum, 33, 174
Soybean Intercropping, 69-77, 156, 159
Soybean Soil, 69, 71, 73, 75, 77
Spodoptera Frugiperda, 114, 119, 133, 139-140, 193
Stacked Insect, 44-45, 47, 49, 51-52
Sustainable Management, 77-79, 81, 83, 85, 161
Systemic Insecticides, 141-142, 145, 184

T
Thiamethoxam, 141-145, 174-181, 183-184
Transgenic Rice, 8-9, 11, 13, 15, 52, 114-119

U
Universal Primer (up), 25-27, 31-32

V
Vector-borne Disease, 35-37, 39, 41-43

W
Weather Extremes, 35-37, 39, 41-43
Weedy Phenotypes, 1
Weedy Relatives, 1, 3-5, 7
Whole-genome Sequencing, 25

Printed in the USA
CPSIA information can be obtained
at www.ICGtesting.com
JSHW051412221024
72173JS00006B/1345

9 781632 397836